Application of Smart Technology and Equipment in Horticulture

Application of Smart Technology and Equipment in Horticulture

Editors

Chenglin Wang
Lufeng Luo

MDPI

Basel • Beijing • Wuhan • Barcelona • Belgrade • Novi Sad • Cluj • Manchester

Editors
Chenglin Wang
Kunming University of Science and Technology
Kunming
China

Lufeng Luo
Foshan University
Foshan
China

Editorial Office
MDPI AG
Grosspeteranlage 5
4052 Basel, Switzerland

This is a reprint of articles from the Special Issue published online in the open access journal *Horticulturae* (ISSN 2311-7524) (available at: https://www.mdpi.com/journal/horticulturae/special_issues/R87IUNW023).

For citation purposes, cite each article independently as indicated on the article page online and as indicated below:

Lastname, A.A.; Lastname, B.B. Article Title. *Journal Name* **Year**, *Volume Number*, Page Range.

ISBN 978-3-7258-1695-8 (Hbk)
ISBN 978-3-7258-1696-5 (PDF)
doi.org/10.3390/books978-3-7258-1696-5

Contents

Preface

As a significant aspect of modern agriculture, horticulture plays a crucial role in both beautifying our surroundings and enriching human nutrition. With the advent of intelligent devices across all facets of agriculture, horticulture—a branch that demands meticulous management and operation—has begun to embrace intelligence and intensification. Consequently, there is a growing need for advanced gardening technologies and intelligent equipment.

To develop intelligent technologies and tools that can aid in gardening, enhance environmental aesthetics, and support the cultivation and breeding of plants, extensive research is necessary. This research aims to boost the adoption of intelligent equipment and enhance the survival rate of cultivated plants. Successful breeding not only diversifies our plant choices but also facilitates urban greening through automated gardening practices. Moreover, the use of intelligent technology and equipment in intensive horticulture can significantly reduce labor costs and enhance the precision and efficiency of management, ultimately leading to increased productivity.

This Special Issue highlights the latest advancements in intelligent technology and equipment designed to beautify the environment, promote agricultural intensification, and support the cultivation and breeding of various plant species.

Chenglin Wang and Lufeng Luo
Editors

horticulturae

MDPI

Editorial

Application of Smart Technology and Equipment in Horticulture

Chenglin Wang [1,*] and Lufeng Luo [2]

1 Faculty of Modern Agricultural Engineering, Kunming University of Science and Technology, Kunming 650500, China
2 School of Mechatronics Engineering and Automation, Foshan University, Foshan 528000, China; luolufeng@fosu.edu.cn
* Correspondence: wangcl86@kust.edu.cn

Horticulture, as an important component of modern agriculture, plays a significant role in beautifying the environment but also in enriching human nutrition [1]. With the widespread application of intelligent devices in various aspects of agriculture, horticulture, which requires more refined management and operations, is also moving towards intelligence and intensification [2]. Therefore, developing intelligent technologies and equipment that can assist in horticulture, beautify the environment, and support plant cultivation and breeding has become a current research focus [3]. This Special Issue, "Application of Intelligent Technology and Equipment in Horticulture", highlights the latest intelligent technologies and equipment in areas such as environmental beautification, agricultural intensification, and plant species cultivation.

Machine learning and deep learning technologies have vast potential in horticulture, and this Special Issue explores some specific application scenarios. By analyzing large amounts of data, machine learning algorithms can help farmers anticipate crop diseases and optimize prevention plans, thereby reducing the use of chemical agents and protecting the ecological environment [4]. Additionally, the application of machine learning in soil and underground structure detection can help improve planting schemes, enhance soil utilization efficiency, and increase crop yields. Through advanced image recognition technology, deep learning algorithms can achieve real-time monitoring and management of key stages such as crop growth, pest and disease control, and flowering periods.

The application of intelligent devices and sensor technology in horticulture allows for real-time monitoring and regulation of environmental parameters and crop growth conditions. These devices can provide precise data support, helping farmers optimize planting strategies, improve resource utilization efficiency, and reduce waste [5]. This Special Issue includes research on the application of sensors in horticulture. For example, wearable sensor technology can monitor the moisture and nutrient status of plants in real-time, allowing for timely adjustments to irrigation and fertilization plans, thereby ensuring the healthy growth of crops.

Plant growth simulation and management optimization technology are additional research focuses of this Special Issue. By establishing mathematical models and simulation algorithms, it is possible to accurately predict and manage various parameters during the crop growth process. This helps farmers better understand the growth patterns of crops and optimize planting density, fertilization amounts, and irrigation frequency, thereby improving crop yield and quality. Additionally, these simulation models can be used for new variety breeding and cultivation technique improvements, promoting the development of horticultural production towards more scientific and precise methods [6].

This Special Issue showcases the multifaceted applications of intelligent technology and equipment in horticulture, covering several cutting-edge fields such as artificial intelligence, deep learning, sensor technology, and plant growth simulation. Through these

Citation: Wang, C.; Luo, L. Application of Smart Technology and Equipment in Horticulture. *Horticulturae* **2024**, *10*, 676. https://doi.org/10.3390/horticulturae10070676

Received: 19 June 2024
Accepted: 20 June 2024
Published: 26 June 2024

studies, various aspects of horticultural production are optimized, not only enhancing production efficiency and crop quality but also promoting the precision and intelligence of horticultural management. In the future, with the continuous advancement of intelligent technology, the horticulture sector will experience more innovations and transformations, providing robust technical support for the sustainable development of modern agriculture. The widespread adoption of intelligent devices and the increased success rate of breeding will further promote the intelligent and intensive development of horticulture [7].

Conflicts of Interest: The author declares no conflicts of interest.

List of Contributions

This Special Issue includes 11 research articles and 1 review article, comprehensively exploring the application of intelligent technology and equipment in horticulture. These contributions have delved into various aspects, such as the application of artificial intelligence in plant pathology, crop management technologies based on machine learning and deep learning, the application of intelligent devices and sensors in environmental monitoring and crop growth, and research on plant growth simulation and management optimization. The Guest Editor would like to thank each author for sharing their knowledge and providing their fascinating research findings to this Special Issue. In addition, they also appreciate the valuable assistance provided by the horticultural industry to make this Special Issue a reality.

1. González-Rodríguez, V.E.; Izquierdo-Bueno, I.; Cantoral, J.M.; Carbú, M.; Garrido, C. Artificial Intelligence: A Promising Tool for Application in Phytopathology. *Horticulturae* **2024**, *10*, 197. https://doi.org/10.3390/horticulturae10030197.
2. Qiu, Z.; Zeng, J.; Tang, W.; Yang, H.; Lu, J.; Zhao, Z. Research on Real-Time Automatic Picking of Ground-Penetrating Radar Image Features by Using Machine Learning. *Horticulturae* **2022**, *8*, 1116. https://doi.org/10.3390/horticulturae8121116.
3. Meng, Y.; Zhai, X.; Zhang, J.; Wei, J.; Zhu, J.; Zhang, T. *HeLoDL*: Hedgerow Localization Based on Deep Learning. *Horticulturae* **2023**, *9*, 227. https://doi.org/10.3390/horticulturae9020227.
4. Zhou, H.; Ou, J.; Meng, P.; Tong, J.; Ye, H.; Li, Z. Reasearch on Kiwi Fruit Flower Recognition for Efficient Pollination Based on an Improved YOLOv5 Algorithm. *Horticulturae* **2023**, *9*, 400. https://doi.org/10.3390/horticulturae9030400.
5. Chen, J.; Ma, A.; Huang, L.; Su, Y.; Li, W.; Zhang, H.; Wang, Z. GA-YOLO: A Lightweight YOLO Model for Dense and Occluded Grape Target Detection. *Horticulturae* **2023**, *9*, 443. https://doi.org/10.3390/horticulturae9040443.
6. Liang, H.; Zhu, J.; Ge, M.; Wang, D.; Liu, K.; Zhou, M.; Sun, Y.; Zhang, Q.; Jiang, K.; Shi, X. A Comparative Analysis of the Grafting Efficiency of Watermelon with a Grafting Machine. *Horticulturae* **2023**, *9*, 600. https://doi.org/10.3390/horticulturae9050600.
7. Zhao, R.; Liao, C.; Yu, T.; Chen, J.; Li, Y.; Lin, G.; Huan, X.; Wang, Z. IMVTS: A Detection Model for Multi-Varieties of Famous Tea Sprouts Based on Deep Learning. *Horticulturae* **2023**, *9*, 819. https://doi.org/10.3390/horticulturae9070819.
8. Guo, S.; Wu, L.; Cao, X.; Sun, X.; Cao, Y.; Li, Y.; Shi, H. Simulation Model Construction of Plant Height and Leaf Area Index Based on the Overground Weight of Greenhouse Tomato: Device Development and Application. *Horticulturae* **2024**, *10*, 270. https://doi.org/10.3390/horticulturae10030270.
9. Huang, Z.; Ou, C.; Guo, Z.; Ye, L.; Li, J. Human-Following Strategy for Orchard Mobile Robot Based on the KCF-YOLO Algorithm. *Horticulturae* **2024**, *10*, 348. https://doi.org/10.3390/horticulturae10040348.
10. Zhang, X.; Kong, L.; Lu, H.; Feng, Q.; Li, T.; Zhang, Q.; Jiang, K. An Original UV Adhesive Watermelon Grafting Method, the Grafting Device, and Experimental Verification. *Horticulturae* **2024**, *10*, 365. https://doi.org/10.3390/horticulturae10040365.

11. Wang, B.; Yang, H.; Li, L.; Zhang, S. Non-Destructive Detection of Cerasus Humilis Fruit Quality by Hyperspectral Imaging Combined with Chemometric Method. *Horticulturae* **2024**, *10*, 519. https://doi.org/10.3390/horticulturae10050519.
12. Zhang, R.; Chai, Y.; Liang, X.; Liu, X.; Wang, X.; Hu, Z. A New Plant-Wearable Sap Flow Sensor Reveals the Dynamic Water Distribution during Watermelon Fruit Development. *Horticulturae* **2024**, *10*, 649. https://doi.org/10.3390/horticulturae10060649.

References

1. Kerig, P. Inaugural Editorial. *J. Trauma. Stress* **2017**, *30*, 5–10. [CrossRef]
2. Das, R.; Bhatt, S.S.; Kathuria, S.; Singh, R.; Chhabra, G.; Malik, P.K. Artificial Intelligence and Internet of Things Based Technological Advancement in Domain of Horticulture 4.0. In Proceedings of the 2023 IEEE Devices for Integrated Circuit (DevIC), Kalyani, India, 7–8 April 2023; pp. 207–211.
3. Singh, R.; Singh, R.; Gehlot, A.; Akram, S.V.; Priyadarshi, N.; Twala, B. Horticulture 4.0: Adoption of Industry 4.0 Technologies in Horticulture for Meeting Sustainable Farming. *Appl. Sci.* **2022**, *12*, 12557. [CrossRef]
4. Yağ, İ.; Altan, A. Artificial Intelligence-Based Robust Hybrid Algorithm Design and Implementation for Real-Time Detection of Plant Diseases in Agricultural Environments. *Biology* **2022**, *11*, 1732. [CrossRef] [PubMed]
5. Jawad, H.M.; Nordin, R.; Gharghan, S.K.; Jawad, A.M.; Ismail, M. Energy-Efficient Wireless Sensor Networks for Precision Agriculture: A Review. *Sensors* **2017**, *17*, 1781. [CrossRef]
6. Rötter, R.P.; Tao, F.; Höhn, J.G.; Palosuo, T. Use of crop simulation modelling to aid ideotype design of future cereal cultivars. *J. Exp. Bot.* **2015**, *66*, 3463–3476. [CrossRef] [PubMed]
7. Harfouche, A.L.J.; Jacobson, D.A.; Kainer, D.; Romero, J.C.; Harfouche, A.H.; Mugnozza, G.S.; Moshelion, M.; Tuskan, G.A.; Keurentjes, J.J.B.; Altman, A. Accelerating Climate Resilient Plant Breeding by Applying Next-Generation Artificial Intelligence. *Trends Biotechnol.* **2019**, *37*, 1217–1235. [CrossRef]

horticulturae

Review

Artificial Intelligence: A Promising Tool for Application in Phytopathology

Victoria E. González-Rodríguez, Inmaculada Izquierdo-Bueno, Jesús M. Cantoral, María Carbú * and Carlos Garrido *

Laboratorio de Microbiología, Departamento de Biomedicina, Biotecnología y Salud Pública, Facultad de Ciencias del Mar y Ambientales, Universidad de Cádiz, 11510 Puerto Real, Spain; victoriaeugenia.gonzalez@uca.es (V.E.G.-R.); inmaculada.izquierdo@uca.es (I.I.-B.); jesusmanuel.cantoral@uca.es (J.M.C.)
* Correspondence: maria.carbu@uca.es (M.C.); carlos.garrido@uca.es (C.G.)

Abstract: Artificial intelligence (AI) is revolutionizing approaches in plant disease management and phytopathological research. This review analyzes current applications and future directions of AI in addressing evolving agricultural challenges. Plant diseases annually cause 10–16% yield losses in major crops, prompting urgent innovations. Artificial intelligence (AI) shows an aptitude for automated disease detection and diagnosis utilizing image recognition techniques, with reported accuracies exceeding 95% and surpassing human visual assessment. Forecasting models integrating weather, soil, and crop data enable preemptive interventions by predicting spatial-temporal outbreak risks weeks in advance at 81–95% precision, minimizing pesticide usage. Precision agriculture powered by AI optimizes data-driven, tailored crop protection strategies boosting resilience. Real-time monitoring leveraging AI discerns pre-symptomatic anomalies from plant and environmental data for early alerts. These applications highlight AI's proficiency in illuminating opaque disease patterns within increasingly complex agricultural data. Machine learning techniques overcome human cognitive constraints by discovering multivariate correlations unnoticed before. AI is poised to transform in-field decision-making around disease prevention and precision management. Overall, AI constitutes a strategic innovation pathway to strengthen ecological plant health management amidst climate change, globalization, and agricultural intensification pressures. With prudent and ethical implementation, AI-enabled tools promise to enable next-generation phytopathology, enhancing crop resilience worldwide.

Keywords: artificial intelligence; phytopathology; emerging disease; climate change; control diseases

Citation: González-Rodríguez, V.E.; Izquierdo-Bueno, I.; Cantoral, J.M.; Carbú, M.; Garrido, C. Artificial Intelligence: A Promising Tool for Application in Phytopathology. *Horticulturae* **2024**, *10*, 197. https://doi.org/10.3390/horticulturae10030197

Academic Editors: Chenglin Wang and Lufeng Luo

Received: 24 January 2024
Revised: 16 February 2024
Accepted: 19 February 2024
Published: 20 February 2024

1. Introduction

Plant diseases have plagued agricultural crops for centuries, presenting a persistent threat to global food security [1,2]. Annually, plant diseases account for an estimated 10–16% of global crop losses, translating into profound economic impacts [3,4]. With the global population projected to reach 9.8 billion by 2050, it is imperative to increase crop yields by 25–70% to meet escalating food demands [5], emphasizing the need for revolutionary advancements in managing plant diseases.

However, the dynamics of plant pathosystems are complex, influenced by genetic and environmental factors, and challenged by the evolution of host–pathogen interactions [6,7]. These interactions have been significantly altered in recent decades due to anthropogenic factors, particularly climate change and modern agricultural practices. Climate change has been a critical driver in the emergence and spread of new plant pathogens, altering the geographical distribution of existing diseases and creating favorable conditions for the emergence of novel pathogens [8,9]. Moreover, the intensification of agricultural practices, including the use of monocultures and high-input farming systems, has reduced crop diversity, making them more susceptible to widespread disease outbreaks [10–12].

These evolving dynamics necessitate innovative solutions to expedite the discovery of knowledge in plant disease dynamics, enhance crop resilience, and understand plant–microbe interactions. Artificial intelligence (AI) offers groundbreaking avenues for deciphering the complexity of plant pathosystems and deriving practical insights for disease management [13,14]. The capacity of AI to analyze large volumes of agricultural data enables the revelation of correlations beyond human cognitive abilities [15,16]. This capability positions AI as a formidable tool in more easily unravelling the nature of plant–disease interactions. These studies involve a substantial amount of data, and AI can identify behavioral patterns in ways that are not readily discernible through purely human analysis. Furthermore, machine learning algorithms can continually self-improve, progressively facilitating the interpretation of new datasets in similar studies, while discarding data that pertain to the inherent variability of the studies [17,18].

This review has three central aims: (1) examine existing and emerging applications of AI supporting plant disease management; (2) identify current challenges and gaps hindering the adoption of AI-driven solutions; and (3) outline a roadmap for stakeholder alignment to mainstream AI in crop protection practices. By realizing these objectives through a detailed literature analysis, this review seeks to catalyze a strategic transition toward AI-enabled plant disease science and agriculture worldwide as a bridge to more sustainable food production, addressing these evolving challenges in plant pathosystems.

2. Overview of Phytopathology

Phytopathology is the scientific discipline dedicated to the study of plant diseases. This field investigates the complex interactions between plants and pathogenic organisms, shedding light on the mechanisms underlying the onset and progression of diseases. The scope of phytopathology encompasses the etiology of diseases, their epidemiology, and the development of integrated strategies for managing them in agricultural and horticultural contexts [2]. It is estimated that over 50,000 species of plant pathogens cause damage to more than 30,000 plant species [1]. These pathogens comprise various taxa, including fungi, bacteria, viruses, viroids, protozoa, and algae. Each pathogen type prompts unique disease manifestations and demands tailored investigative approaches. Furthermore, the effects of climate change, globalization, and crop intensification add complexity to deciphering modern plant disease epidemiology [8].

As a discipline so integral to food security and agricultural sustainability, the importance of phytopathology cannot be overstated. As mentioned earlier, plant diseases result in substantial economic losses in major staple crops worldwide, amounting to USD 220 billion in annual economic damages globally [3,4]. For instance, Fusarium wilt disease alone results in approximately USD 410 million in annual banana crop damages, while cassava brown streak disease incited over USD 100 million in crop damages across eastern Africa in the early 1990s [19,20]. By elucidating plant–pathogen interactions and disease epidemiology, phytopathology enables breeding disease-resistant varieties, optimizing cultural practices, and implementing integrated pest management interventions that minimize disease impacts and crop loss [6]. The development of resistant cultivars alone has saved certain crops from near extinction, as exemplified by saving papaya production in Hawaii from papaya ringspot virus in the mid-20th century [21]. A recent example of success in phytopathology is the management of coffee rust disease in Central America. Since 2012, coffee rust has significantly threatened coffee production, but the implementation of resistant varieties and improved agronomic practices has resulted in a notable recovery in affected regions [22]. Another case is the management of citrus tristeza virus in Florida, where the use of tolerant rootstocks and vector control has helped mitigate the impacts of the disease [23].

However, current disease management strategies often provide incomplete and temporary solutions in the face of an evolving pathogen landscape. In Figure 1, we present a conceptual framework that lists some of the major challenges in contemporary phytopathology, including emerging diseases, climate change, global trade and pathogen dissemination, breakdown of resistance, and data analysis and integration. The examples cited illustrate how phytopathological science must respond to specific diseases with innovations and adaptive strategies, highlighting its relevance in an ever-changing agricultural world.

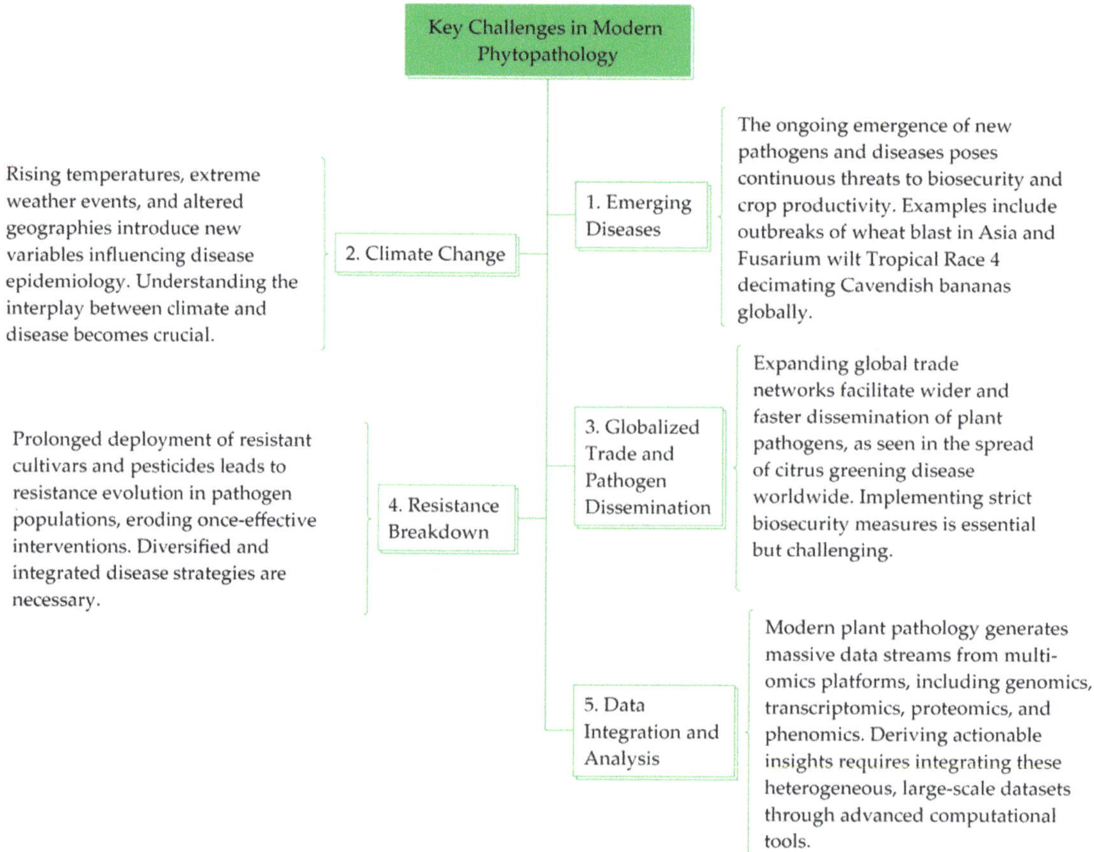

Figure 1. Several key challenges for innovation in modern phytopathology.

3. Role of Technology in Phytopathology

Historically, phytopathologists predominantly relied on conventional methods, such as visual inspection, symptomatology characterization, and pathogen isolation for plant disease diagnosis and management [2]. While these traditional techniques are valuable, they have inherent limitations, especially when considering the emerging agricultural challenges of the modern world. For instance, visual disease symptoms often do not manifest until infections are well-established, leading to delayed intervention and unchecked pathogen spread [24]. Reliance on visual symptoms alone also poses challenges in distinguishing between diseases with similar outward manifestations [25].

Traditional methods, such as pathogen isolation and culture, remain cornerstones in diagnostics. They require time-consuming processes, and obtaining pure cultures can be technically challenging [26]. Furthermore, many phytopathogenic microbes exhibit complex life cycles, switching between morphological forms, which traditional techniques often fail to detect at low pathogen levels or in identifying novel strains [20,27]. This limits their reliability and applicability in the dynamic agricultural ecosystems of today.

3.1. Advent of Emerging Technologies in Agriculture

The advent of emerging technologies and advanced analytical tools has significantly altered the agricultural landscape. Next-generation high-throughput DNA sequencing platforms, for instance, have revolutionized plant–microbiome studies, enabling the rapid genomic characterization of plant-associated microbiota and pathogens [28,29]. Metagenomic approaches have elucidated complex plant–microbe interactions, identified novel pathogens, and assessed microbiome shifts correlating with health–disease transitions. Additionally, ultra-sensitive quantitative DNA and RNA diagnostic tests now facilitate the detection of exceedingly low pathogen levels at early infection stages [25,30].

Remote sensing technologies and high-resolution spectral imaging through satellites, planes, and unmanned aerial vehicles offer large-scale capabilities in monitoring crop health and stress levels [31,32]. These tools enable the real-time, non-invasive assessment of plant vigor and the detection of disease outbreak locations in the field, facilitating timely and precise management interventions [33]. Recent advancements in nano-biosensors and lab-on-chip devices have allowed for the continuous monitoring of environmental parameters influencing disease development, such as temperature, humidity, soil water content, and microclimate conditions [34]. The integration of these sensors in agricultural ecosystems generates comprehensive datasets, shedding light on the crop–climate–disease interplay [35].

Big data analytics, automation, robotics, and artificial intelligence (AI) are accelerating a paradigm shift towards data-driven precision agriculture systems [36–38]. Phytopathology, transitioning into a highly interdisciplinary and technology-intensive science, integrates diverse data streams. Advanced computational methods offer immense promise in deriving actionable insights from the wealth of agricultural big data for efficient disease management [39].

3.2. Need for Advanced Data-Driven Solutions

While emerging technologies provide promising avenues, significant challenges persist in effectively managing diseases within the highly complex and dynamic agricultural ecosystems of today. Globalization, climate change, and intensive farming systems facilitate the increased emergence and faster evolution of plant pathogens [8,40]. Many conventional disease management approaches now face diminishing effectiveness due to rising pathogen resistance, alongside serious environmental and health concerns [11,12,41].

The complexity characterizing plant–pathogen interactions and disease epidemiology necessitates a paradigm shift towards sophisticated, integrated solutions. In this context, AI and advanced machine learning algorithms emerge as potentially transformative tools in modern data-driven phytopathology. Machine learning models can analyze vast, disparate datasets, including weather, soil, plant omics, microbiome, and pathogen genomic information [42]. These models discern subtle multivariate relationships, predict disease outbreak risks, and enable targeted intervention strategies undetectable via conventional approaches [13,14,42]. Continually learning from accumulating agricultural data streams, such AI-based systems progressively improve their predictive capabilities and decision support functionalities. Therefore, harnessing modern technology and computational innovation is imperative for developing dynamic, ecologically balanced, and economically viable plant disease management regimes, crucial in addressing the pressing food security challenges of the future [43].

4. Introduction to Artificial Intelligence (AI)

Artificial intelligence (AI) represents a transformative paradigm in computing, revolutionizing how machines perform tasks that typically require human intelligence [44,45]. In this section, we delve into the fundamental aspects of AI, tracing its evolution, understanding its basic principles, and exploring its relevance to the field of phytopathology.

4.1. Definition and Basics of AI

Artificial intelligence (AI) is the capacity of computer systems to undertake tasks that usually require human cognition, such as learning, reasoning, perception, prediction, and decision-making [46]. At its foundation, AI is about developing algorithms allowing machines to emulate aspects of human intellect, like processing and adapting to information over time. AI spans several sub-domains: machine learning uncovers data patterns without direct programming [47], computer vision gives machines sight [48], and natural language processing enables them to understand and produce human language [49]. Expert systems represent human knowledge in structured domains [50], and robotics combines these AI capabilities for environmental interaction [51–53]. AI can be general, with human-like cognitive abilities, or narrow, focused on specific tasks, where most advancements occur, revolutionizing industries with applications like smart assistants and fraud detection [54,55]. Deep learning, a machine learning subset using multi-layer neural networks, exemplifies AI's core research areas, alongside computer vision and natural language processing, pushing autonomous pattern recognition and decision-making further [47]. Table 1 presents different AI models, detailing their definitions and pivotal development dates, illuminating AI's diverse landscape. Furthermore, the integration of deep learning algorithms optimizes neural network performance, as highlighted in [56], underscoring ongoing innovations in AI's algorithmic framework.

Table 1. Different artificial intelligence models. Definition and significant dates of development are included.

Artificial Intelligence Models	Definition and Significant Dates	Reference
LLM—Large Language Model	These are systems that use large-scale neural networks to understand and generate human-like language. They excel in natural language processing tasks, such as text completion and language translation. Notable developments in large language models, especially the introduction of GPT-3, occurred around 2020–2021.	[57]
CNN—Convolutional Neural Network	A type of neural network designed for image processing and recognition. It uses convolutional layers to automatically and adaptively learn spatial hierarchies of features from input images. Proposed by Yann LeCun in the early 1990s, CNNs gained prominence in the mid-2010s with breakthroughs in image recognition tasks.	[58,59]
RNN—Recurrent Neural Network	A type of neural network architecture designed to recognize patterns in sequences of data. RNNs are well suited for tasks involving sequential data, such as time series analysis and natural language processing. While the concept of RNNs dates back to the 1980s, their resurgence and success in various applications, especially in natural language processing, gained momentum in the mid-2010s.	[60]
GAN—Generative Adversarial Network	GANs consist of two neural networks, a generator and a discriminator, which are trained simultaneously through adversarial training. GANs are used for generating new, realistic data instances, such as images. Introduced by Ian Goodfellow and his colleagues in 2014, GANs have since become a revolutionary concept in the generation of realistic data.	[61]
Decision Tree and XGBoost (eXtreme Gradient Boosting)	They are powerful models for classification, regression, and ranking tasks. Decision Trees are simple yet effective models that partition data into subsets based on feature values, using a tree-like structure of decisions and their possible consequences. XGBoost, an implementation of gradient boosted decision trees designed for speed and performance, significantly improves model accuracy by combining multiple decision trees to correct the errors of predecessors. Introduced by Chen and Guestrin in 2016, XGBoost has become a dominant force in machine learning competitions due to its efficiency and effectiveness.	[62,63]

Table 1. *Cont.*

Artificial Intelligence Models	Definition and Significant Dates	Reference
ElasticNet, Lasso, and Ridge Regression	They are regularization techniques in linear regression that address overfitting by penalizing the size of coefficients. ElasticNet combines the properties of both Lasso (Least Absolute Shrinkage and Selection Operator) and Ridge Regression by integrating their penalty terms; it is particularly effective when dealing with highly correlated data. Lasso contributes to feature selection by reducing the coefficients of less important features to zero, while Ridge Regression shrinks the coefficients but does not set them to zero. These methods were developed in the early 21st century, with ElasticNet introduced by Zou and Hastie in 2005, offering a bridge between Lasso's feature selection and Ridge's coefficient shrinkage.	[64]
Random Forest	It is an ensemble learning method renowned for its versatility and accuracy in classification and regression tasks. By constructing multiple decision trees at training time and outputting the mode of the classes (for classification) or mean prediction (for regression) of the individual trees, Random Forest mitigates the overfitting problem common to single decision trees. This model's significant development dates back to the early 2000s, with Breiman's seminal paper in 2001 laying the foundational framework for Random Forest algorithms.	[65]
SVM—Support Vector Machine	A supervised machine learning algorithm used for classification and regression analysis. SVMs are effective in high-dimensional spaces and are particularly useful in tasks like image classification and handwriting recognition. Proposed by Vladimir Vapnik and Corinna Cortes in the 1990s, SVMs gained popularity in the early 2000s and became a staple in machine learning applications.	[66,67]
KNN—k-Nearest Neighbors	A simple and effective algorithm used for classification and regression tasks. KNN makes predictions based on the majority class or average of the k-nearest data points in the feature space. KNN is a classical algorithm, and its principles have been known for decades. It is widely applied in various fields since the 1960s.	[68,69]
DNN—Deep Neural Network	A neural network with three or more layers, including an input layer, one or more hidden layers, and an output layer. Deep neural networks are capable of learning intricate representations and are used in various applications. While the concept of deep neural networks has roots in the 1960s, their resurgence and practical success came in the mid to late 2000s with advancements in training algorithms and hardware.	[70]
MLP—Multilayer Perceptron	MLP is an artificial neural network model consisting of an input layer, multiple hidden layers, and an output layer, with each layer fully connected to the next. It employs backpropagation for learning, allowing it to model complex non-linear relationships. Developed in the 1980s, MLPs are versatile in applications ranging from pattern recognition to classification and regression tasks, marking a significant advance in the field of deep learning.	[71]
SGD—Stochastic Gradient Descent	It is an optimization algorithm pivotal for training a broad spectrum of artificial intelligence models, notably in deep learning. It optimizes model parameters by calculating gradients based on randomly selected data subsets, enhancing training efficiency across large datasets. Introduced in the context of machine learning in the late 20th century, its conceptual roots trace back to Robbins and Monro's stochastic approximation method in 1951, laying the theoretical groundwork for iterative stochastic optimization techniques in AI.	[72]

Table 1. *Cont.*

Artificial Intelligence Models	Definition and Significant Dates	Reference
LSTM—Long Short-Term Memory	A type of recurrent neural network architecture designed to overcome the limitations of traditional RNNs in capturing long-term dependencies in sequential data. LSTMs are widely used in natural language processing and speech recognition. Proposed by Sepp Hochreiter and Jürgen Schmidhuber in 1997, LSTMs became popular in the mid-2010s, addressing challenges in capturing long-term dependencies.	[60,73]
RL—Reinforcement Learning	An area of machine learning where an agent learns to make decisions by interacting with an environment. The agent receives feedback in the form of rewards or penalties, allowing it to learn optimal strategies over time. RL has a history dating back to the 1950s and 1960s, but recent advancements, especially in deep reinforcement learning, have gained prominence since the mid-2010s.	[74,75]
BERT—Bidirectional Encoder Representations from Transformers	A pre-trained natural language processing model based on transformer architecture. BERT is particularly effective in understanding the context of words in a sentence and is used for various language-related tasks. Introduced by Google AI in 2018, BERT brought a breakthrough in natural language processing by capturing contextual information bidirectionally.	[76]

4.2. *Evolution of AI*

The evolution of AI began with early conceptualizations by figures like Ada Lovelace and Alan Turing, progressing through the 1950s with attempts at creating intelligent machines [46] (Figure 2). Despite initial successes, challenges led to periods of stagnation, known as "AI winters" [77]. The 21st century marked a resurgence, fueled by advancements in computational power, data generation, and machine learning, leading to breakthroughs in areas like vision and speech [47]. Recent developments in natural language processing indicate significant advancements in AI's ability to understand and generate human-like language [78], hinting at future communicative AI systems' potential [79–81] and supported by large-scale data and computational resources [82]. Rapid innovation continues toward safer and more robust language models aligned with human values [83]. The historical context of AI's development underscored by data-driven approaches is detailed in Figure 2.

4.3. *Relevance of AI in Various Fields*

In the realm of artificial intelligence (AI), its transformative impact extends far beyond theoretical frameworks, finding tangible applications in diverse fields. AI is making significant strides, revolutionizing industries and scientific endeavors.

4.3.1. Healthcare: Enhanced Diagnostics and Personalized Medicine

In the healthcare sector, AI is significantly improving clinical diagnostics and personalizing medicine. It utilizes advanced algorithms for medical image analysis and genomic pattern recognition, enabling the early detection of conditions not visible through conventional methods [84]. In digital pathology, AI predicts the risk of cancer metastasis earlier than traditional clinical indicators [85–87]. For drug discovery, deep learning has shortened the timeline for developing new medications [88]. AI also customizes treatment plans to individual genetic profiles, enhancing patient outcomes [89]. This integration of AI is transforming healthcare diagnostics, treatments, and drug development processes.

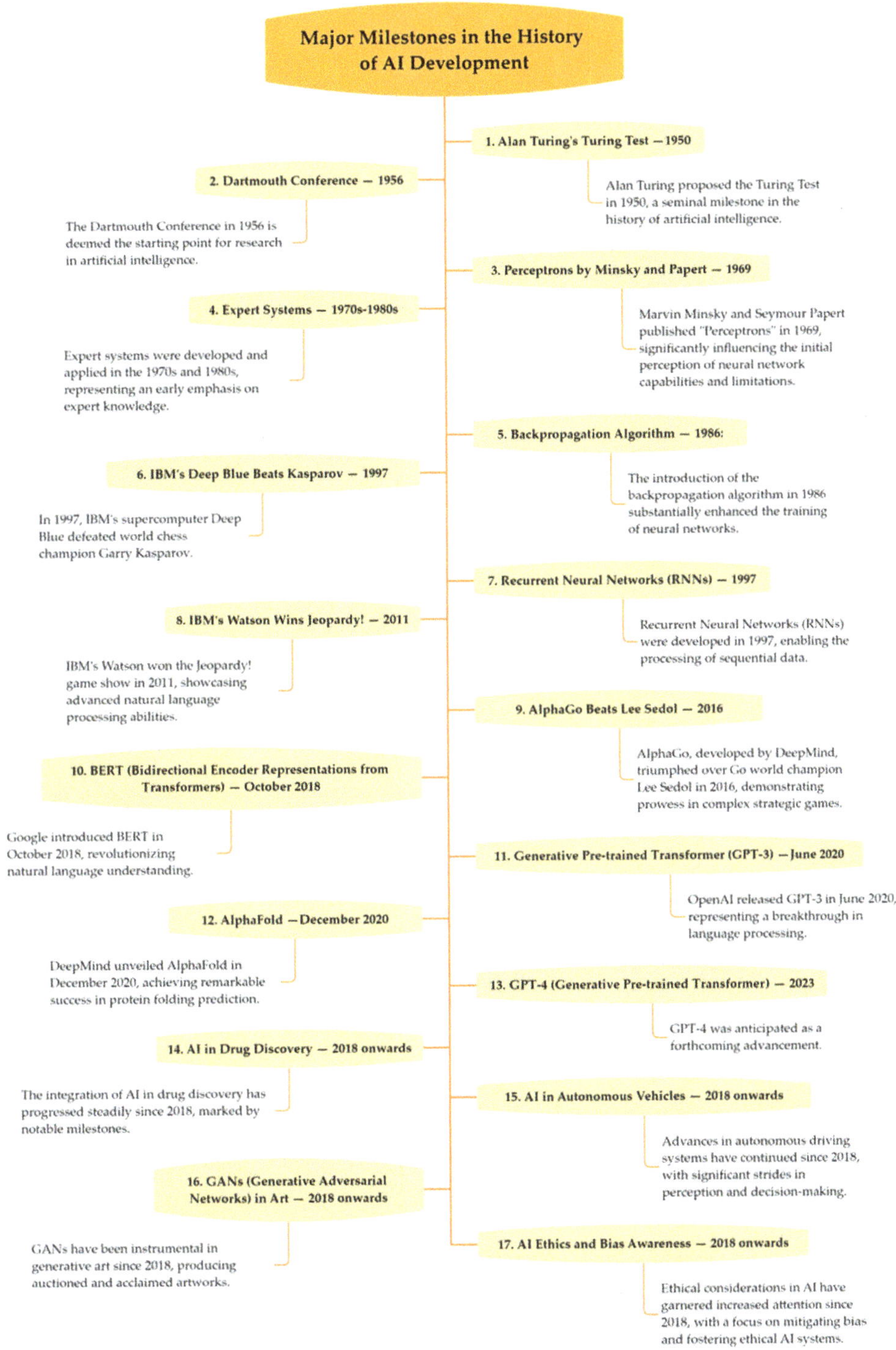

Figure 2. Major milestones in the history of AI development.

4.3.2. Engineering: Optimizing Complex Systems

In the field of engineering, AI plays a crucial role in optimizing complex systems through autonomous design enhancements, predictive maintenance, and self-adaptive mechanisms. It has made significant strides in improving aerodynamic designs, reducing the development time for aircraft wing prototypes [90], and employing AI vision systems for the early detection of structural weaknesses, such as bridge cracks, to facilitate timely repairs [91]. Moreover, neural networks allow electronics to dynamically reconfigure circuitry, enhancing resilience and operational efficiency despite component damages [92,93], underscoring AI's profound impact on fostering innovation and efficiency in engineering.

4.3.3. Business: Data-Driven Decision Automation

In the business sector, AI is revolutionizing decision-making by automating complex analyses previously dependent on human judgment. It leverages advanced algorithms to process diverse datasets, from historical records to market trends, enhancing risk management and optimizing processes [94]. In finance, AI, through deep reinforcement learning, outperforms human stock trading strategies [95]. Additionally, AI rapidly analyzes complex legal documents and drives business analytics and marketing decisions, offering strategies that significantly improve efficiency and outcomes [96,97]. This marks AI's significant role in advancing business practices through data-driven insights.

4.3.4. Transportation: Optimized Mobility

In transportation, AI enhances urban mobility and develops autonomous driving technologies. Adaptive traffic systems, using deep learning, coordinate signals to reduce congestion, significantly improving travel times [98]. Autonomous vehicles combine sensor data for real-time environmental perception, crucial for safe navigation, including cameras, LiDAR, radar, GPS, wheel odometry, and IMUs [99,100]. The integration of sensors, software, and AI computational power is indeed enhancing the safety of autonomous vehicles [101]. For instance, Waymo's autonomous cars have covered over 20 million miles, demonstrating sophisticated navigation capabilities [102]. Recent studies focus on integrating AI into automotive manufacturing and navigation systems, promising safer, more efficient transportation solutions [103,104].

4.3.5. Space Exploration: Autonomous Exploration and Data Analysis

AI is transforming space exploration by enabling autonomous operations and advanced data analysis. Machine learning equips spacecraft and rovers, like NASA's Mars rovers, to navigate and collect data independently, optimizing paths in real time. Research expands AI's role in autonomous spacecraft technology and space law [105,106], with a focus on developing trusted AI systems for mission autonomy [107]. These AI advancements promise to significantly enhance the efficiency, autonomy, and intelligence of space missions, marking a new era in space exploration and analysis.

4.3.6. Education: Personalized Learning and Student Support

AI is revolutionizing education sector by facilitating personalized learning and supporting students. It employs algorithms in adaptive platforms to customize educational content for individual learners, enhancing their educational journey. AI-driven chatbots offer instant assistance and guidance. For example, Duolingo uses AI to adjust language lessons based on user progress [108]. Research in this area includes analyzing AI's impact on personalized learning [109], studying AI's role in learning methodologies [110], and employing machine learning to identify learning styles [111], illustrating AI's transformative potential in education.

5. Applications of AI in Phytopathology

Artificial intelligence (AI) is transforming approaches in phytopathology, catalyzing innovations in understanding, managing, and mitigating plant diseases. AI's capacity to analyze vast datasets reveals subtle correlations in plant–pathogen interactions, granting key insights for disease control [16]. This section surveys prominent applications of AI across major facets of phytopathology.

5.1. Disease Detection and Diagnosis

Artificial intelligence (AI) enables rapid and precise disease detection and diagnosis, overcoming the limitations of techniques reliant on visual inspection. Numerous studies demonstrate the efficacy of AI in accurately diagnosing complex diseases. In an early example, Ramcharan et al. [14] applied deep learning techniques for detecting and diagnosing cassava diseases through image analysis. Using a convolutional neural network, they achieved diagnostic accuracy above 90%, demonstrating deep learning's effectiveness in identifying various cassava diseases. This approach not only surpassed traditional methods in terms of accuracy and speed but also enabled the implementation of these models on mobile devices, facilitating diagnosis in the field.

Similarly, Fuentes et al. [112] implemented three artificial intelligence architectures—Faster R-CNN, SSD, and R-FCN—to detect and diagnose diseases and pests in tomato plants. These architectures fall within the broader context of convolutional neural networks (CNNs), which are particularly suited for image recognition tasks due to their ability to learn spatial hierarchies of features from input images (see information in Table 1). The application of these models in the study marked a significant advancement in the application of CNNs in image recognition tasks since their proposal in the 1990s. The authors of [112] used images captured by cameras at various resolutions, both of healthy plants and plants with symptoms. With these images, they trained the artificial processing models, which significantly improved the accuracy in disease and pest recognition and reduced false positives during the training phase. This systematic approach allowed the AI system to effectively recognize nine different types of diseases and pests in tomato plants, demonstrating the capability of these models to handle complex environmental variables present in a plant's surroundings. Following in the footsteps of these works, but not focused on a specific plant species, Sladojevic et al. [113] also used deep convolutional neural networks (CNNs), training the artificial model with an extensive database, which allowed it to distinguish between different types of diseases in the leaves of various genera and species. The novelty and advancement of the developed model lie in its simplicity, where healthy leaves and background images are aligned with other classes, allowing the model to distinguish between diseased and healthy leaves or their surroundings using deep CNNs. The experimental results showed an accuracy of between 91% and 99% in separate class tests and an overall accuracy of 95.8% in the trained model. These studies are a clear example of CNNs' ability to handle the complexity of visual data and improve the accuracy of automated diagnosis [113].

Recent studies have continued to demonstrate AI's potential in plant disease detection and diagnosis using more modern, precise, and powerful models thanks to the development of new AI capabilities. In 2022, Arinichev [114] explored the use of artificial intelligence technologies for diagnosing fungal diseases in cereals, specifically in wheat and rice, through methods of vision and automated recognition. This analysis revealed that artificial neural networks have the capability to detect and classify disease patterns, such as yellow spots, yellow and brown rust, and brown spots, with classification metrics ranging between 0.95 and 0.99. To advance in this line of research, Arinichev examined four well-established and relatively light convolutional neural network (CNN) architectures, namely, GoogleNet, ResNet-18, SqueezeNet-1.0, and DenseNet-121, with the DenseNet-121 model particularly standing out for its optimal combination of high precision and operational efficiency. Characterized by a relatively low number of parameters and a file size suitable for mobile devices, this model achieved exceptionally high classification accuracy, surpassing the other evaluated models. Similar to previous research, such as that of Ramcharan et al., the implementation of a light neural network like DenseNet-121 facilitated its application in the field on mobile devices, allowing for quick and accurate diagnostics [114].

In the case of the study carried out by Feng et al. [115], the authors developed a convolutional neural network model for potato late blight detection method using deep learning, with high accuracy and fast inference speed, using a dataset of potato leaf disease images in single and complex backgrounds. Feng et al. used the ShuffleNetV2 2× model, characterized by its high classification accuracy, while also having a larger parameter scale and memory space compared to other models with equal accuracy. The authors improved the model through strategies that included introducing an attention module, reducing network depth, and minimizing the number of 1×1 convolutions. This resulted in an enhancement of classification accuracy while simultaneously maintaining efficient inference speed on CPUs in the devices used for its application. In the same line of work, Bracino et al. [116] carried out a study focus on the non-destructive classification of paddy rice leaf diseases using deep learning algorithms such as EfficientNet-b0, MobileNet-v2, and Places365-GoogLeNet. They aim to identify whether the rice paddy leaf is normal or infected with various diseases including bacterial leaf blight (BLB), bacterial leaf streaks (BLS), bacterial panicle blight (BPB), heart, downy mildew, hispa, or rice tungro disease (RTD). Of the models used, EfficientNet-b0 was identified as the most effective, achieving an average accuracy of 97.74%. This model is distinguished by its focus on maximizing efficiency, optimally balancing network depth, width, and the resolution of input images through a compound scaling technique, resulting in superior performance with minimal memory requirements and floating-point operations per second (FLOPS). This efficiency and precision capability distinguish it from the model used by Feng et al., the ShuffleNetV2 2×, which, although highly precise, focuses on improving inference speed and reducing parameter size through the introduction of an attention module and the optimization of the network architecture. Bracino et al.'s significant contribution lies in providing a precise and non-destructive diagnostic method for rice diseases, thereby supporting the prevention of product loss and improving crop quality through the application of advanced and efficient AI technologies.

A deep convolutional neural network model was also developed by Jouini et al. [117] to detect wheat leaf rust. The authors advanced the application of a CNN by developing a model specifically designed for the detection of wheat leaf diseases using hyperspectral images, achieving an impressive testing accuracy of 94%. This study showed the feasibility of real-time disease detection in wheat, a critical advancement for resource-constrained environments where timely and effective disease management is vital [117]. In a related study, Zhou et al. [118] introduced a novel spectral feature pseudo-graph-based residual network (SFPGRN) for the spectral analysis of plant diseases. Their method innovatively constructs a residual network model using a characteristic surface derived from natural neighborhood interpolation based on preprocessed near-infrared spectral reflection signals and first-order differential spectral index, achieving a classification accuracy of 93.21% on a dataset of apple leaf diseases and insect pests [118]. Complementing these developments, Shi et al. [93] introduced a novel fast Fourier convolutional deep neural network (FFCDNN) designed for the accurate and interpretable detection of wheat yellow rust and nitrogen deficiency from Sentinel-2 time series data. The FFCDNN model stands out for its innovative use of a fast Fourier convolutional block and a capsule feature encoder, significantly enhancing computing efficiency and model interpretability. This approach not only achieves high classification accuracy but also provides insights into the host–stress interaction, marking a significant advancement over previous studies by integrating spatial-temporal information for global feature extraction [93].

In recent times, the research group of Hassan et al. [119] introduced a groundbreaking CNN architecture for plant disease identification, leveraging inception layers and residual connections to enhance feature extraction, while employing depth wise separable convolution to significantly reduce computational complexity. This model is distinct in its ability to achieve high accuracy across various plant disease datasets with a markedly lower parameter count, illustrating a significant advancement in AI's application to phytopathology. By optimizing the model to require fewer computational resources, this work facilitates the deployment of AI technologies on devices with limited processing capabilities, making sophisticated disease diagnosis tools more accessible to a broader range of users and applications [119].

A new evolution of CNN is the Siamese convolutional neural network (SNN). Narain et al. [120] introduced an enhanced approach to detection systems by implementing a SNN for identifying diseases in tomato leaves. Siamese neural networks stand out from conventional CNN models due to their unique structure, designed to learn to differentiate between pairs of inputs, making them exceptionally suitable for comparison and differentiation tasks. By evaluating similarities or differences between pairs of images, SNNs can offer notable accuracy in disease classification, often overcoming challenges faced by traditional CNNs in terms of intraclass variability and the scarcity of labeled data (Figure 3). In this work, Narain et al. developed a customized SNN by training with a specially collected dataset of 155 tomato leaf images, and the system demonstrated high efficacy, achieving an accuracy of 83.749% in training and 80.4% in testing. This improvement in disease classification represents a significant advancement over more classic CNN models. The implementation of Siamese networks signifies an optimization in the accuracy and efficiency of disease detection in crops, allowing for the application of appropriate management measures more quickly and accurately by providing a more robust and adaptable mechanism for recognizing complex patterns associated with various plant diseases [120].

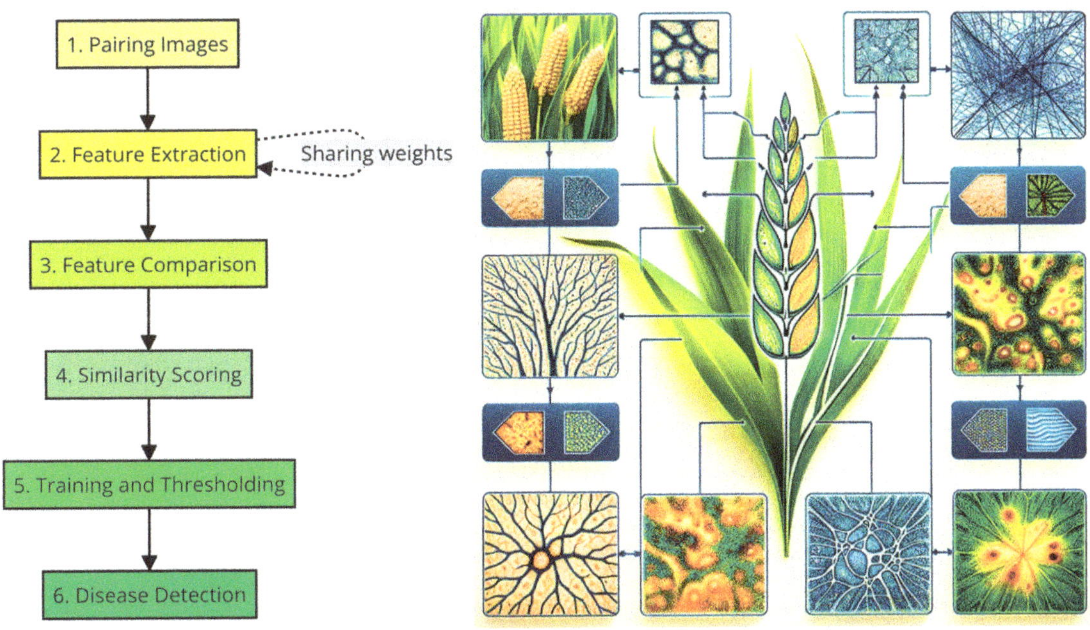

Figure 3. Graph diagram illustrating the operation of Siamese convolutional neural networks (SNNs) for detecting plant diseases from an image dataset.

Siamese convolutional neural networks operate into a sequence of steps illustrated in Figure 3: (a) Pairing Images: a set of image pairs of plants is created, where each pair consists of two images: it could be one of a healthy plant and one of a diseased plant, or two healthy plants, or two diseased plants. (b) Feature Extraction: each image in the pair is fed through a convolutional network that acts as a feature extractor. The key here is that both images go through the same network (sharing weights), ensuring that features are extracted uniformly. (c) Feature Comparison: the features extracted from each image are combined and fed into a layer that compares the two images. This comparison could be an absolute difference, a concatenation operation followed by dense layers, or even a more complex metric. (d) Similarity Scoring: the network produces a score that reflects the similarity between the two images. In the context of plant disease detection, a high score might indicate that both images are of plants in the same condition (both healthy or both diseased), while a low score might suggest one is healthy and the other is diseased. (e) Training and Thresholding: during training, the network learns what features are important for distinguishing between healthy and diseased plants. A similarity threshold is adjusted that best separates pairs of healthy plant images from pairs with at least one diseased plant. (f) Disease Detection: once trained, the network can take a pair of images, process them through the network to obtain the similarity score, and using the learned threshold, determine if the plants are healthy or diseased.

Other authors are making significant advances in developing models which are more advanced in capabilities and simpler in their handling, thanks to the evolution that vision systems and their conjunction with large language modeling systems are undergoing in recent months. In this line, Tabbakh and Barpanda [121] introduced an innovative hybrid model for the classification of plant diseases, through the integration of Transfer Learning with a Vision Transformer (TLMViT). This hybrid approach stands out for its unique ability to extract and analyze deep features of plant leaf images, achieving exceptionally high accuracy in the evaluated datasets. The TLMViT is a key innovation in this study, leveraging the architecture of transformers, which has revolutionized natural language processing, to apply it in the realm of computer vision. Vision transformers adapt the concept of attention, allowing the model to focus on the most relevant parts of the image for the classification task, significantly improving accuracy and efficiency in disease identification. Tabbakh and Barpanda used a dataset freely available in the PlantVillage project, as the authors comment. This dataset encompasses more than 54,000 images of more than 38 different crop species, with a particular focus on cassava, tomato, pepper, and potato. Each image within the dataset is labeled with the plant species and, if present, the disease. This resource is freely available for computer vision and deep learning tasks, such as image classification, object detection, and semantic segmentation. In the specific research of Tabbakh et al., three different crops from the PlantVillage dataset (pepper, potato, and tomato) were used, which include 20,638 images of diseased and healthy leaves. The application of their model managed to achieve identification accuracies above 98%. This hybrid model, combining transferred learning with the power of vision transformers, illustrates a qualitative leap in the detection and classification of plant diseases, offering new perspectives for precision agriculture and sustainable crop management [121].

The integration of advanced AI models, from deep convolutional neural networks to Siamese networks and vision transformers, underscores a transformative period in the field of phytopathology. These studies collectively represent a leap forward in precision phytopathology, offering not just higher accuracy in disease diagnosis but also a model for future research to build upon. Particularly, the adoption of vision transformers marks a novel approach, leveraging the strengths of AI to address complex agricultural challenges. This evolution of AI methodologies, characterized by increased model sophistication and adaptability, promises to significantly enhance disease detection capabilities, paving the way for more targeted and effective disease management strategies.

5.2. *Advancements in Plant Disease Propagation Modeling*

The field of plant disease propagation modeling has witnessed transformative growth through the incorporation of artificial intelligence (AI) and machine learning techniques, opening new vistas in pathogen prediction and management. A pivotal approach in this field is the application of machine learning models for disease prediction based on symptoms and environmental data. In this context, the existing scientific literature encompasses a variety of meticulously developed strategies that significantly contribute to the advancement of predictive model development in the field of plant pathology.

During 2022 several studies were published to apply different algorithms and predictive models using AI. For example, a very interesting work is that published by Garrett et al. [39] in which the authors utilized Random Forest and Support Vector Machines (SVMs) to analyze the complex interplay between climate change and pathogen emergence. Random Forest is an ensemble learning method for classification, regression, and other tasks. It operates by constructing a multitude of decision trees during training for more accurate and robust predictions (consult Table 1 and Figure 4). SVMs, in contrast, are powerful supervised machine learning models used for classification and regression challenges, effectively handling high-dimensional spaces and complex datasets. Employing these algorithms, Garrett et al. aimed to capture and model the nuanced relationships between environmental factors and the likelihood of pathogen spread, underscoring the potential of AI to offer predictive insights into plant disease dynamics influenced by climate variables. Their methodology showcases the strengths of combining multiple AI approaches to enhance predictive accuracy and provide actionable insights into pathogen management strategies [39]. These modeling approaches were similarly utilized by Otero et al. [122], who delved into the creation of data-driven predictive models utilizing artificial intelligence to anticipate the occurrence of *Plasmopara viticola* and *Uncinula necator* in the viticultural regions of Southern Europe. Otero et al. employed a variety of machine and deep learning algorithms, including Logistic Regression, Decision Trees, Random Forest, Gradient Boosting, K-Nearest Neighbors, Naïve Bayes, Support Vector Machines, and Deep Neural Networks. Logistic Regression provides a probabilistic approach for binary outcomes, making it suitable for disease presence predictions. Decision Trees offer clear, interpretable decisions. Random Forest improves on Decision Trees by combining multiple trees to reduce overfitting. Gradient Boosting sequentially corrects errors, enhancing model performance. K-Nearest Neighbors classifies based on the majority vote of nearest data points, offering simplicity and effectiveness. Naïve Bayes, based on Bayes' theorem, excels in classification with an assumption of feature independence. Support Vector Machines efficiently handle high-dimensional data, optimizing margins between classes for clear decision boundaries. Notably, the models employed by Otero et al. achieved over 90% accuracy for infection risk and over 80% for treatment recommendations, highlighting the potential of AI in enhancing disease management strategies in vineyards across Southern Europe [122].

Collaboration and innovation in AI and cloud-based platforms are charting new paths in the monitoring and forecasting of plant diseases. The study published by Lavanya and Krishna [123] has developed a collaborative AI and cloud-based platform for plant disease identification, tracking, and forecasting. This innovative approach merges a mobile application with AI algorithms, providing real-time disease diagnostics and disease density mapping. This collaborative and technology-driven approach reflects a shift towards more integrated and interconnected systems for plant disease management, akin to the initiatives by Otero et al. [122] and Zen et al. [124]. The study by Zen et al. focuses on developing an AI-based mobile application for detecting plant diseases with high accuracy, utilizing CNN and RNN with TensorFlow.js. TensorFlow.js is an open-source library developed by Google for machine learning in JavaScript. It enables the training and deployment of machine learning models directly in the browser or on Node.js. TensorFlow.js provides a flexible and efficient platform for building and executing machine learning algorithms on web-based applications, allowing for interactive and real-time applications of AI technologies without the need for backend servers. The application in this study was tested on tomato plant diseases, achieving prediction accuracies of 100% for early blight, 90% for bacterial spot, and 100% for both healthy and late blight conditions. This research showcases the application's capability to recommend treatment options based on image analysis, offering a significant tool for farmers to identify and manage plant diseases effectively [122–124].

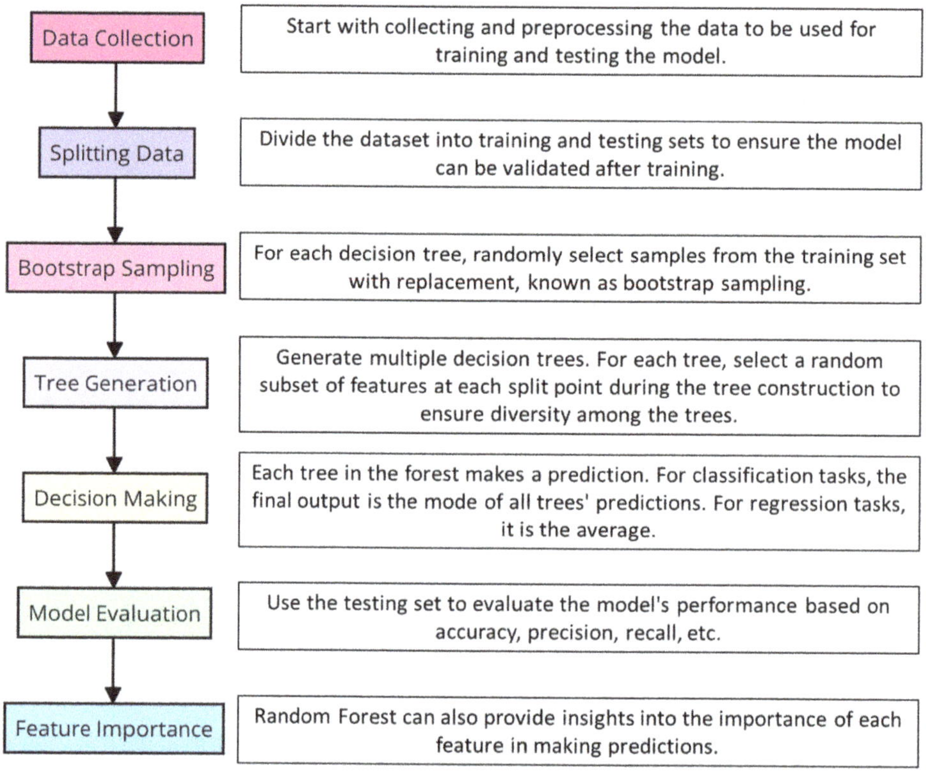

Figure 4. Graph diagram illustrating how Random Forest system works.

During 2023, in the realm of early disease detection and prediction in plants, technological advancements, particularly in Enhanced Data rates for GSM Evolution (EDGE) and deep learning, have played a pivotal role. Marco-Detchart et al. [125] focuses on the development of a robust, multi-sensor consensus approach for plant disease detection using the Choquet Integral. The Choquet Integral is a mathematical concept used in decision-making and information aggregation integrated in an Edge-AI device. Edge-AI refers to the deployment of AI applications directly on devices located at the "edge" of the network, rather than relying on centralized cloud services. This means that computations are performed close to where data are generated, such as in smartphones, surveillance cameras, or IoT devices, which allows for faster processing times, reduced bandwidth costs, and improved data privacy. This device was designed to improve disease classification by capturing multiple images of plant leaves and applying data fusion techniques. The system demonstrated increased robustness in classification responses to potential plant diseases, leveraging deep learning models for better analysis and classification. This innovatively implemented a multi-sensor consensus approach for plant disease detection, monitoring and prediction demonstrating efficacy surpassing traditional single-camera setups. Complementing this, Ojo and Zahid [126] have refined deep learning classifiers for plant disease detection by adeptly applying image preprocessing techniques and addressing class imbalance issues. Ojo and Zahid focused on enhancing deep learning classifiers for plant disease detection by addressing data imbalances and applying image preprocessing techniques. They used techniques like Adaptive Histogram Equalization (AHE), Contrast Limited Adaptive Histogram Equalization (CLAHE), and image sharpening to improve image quality. They tackled class imbalance with methods like Synthetic Minority Oversampling Technique (SMOTE), Major-to-Minor Translation (M2m), and Generative Adversarial Networks (GANs). These studies collectively highlight the critical role of cutting-edge technology in efficient disease management and resource optimization in agricultural sectors [125,126].

The study by Vardhan et al. [127] takes an innovative approach to plant disease detection and monitoring using drone-captured imagery. They developed a comprehensive database from online sources, categorizing various plant species and diseases for analysis. This database was crucial for testing the accuracy and reliability of their CNN-based model. Their methodology emphasizes the use of CNNs due to their effectiveness in complex categorization and detection challenges, especially under varied imaging conditions. Additionally, they introduced a prototype drone equipped with a high-resolution camera for live field monitoring, showcasing the practical application of their research in real agricultural settings. This integration of advanced imaging techniques and AI algorithms represents a significant step forward in agricultural technology, offering a more efficient and accurate method for plant health assessment and disease management. In relation, the study published by Dagwale et al. [128] showed the YOLOv5 model, an advanced neural network architecture for real-time object detection, to accurately predict leaf species and diseases across various plant types using the PlantDoc dataset. YOLOv5 (You Only Look Once version 5) is designed for rapid image processing, identifying, and classifying multiple objects simultaneously with high precision. This integration showcases the potential of leveraging cutting-edge AI technologies like YOLOv5 to enhance disease detection accuracy in agriculture, marking a significant advancement in plant pathology diagnostics and disease spread and monitoring [128].

It is notably how neural networks, especially convolutional neural networks (CNNs), are emerging as the predominant technique for classifying plant diseases, thanks to their inherent flexibility and automatic feature extraction capabilities [124,127]. The developments shown in this subsection mark a milestone at the confluence of advanced technology and agronomy, heralding a new era in plant disease management. The fusion of machine learning techniques with cloud-based collaborative platforms is redefining the approach of farmers and scientists to plant disease challenges. These advancements not only enhance accuracy in disease detection and management but also facilitate a prompter and effective response, crucial for global sustainability and food security.

5.3. Comprehensive Evaluation and Prospects of AI Technologies in Phytopathological Applications

In the rapidly evolving domain of AI-assisted plant disease management, the integration of pretrained models, such as ResNet-18 and ResNet-50, marks a significant leap towards refining disease detection and diagnostic accuracy. These models, part of the Residual Networks (ResNets) introduced to mitigate the vanishing gradient problem in deep convolutional neural networks (CNNs), incorporate "shortcut connections" that allow gradients to flow through the network without undergoing non-linear transformations, thereby facilitating the training of much deeper networks. ResNet-50, a 50-layer CNN comprising 48 convolutional layers, one MaxPool layer, and one average pool layer, employs a "bottleneck" design in each residual block to reduce the number of parameters and accelerate layer training. This bottleneck design, featuring a stack of three layers instead of two, utilizes 1×1 convolutions to compress and then expand the number of channels, significantly lowering computational complexity while maintaining or enhancing model performance [129].

Originally trained on the expansive ImageNet dataset, these models exhibit exceptional prowess in feature recognition, offering tailored solutions for the nuanced challenges of phytopathology. Their capability to discern complex image characteristics with remarkable precision positions them as indispensable tools for identifying plant pathologies, often surpassing traditional visual inspection methods with accuracies ranging between 95% and 97%. Furthermore, the application of these pretrained models extends beyond disease identification to encompass broader spatial analyses, as evidenced by their deployment within the ArcGIS ecosystem for tasks like land cover classification and aerial feature extraction, underscoring their potential to revolutionize the monitoring and management of plant health on a large scale [130].

The comparative analysis of AI models, including CNNs, YOLOv5, and MobileNet, illuminates the diverse applicability and efficacy of these technologies in phytopathology. Each model, with its unique strengths (be it CNNs for their image processing capabilities, YOLOv5 for its rapid processing speed facilitating timely interventions, or MobileNet for offering an efficient solution on low-power devices), advances our capacity to manage plant disease spread through predictive analyses that integrate environmental and symptomatic data. This synthesis not only augments diagnostic precision but also enhances proactive disease management strategies. Nevertheless, the practical application of these models encounters challenges such as the need for image preprocessing and handling unbalanced datasets, propelling the pursuit of technological innovations, especially in the realm of Edge-AI devices. These advancements promise a transformative impact on disease monitoring, enabling more accurate and accessible diagnostics.

As AI technologies continue to evolve, alongside breakthroughs in sensor technologies, we are ushered towards a new era of integrated and automated plant disease management. This journey is not without its hurdles, necessitating innovative approaches like transfer learning and the development of multisensorial detection systems to overcome current limitations. The ongoing exploration and refinement of AI models in phytopathology not only pave the way for future research directions but also highlight the pivotal role of AI in crafting sustainable, precision-based solutions for global agricultural challenges.

6. Applications of AI in Precision Agriculture and Management

The advent of artificial intelligence (AI) in precision agriculture marks a significant milestone in the evolution of precision farming, offering a promising avenue for enhancing yields while minimizing ecological impacts. The advantages of AI are not only related to phytopathology; groundbreaking developments in AI, such as advanced robotic weeders equipped with computer vision, have substantially reduced herbicide usage, exemplifying a move towards more sustainable farming practices [131]. Machine learning models that integrate weather, soil, and crop data have become increasingly sophisticated, aiding farmers in making well-informed decisions about irrigation, fertilization, and harvesting [132]. The democratization of AI through affordable solutions is further expanding the accessibility and effectiveness of precision farming [133].

The role of AI in agriculture, as outlined by Buja et al. [134], emphasizes the importance of early detection and rapid, accurate diagnostics for managing phytopathological challenges. This progress, marked by the application of nanotechnologies and the integration of the Internet of Things (IoT), is revolutionizing preventive strategies in combatting phytopathogens and precision agriculture. Liakos et al. [15] provide a comprehensive review of machine learning applications in agriculture, demonstrating how AI, combined with sensor data, is transforming farm management systems into real-time, intelligent platforms. These platforms offer insightful recommendations, significantly aiding in farmer decision-making across various aspects of agriculture, including crop, livestock, water, and soil management.

Kumar et al. [135] introduced DeepMC, a deep learning-based microclimate prediction framework utilizing IoT data, which exemplifies the potential of AI in enhancing precision agriculture. DeepMC's innovative approach to predicting a range of climatic parameters, including soil moisture, humidity, and temperature, offers accurate forecasts crucial for agricultural decision-making. The integration of AI in image processing has made significant contributions to precision agriculture. Studies by G S. and Rajamohan [136] and Sasikala D. and Sharma K. [137] demonstrate how AI-driven image processing technologies improve crop monitoring and management, further bolstering the efficiency and accuracy of agricultural practices.

Furthermore, the work of Joseph R.B. et al. [138] and Arokia Raj V.H. and Xavier de Carvalho C. [139] highlights the integration of AI in agricultural automation and agrometeorology, respectively. These studies underscore AI's potential in enhancing the efficiency of agricultural products and in offering model-based decision support systems that unite AI with precision agriculture.

Lastly, Williams et al. [140] developed the AI2Farm model, a machine learning-based approach that analyzes the impact of global and domestic events on agricultural production, consumption, and pricing. This model represents a significant advancement in precision agriculture by providing farmers with tools to adapt to both conventional and unconventional challenges in agriculture.

In summary, the integration of AI into precision agriculture and management marks a transformative shift in modern farming encompassing sustainable practices, advanced diagnostics, data-driven decisions, and innovative technologies. These developments are crucial to meet escalating food demands while maintaining ecological balance. Concurrently, manifold AI applications in detection, forecasting, precision management, and monitoring are transforming phytopathology. As these techniques mature, AI-enabled tools promise to strengthen global food security and agricultural sustainability amidst evolving plant disease challenges. Taken together, the advent of precision agriculture powered by AI constitutes a strategic innovation pathway for next-generation phytopathology and plant protection practices worldwide. With prudent and ethical implementation, data-driven smart farming technologies can enable the sustainable intensification of crop productivity to feed rising populations in the face of climate change and agricultural pressures.

7. Integration Challenges and Ethical Considerations

7.1. Technical Barriers to AI Implementation

While artificial intelligence promises transformative phytopathology innovations, prudent precautions are necessary for its successful integration into agricultural systems. Technical barriers persist in developing robust, reliable AI solutions for real-world plant disease environments [47]. A key limitation of many current machine learning models is their narrow focus on specific crops, pathogens, and controlled settings [141]. Algorithms trained on limited datasets often fail to generalize across diverse agricultural contexts. The myriad variations in crop cultivars, growth stages, climates, soil conditions, and pathogen strains pose challenges in creating AI tools with sufficient flexibility for in situ usage [13,142].

Progress is also impeded by a lack of coordination across data collection efforts and an unwillingness to openly share datasets between research groups and private entities. Most available plant disease datasets remain relatively small-scale and sparse [14]. Such fragmented data restrict the training and performance scope of AI systems. While emerging sensor, imagery, and genomic technologies offer copious agricultural data streams, integrating such disparate formats for AI utilization is non-trivial and requires dedicated preprocessing pipelines [15].

Researchers have outlined frameworks to methodically address these technical barriers through good data practices and coordinated action [141,143]. Recommendations include collaborative open-access data platforms, standardized collection protocols, and emphasis on creating shareable datasets with diversity. Transfer learning methods that leverage models pre-trained on large natural image repositories are also being explored to improve generalization despite limited domain-specific agricultural data [144].

In additions, while AI holds the potential to drive sustainable agricultural practices, such as optimizing resource use and minimizing chemical inputs, it is also essential to consider the environmental footprint of the AI technology itself. This includes the carbon footprint associated with data centers powering AI applications and the environmental impact of manufacturing AI hardware. Sustainable AI in agriculture should strive for a balance where the ecological benefits of its application significantly outweigh its environmental costs [145].

7.2. Ethical Issues in AI-Driven Phytopathology

The integration of AI into phytopathology or precision agriculture also raises pressing ethical concerns regarding data privacy, accountability, labor impacts, and environmental sustainability that warrant scrutiny [146,147]. Critics caution that AI-enabled crop management regimes could reinforce unsustainable industrial farming at the cost of rural livelihoods, localized knowledge, and the food sovereignty of smallholder farmers [148,149]. Therefore, phytopathology AI systems must be designed through inclusive stakeholder participation, centering human needs and values.

There is apprehension surrounding the data privacy and consent procedures involved in collecting large agricultural datasets for training AI models, which could include farmer proprietary information alongside field images or soil data [36]. The onus is on researchers to implement ethical data management practices that protect farmer interests and anonymity. Moreover, the proprietary black-box nature of some commercial AI technologies obscures model biases and prevents oversight into decision-making rationales [150] (Ribeiro et al., 2016). Such opacity becomes ethically problematic for AI systems deployed in social realms including agriculture [151].

Broader concerns also exist around delegating data-driven farming fully to AI, potentially marginalizing rural communities and eroding farmers' autonomy, knowledge, and sense of place [148]. Hence, human-centered design considerations must shape responsible AI integration in phytopathology, serving to augment, not replace, agricultural expertise and intuition. Ongoing farmer education and upskilling initiatives are imperative to democratize AI access, allowing rural communities to reap the benefits equitably and partake in co-developing solutions attuned to local needs [43,152].

7.3. Regulatory Frameworks and Standards

Realizing ethical AI for agriculture requires establishing regulatory frameworks and technological standards guiding development and deployment [153]. At present, there is a lack of governance surrounding the creation, sales, and monitoring of AI phytopathology technologies. Policy interventions are required at national and global levels to regulate the quality control, risk assessments, and liability attribution of agricultural AI systems. Such oversight can mitigate dangers of hastily implemented tools with unreliable real-world performances or unexamined biases causing harm [154,155].

Global agreements are also needed to align technological approaches, architecture choices, data formats, curation protocols, and performance benchmarks across the emerging field of AI phytopathology [156]. Common technology standards will support collaboration, open data sharing, and interoperability, accelerating innovation. Furthermore, voluntary professional codes of ethics around topics such as model transparency, auditability, and farmer privacy could guide institutional research and industry product design [157]. Overall, multi-tiered governance combining binding regulations and soft-law ethics codes can steer agricultural AI progress along responsible trajectories.

In summary, realizing AI's promise in transforming 21st century phytopathology necessitates prudently navigating the accompanying integration barriers and ethical tensions. Only through holistic frameworks considering all stakeholder needs can AI technologies serve humanity in enabling sustainable plant disease management worldwide. The path forward lies in an inclusive and value-based co-development of agricultural AI tools, supported by emergent policy regimes governing these evolving technologies for societal benefit.

8. Conclusions

8.1. Summary of Key Findings

This comprehensive review highlights the immense potential of artificial intelligence to transform modern approaches in plant disease management and phytopathological research. Through an extensive analysis of the existing literature, manifold AI applications across major facets of phytopathology have been delineated.

Notable successes have been demonstrated in employing AI for automated disease detection and diagnosis using image recognition techniques like convolutional neural networks. Studies indicate that AI can identify plant diseases, often with 95–97% accuracy, exceeding human visual inspection. AI also shows an aptitude for data-driven disease spread forecasting, integrating weather, soil, and crop parameters to predict outbreak risks up to 3 weeks prior at 81–95% precision. This enables preemptive and targeted protection strategies minimizing pesticide usage.

Furthermore, AI is optimizing precision agriculture through site-specific interventions tailored to local conditions based on integrated crop data analysis. These holds promise to boost yields while protecting ecosystems. The AI monitoring of plant and environmental cues also facilitates pre-symptomatic disease alerts for early action. Ongoing research on explainable and transparent AI can mitigate issues surrounding model opacity.

Overall, real-world evidence affirms that AI-enabled tools can strengthen disease control, enhance crop resilience, and unlock novel phytopathological insights from increasingly complex agricultural data streams [158]. AI's self-improving and generalizable capabilities make it well suited to address evolving plant health challenges amidst climate change, globalization, and intensified farming systems.

8.2. Implications for the Future of Phytopathology

The integration of AI portends a paradigm shift in phytopathology and plant protection strategies worldwide. As algorithms become more robust and tailored for agricultural settings, AI's role is poised to expand from assisting tasks to autonomous in-field decision-making around disease management. With sufficient training data encompassing diverse cropping contexts, AI systems can attain the flexibility and adaptiveness required for broad deployment.

Advances in sensors, automation, and robotics will enable expansive data generation on crop status, disease progression, and environmental influences. AI's capacity to assimilate such big data and discern correlations can illuminate plant–microbe interactions, evolutionary dynamics, and epidemiology at unprecedented resolution. These insights promise to accelerate knowledge discovery and innovation in phytopathology, seeding 21st century breakthroughs.

Overall, the advent of data-driven smart farming powered by AI algorithms marks a historic juncture in tackling plant disease burdens. As phytopathology transitions into an interdisciplinary, technology-intensive science, AI will catalyze a strategic shift towards precision and sustainable agriculture. This new paradigm seeks ecological disease prevention over chemical controls, supporting global food security and environmental objectives.

8.3. Call to Action for Researchers and Stakeholders

Realizing the immense promise of AI in enabling next-generation phytopathology necessitates focused efforts by researchers worldwide alongside multi-stakeholder participation. Key priorities include assembling diverse open-access datasets, advancing collaborative models, strengthening farmer education, and developing supportive policies.

Researchers must coordinate shared protocols and create expansive training datasets encompassing various crops, cultivars, pathogens, and agricultural environments. This will improve AI model robustness, avoiding dataset limitations. Committing to open data access and developing regional repositories are critical to accelerated innovation. Advancing participatory models where farmers help co-design context-specific AI tools is vital to democratize benefits equitably. Capacity building to equip farming communities in adopting smart technologies responsibly is imperative. Policymakers must also implement updated regulations governing agricultural AI development and deployment for the public good.

In summary, the promising future of AI in plant disease management calls for collective action by stakeholders worldwide. Through ethically aligned, inclusive efforts that put farmers first, AI can help secure the productivity and sustainability of agricultural systems globally in the face of rising pressures. This necessitates bridging disciplinary divides and steering agricultural AI progress along humanistic trajectories.

Author Contributions: All the authors collaborated in the research and review described in this article. Conceptualization, V.E.G.-R. and I.I.-B.; investigation, V.E.G.-R., I.I.-B., J.M.C., M.C. and C.G.; writing—original draft preparation, V.E.G.-R. and I.I.-B.; writing—review and editing, J.M.C., M.C., and C.G.; supervision, M.C. and C.G.; funding acquisition, J.M.C., M.C. and C.G. All authors have read and agreed to the published version of the manuscript.

Funding: This research was supported by grants from MCIN PID2021-122899O-B-C22 MCIN/AEI/ FEDER, EU) and the University of Cádiz through the "Programa de Fomento e Impulso de la actividad de Investigación y Transferencia de la Universidad de Cádiz".

Data Availability Statement: Not applicable.

Conflicts of Interest: The authors declare no conflicts of interest.

References

1. Strange, R.N.; Scott, P.R. Plant Disease: A Threat to Global Food Security. *Annu. Rev. Phytopathol.* **2005**, *43*, 83–116. [CrossRef]
2. Agrios, G.N. *Plant Pathology*, 5th ed.; Academic Press: Amsterdam, The Netherlands, 2004; ISBN 9780120445653.
3. Oerke, E.C. Crop Losses to Pests. *J. Agric. Sci.* **2005**, *144*, 31–43. [CrossRef]
4. Savary, S.; Willocquet, L.; Pethybridge, S.J.; Esker, P.; McRoberts, N.; Nelson, A. The Global Burden of Pathogens and Pests on Major Food Crops. *Nat. Ecol. Evol.* **2019**, *3*, 430–439. [CrossRef] [PubMed]
5. Ray, D.K.; Mueller, N.D.; West, P.C.; Foley, J.A. Yield Trends Are Insufficient to Double Global Crop Production by 2050. *PLoS ONE* **2013**, *8*, e66428. [CrossRef] [PubMed]
6. Lucas, J.A.; Hawkins, N.J.; Fraaije, B.A. The Evolution of Fungicide Resistance. *Adv. Appl. Microbiol.* **2015**, *90*, 29–92. [CrossRef] [PubMed]
7. Parnell, S.; Gottwald, T.R.; Gilligan, C.A.; Cunniffe, N.J.; Van Den Bosch, F. The Effect of Landscape Pattern on the Optimal Eradication Zone of an Invading Epidemic. *Phytopathology* **2010**, *100*, 638–644. [CrossRef] [PubMed]
8. Anderson, P.K.; Cunningham, A.A.; Patel, N.G.; Morales, F.J.; Epstein, P.R.; Daszak, P. Emerging Infectious Diseases of Plants: Pathogen Pollution, Climate Change and Agrotechnology Drivers. *Trends Ecol. Evol.* **2004**, *19*, 535–544. [CrossRef] [PubMed]
9. Bebber, D.P. Range-Expanding Pests and Pathogens in a Warming World. *Annu. Rev. Phytopathol.* **2015**, *53*, 335–356. [CrossRef] [PubMed]
10. McDonald, B.A.; Linde, C. Pathogen Population Genetics, Evolutionary Potential, And Durable Resistance. *Annu. Rev. Phytopathol.* **2003**, *40*, 349–379. [CrossRef]
11. Collinge, D.B.; Jensen, D.F.; Rabiey, M.; Sarrocco, S.; Shaw, M.W.; Shaw, R.H. Biological Control of Plant Diseases—What Has Been Achieved and What Is the Direction? *Plant Pathol.* **2022**, *71*, 1024–1047. [CrossRef]
12. Ferguson, M.; Hsu, C.K.; Grim, C.; Kauffman, M.; Jarvis, K.; Pettengill, J.B.; Babu, U.S.; Harrison, L.M.; Li, B.; Hayford, A.; et al. A Longitudinal Study to Examine the Influence of Farming Practices and Environmental Factors on Pathogen Prevalence Using Structural Equation Modeling. *Front. Microbiol.* **2023**, *14*, 1141043. [CrossRef]
13. Singh, A.; Ganapathysubramanian, B.; Singh, A.K.; Sarkar, S. Machine Learning for High-Throughput Stress Phenotyping in Plants. *Trends Plant Sci.* **2016**, *21*, 110–124. [CrossRef] [PubMed]
14. Ramcharan, A.; Baranowski, K.; McCloskey, P.; Ahmed, B.; Legg, J.; Hughes, D.P. Deep Learning for Image-Based Cassava Disease Detection. *Front. Plant Sci.* **2017**, *8*, 1852. [CrossRef] [PubMed]
15. Liakos, K.G.; Busato, P.; Moshou, D.; Pearson, S.; Bochtis, D. Machine Learning in Agriculture: A Review. *Sensors* **2018**, *18*, 2674. [CrossRef] [PubMed]
16. Zhao, L.; Walkowiak, S.; Fernando, W.G.D. Artificial Intelligence: A Promising Tool in Exploring the Phytomicrobiome in Managing Disease and Promoting Plant Health. *Plants* **2023**, *12*, 1852. [CrossRef] [PubMed]
17. Shoaib, M.; Shah, B.; EI-Sappagh, S.; Ali, A.; Ullah, A.; Alenezi, F.; Gechev, T.; Hussain, T.; Ali, F. An Advanced Deep Learning Models-Based Plant Disease Detection: A Review of Recent Research. *Front. Plant Sci.* **2023**, *14*, 1158933. [CrossRef] [PubMed]
18. Das, S.; Pattanayak, S.; Behera, P.R. Application of Machine Learning: A Recent Advancement in Plant Diseases Detection. *J. Plant Prot. Res.* **2022**, *62*, 122–135. [CrossRef]
19. Legg, J.P.; Lava Kumar, P.; Makeshkumar, T.; Tripathi, L.; Ferguson, M.; Kanju, E.; Ntawuruhunga, P.; Cuellar, W. Cassava Virus Diseases: Biology, Epidemiology, and Management. *Adv. Virus Res.* **2015**, *91*, 85–142. [CrossRef]
20. Ploetz, R.C. Fusarium Wilt of Banana. *Phytopathology* **2015**, *105*, 1512–1521. [CrossRef]
21. Gonsalves, D. Control Of Papaya Ringspot Virus In Papaya: A Case Study. *Annu. Rev. Phytopathol.* **1998**, *36*, 415–452. [CrossRef]
22. Avelino, J.; Cristancho, M.; Georgiou, S.; Imbach, P.; Aguilar, L.; Bornemann, G.; Läderach, P.; Anzueto, F.; Hruska, A.J.; Morales, C. The Coffee Rust Crises in Colombia and Central America (2008–2013): Impacts, Plausible Causes and Proposed Solutions. *Food Secur.* **2015**, *7*, 303–321. [CrossRef]

23. Gottwald, T.R. Current Epidemiological Understanding of Citrus Huanglongbing*. *Annu. Rev. Phytopathol.* **2010**, *48*, 119–139. [CrossRef] [PubMed]
24. Pethybridge, S.J.; Nelson, S.C. Leaf Doctor: A New Portable Application for Quantifying Plant Disease Severity. *Plant Dis.* **2015**, *99*, 1310–1316. [CrossRef]
25. Mahlein, A.K. Plant Disease Detection by Imaging Sensors—Parallels and Specific Demands for Precision Agriculture and Plant Phenotyping. *Plant Dis.* **2016**, *100*, 241–254. [CrossRef] [PubMed]
26. Ahmad, A.; Hettiarachchi, R.; Khezri, A.; Singh Ahluwalia, B.; Wadduwage, D.N.; Ahmad, R. Highly Sensitive Quantitative Phase Microscopy and Deep Learning Aided with Whole Genome Sequencing for Rapid Detection of Infection and Antimicrobial Resistance. *Front. Microbiol.* **2023**, *14*, 1154620. [CrossRef] [PubMed]
27. Proaño-Cuenca, F.; Espindola, A.S.; Garzon, C.D. Detection of Phytophthora, Pythium, Globisporangium, Hyaloperonospora, and Plasmopara Species in High-Throughput Sequencing Data by In Silico and In Vitro Analysis Using Microbe Finder (MiFi). *PhytoFrontiers* **2023**, *3*, 124–136. [CrossRef]
28. Kemen, E. Microbe–Microbe Interactions Determine Oomycete and Fungal Host Colonization. *Curr. Opin. Plant Biol.* **2014**, *20*, 75–81. [CrossRef]
29. Joshi, A.; Song, H.G.; Yang, S.Y.; Lee, J.H. Integrated Molecular and Bioinformatics Approaches for Disease-Related Genes in Plants. *Plants* **2023**, *12*, 2454. [CrossRef]
30. Venbrux, M.; Crauwels, S.; Rediers, H. Current and Emerging Trends in Techniques for Plant Pathogen Detection. *Front. Plant Sci.* **2023**, *14*, 1120968. [CrossRef]
31. Qin, J.; Monje, O.; Nugent, M.R.; Finn, J.R.; O'Rourke, A.E.; Wilson, K.D.; Fritsche, R.F.; Baek, I.; Chan, D.E.; Kim, M.S. A Hyperspectral Plant Health Monitoring System for Space Crop Production. *Front. Plant Sci.* **2023**, *14*, 1133505. [CrossRef]
32. Schirrmann, M. Advances in Remote Sensing Technologies for Assessing Crop Health. In *Advances in Sensor Technology for Sustainable Crop Production*; Burleigh Dodds Science Publishing Limited: Cambridge, UK, 2023; pp. 43–64. ISBN 9781786769770.
33. Abbas, A.; Zhang, Z.; Zheng, H.; Alami, M.M.; Alrefaei, A.F.; Abbas, Q.; Naqvi, S.A.H.; Rao, M.J.; Mosa, W.F.A.; Abbas, Q.; et al. Drones in Plant Disease Assessment, Efficient Monitoring, and Detection: A Way Forward to Smart Agriculture. *Agronomy* **2023**, *13*, 1524. [CrossRef]
34. Hu, H.; Wang, N.; Liao, J.; Tovar-Lopez, F.J. Recent Progress in Micro- and Nanotechnology-Enabled Sensors for Biomedical and Environmental Challenges. *Sensors* **2023**, *23*, 5406. [CrossRef]
35. Arshad, F.; Deliorman, M.; Sukumar, P.; Qasaimeh, M.A.; Olarve, J.S.; Santos, G.N.; Bansal, V.; Ahmed, M.U. Recent Developments and Applications of Nanomaterial-Based Lab-on-a-Chip Devices for Sustainable Agri-Food Industries. *Trends Food Sci. Technol.* **2023**, *136*, 145–158. [CrossRef]
36. Kamilaris, A.; Kartakoullis, A.; Prenafeta-Boldú, F.X. A Review on the Practice of Big Data Analysis in Agriculture. *Comput. Electron. Agric.* **2017**, *143*, 23–37. [CrossRef]
37. Glass, C.R.; Gonzalez, F.J.E. Developing of New Technologies Driving Advances in Precision Agriculture to Optimise Inputs and Reduce Environmental Footprint. *C3-BIOECONOMY Circ. Sustain. Bioeconomy* **2022**, 69–75. [CrossRef]
38. Kotpalliwar, P.; Barmate, M.; Satpute, P.; Manapure, D.; Hassan, M.; Student, U.G.; Prof, A. Agro Analysis System for Precision Agriculture. *Int. J. Res. Appl. Sci. Eng. Technol. (IJRASET)* **2023**, *11*, 960–963. [CrossRef]
39. Garrett, K.A.; Bebber, D.P.; Etherton, B.A.; Gold, K.M.; Sulá, A.I.P.; Selvaraj, M.G. Climate Change Effects on Pathogen Emergence: Artificial Intelligence to Translate Big Data for Mitigation. *Annu. Rev. Phytopathol.* **2022**, *60*, 357–378. [CrossRef] [PubMed]
40. Tatineni, S.; Hein, G.L. Plant Viruses of Agricultural Importance: Current and Future Perspectives of Virus Disease Management Strategies. *Phytopathology* **2023**, *113*, 117–141. [CrossRef] [PubMed]
41. Miller, S.A.; Ferreira, J.P.; Lejeune, J.T. Antimicrobial Use and Resistance in Plant Agriculture: A One Health Perspective. *Agriculture* **2022**, *12*, 289. [CrossRef]
42. Banerjee, S.; Mondal, A.C. An Intelligent Approach to Reducing Plant Disease and Enhancing Productivity Using Machine Learning. *Int. J. Recent Innov. Trends Comput. Commun.* **2023**, *11*, 250–262. [CrossRef]
43. Tzachor, A.; Devare, M.; King, B.; Avin, S.; Héigeartaigh, S.Ó. Responsible Artificial Intelligence in Agriculture Requires Systemic Understanding of Risks and Externalities. *Nat. Mach. Intell.* **2022**, *4*, 104–109. [CrossRef]
44. Gruetzemacher, R.; Whittlestone, J. The Transformative Potential of Artificial Intelligence. *Futures* **2022**, *135*, 102884. [CrossRef]
45. Pal, S. A Paradigm Shift in Research: Exploring the Intersection of Artificial Intelligence and Research Methodology. *Int. J. Innov. Res. Eng. Multidiscip. Phys. Sci.* **2023**, *11*, 230125. [CrossRef]
46. Russell, S.; Norvig, P. *Artificial Intelligence a Modern Approach*, 3rd ed.; Prentice Hall: New Jersey, NJ, USA, 2010; ISBN 9780136042594.
47. Jordan, M.I.; Mitchell, T.M. Machine Learning: Trends, Perspectives, and Prospects. *Science* **2015**, *349*, 255–260. [CrossRef]
48. Lecun, Y.; Bengio, Y.; Hinton, G. Deep Learning. *Nature* **2015**, *521*, 436–444. [CrossRef] [PubMed]
49. Hirschberg, J.; Manning, C.D. Advances in Natural Language Processing. *Science* **2015**, *349*, 261–266. [CrossRef]
50. Jackson, P. *Introduction to Expert Systems*, 3rd ed.; Addison-Wesley Longman Publishing Co., Inc.: Upper Saddle River, NJ, USA, 1998; ISBN 0201876868.
51. Kober, J.; Bagnell, J.A.; Peters, J. Reinforcement Learning in Robotics: A Survey. *Int. J. Robot. Res.* **2013**, *32*, 1238–1274. [CrossRef]

52. Arockia Venice, J.; Thoti, K.K.; Mary Henrietta, H.; Elangovan, M.; Anusha, D.J.; Zhakupova, A. Artificial Intelligence Based Robotic System with Enhanced Information Technology. In Proceedings of the 2022 Sixth International Conference on I-SMAC (IoT in Social, Mobile, Analytics and Cloud) (I-SMAC), Dharan, Nepal, 10–12 November 2022; Institute of Electrical and Electronics Engineers Inc.: New York, NY, USA, 2022; pp. 705–714.
53. Schraagen, J.M.; van Diggelen, J. A Brief History of the Relationship between Expertise and Artificial Intelligence. In *Expertise at Work: Current and Emerging Trends*; Springer: Berlin/Heidelberg, Germany, 2021; pp. 149–175. ISBN 9783030643713.
54. Long, L.N.; Cotner, C.F. A Review and Proposed Framework for Artificial General Intelligence. In Proceedings of the 2019 IEEE Aerospace Conference, Big Sky, MT, USA, 2–9 March 2019. [CrossRef]
55. Beaulac, C.; Larribe, F. Narrow Artificial Intelligence with Machine Learning for Real-Time Estimation of a Mobile Agent's Location Using Hidden Markov Models. *Hindawi Int. J. Comput. Games Technol.* **2017**, *2017*, 4939261. [CrossRef]
56. Keisler, J.; Talbi, E.-G.; Claudel, S.; Cabriel, G. An Algorithmic Framework for the Optimization of Deep Neural Networks Architectures and Hyperparameters. *arXiv* **2023**, arXiv:2303.12797. [CrossRef]
57. Brown, T.B.; Mann, B.; Ryder, N.; Subbiah, M.; Kaplan, J.; Dhariwal, P.; Neelakantan, A.; Shyam, P.; Sastry, G.; Askell, A.; et al. Language Models Are Few-Shot Learners. *Adv. Neural Inf. Process Syst.* **2020**, *33*, 1877–1901.
58. LeCun, Y.; Bottou, L.; Bengio, Y.; Haffner, P. Gradient-Based Learning Applied to Document Recognition. *Proc. IEEE* **1998**, *86*, 2278–2323. [CrossRef]
59. Celeghin, A.; Borriero, A.; Orsenigo, D.; Diano, M.; Méndez Guerrero, C.A.; Perotti, A.; Petri, G.; Tamietto, M. Convolutional Neural Networks for Vision Neuroscience: Significance, Developments, and Outstanding Issues. *Front. Comput. Neurosci.* **2023**, *17*, 1153572. [CrossRef]
60. Hochreiter, S.; Schmidhuber, J. Long Short-Term Memory. *Neural Comput.* **1997**, *9*, 1735–1780. [CrossRef] [PubMed]
61. Goodfellow, I.J.; Pouget-Abadie, J.; Mirza, M.; Xu, B.; Warde-Farley, D.; Ozair, S.; Courville, A.; Bengio, Y. Generative Adversarial Nets. *Adv. Neural Inf. Process Syst.* **2014**, *27*, 1–9.
62. Chen, T.; Guestrin, C. XGBoost: A Scalable Tree Boosting System. In Proceedings of the KDD '16: 22nd ACM SIGKDD International Conference on Knowledge Discovery and Data Mining, San Francisco, CA, USA, 13–17 August 2016; pp. 785–794. [CrossRef]
63. Niedbała, G.; Kurek, J.; Świderski, B.; Wojciechowski, T.; Antoniuk, I.; Bobran, K. Prediction of Blueberry (*Vaccinium Corymbosum* L.) Yield Based on Artificial Intelligence Methods. *Agriculture* **2022**, *12*, 2089. [CrossRef]
64. Zou, H.; Hastie, T. Regularization and Variable Selection Via the Elastic Net. *J. R. Stat. Soc. Series B Stat. Methodol.* **2005**, *67*, 301–320. [CrossRef]
65. Breiman, L. Random Forests. *Mach. Learn.* **2001**, *45*, 5–32. [CrossRef]
66. Cortes, C.; Vapnik, V.; Saitta, L. Support-Vector Networks Editor. *Mach. Leam.* **1995**, *20*, 273–297. [CrossRef]
67. Suzuki, T.; Hasebe, T.; Miyazaki, T. Quantum Support Vector Machines for Classification and Regression on a Trapped-Ion Quantum Computer. *arXiv* **2023**, arXiv:2307.02091. [CrossRef]
68. Daulay, R.S.A.; Efendi, S. Suherman Review of Literature on Improving the KNN Algorithm. *Trans. Eng. Comput. Sci.* **2023**, *11*, 63–72. [CrossRef]
69. Cover, T.M.; Hart, P.E. Nearest Neighbor Pattern Classification. *IEEE Trans. Inf. Theory* **1967**, *13*, 21–27. [CrossRef]
70. Bengio, Y.; Courville, A.; Vincent, P. Representation Learning: A Review and New Perspectives. *IEEE Trans. Pattern Anal. Mach. Intell.* **2012**, *35*, 1798–1828. [CrossRef]
71. Frean, M. The Upstart Algorithm: A Method for Constructing and Training Feedforward Neural Networks. *Neural Comput.* **1990**, *2*, 198–209. [CrossRef]
72. Robbins, H.; Monro, S. A Stochastic Approximation Method. *Ann. Math. Stat.* **1951**, *22*, 400–407. [CrossRef]
73. Liu, F.; Li, J.; Wang, L. PI-LSTM: Physics-Informed Long Short-Term Memory Network for Structural Response Modeling. *Eng. Struct.* **2023**, *292*, 116500. [CrossRef]
74. Sutton, R.S.; Barto, A.G. Reinforcement Learning: An Introduction. *IEEE Trans. Neural Netw.* **1998**, *9*, 1054. [CrossRef]
75. Liu, Z.; Xu, B. Researches Advanced in the Application of Reinforcement Learning. In Proceedings of the 2nd 928 International Conference on Artificial Intelligence, Automation, and High-Performance Computing (AIAHPC 2022), Zhuhai, China, 25–27 February 2022; p. 134.
76. Devlin, J.; Chang, M.-W.; Lee, K.; Google, K.T.; Language, A.I. BERT: Pre-Training of Deep Bidirectional Transformers for Language Understanding. In *Proceedings of the 2019 Conference of the North*; Association for Computational Linguistics: Stroudsburg, PA, USA, 2019; pp. 4171–4186.
77. Toosi, A.; Bottino, A.G.; Saboury, B.; Siegel, E.; Rahmim, A. A Brief History of AI: How to Prevent Another Winter (A Critical Review). *PET Clin.* **2021**, *16*, 449–469. [CrossRef] [PubMed]
78. Auda, Z.M.; Radhi, S.J. Artificial Intelligence and Evolution of the Global System. *IPRI J.* **2022**, *22*, 87–105. [CrossRef]
79. Magruder, M.L.; Delanois, R.E.; Nace, J.; Mont, M.A. ChatGPT and Other Natural Language Processing Artificial Intelligence Models in Adult Reconstruction. *J. Arthroplast.* **2023**, *38*, 2191–2192. [CrossRef] [PubMed]
80. Nguyen, J.; Pepping, C.A. The Application of ChatGPT in Healthcare Progress Notes: A Commentary from a Clinical and Research Perspective. *Clin. Transl. Med.* **2023**, *13*, e1324. [CrossRef] [PubMed]
81. Shirodkar, S.S.; Raotole, A.N.; Gajbhiye, C.R. The AI Galaxy: A Comparative Study of Cutting-Edge AI Technology. *Int. J. Res. Appl. Sci. Eng. Technol.* **2023**, *11*, 1010–1017. [CrossRef]

82. Kaplan-Marans, E.; Khurgin, J. ChatGPT Wrote This Article. *Urology* **2023**, *179*, 1–4. [CrossRef]
83. Doshi, R.H.; Bajaj, S.S.; Krumholz, H.M. ChatGPT: Temptations of Progress. *Am. J. Bioeth.* **2023**, *23*, 6–8. [CrossRef] [PubMed]
84. Ahmad, S.S.; Meehan, A.; Crispin-Ortuzar, M.; Weckman, N.E. Personalized, Connected Health Enabled by AI and Home-Based Diagnostics. *Trends Biotechnol.* **2023**, *41*, 982–983. [CrossRef] [PubMed]
85. Bejnordi, B.E.; Veta, M.; Van Diest, P.J.; Van Ginneken, B.; Karssemeijer, N.; Litjens, G.; Van Der Laak, J.A.W.M.; Hermsen, M.; Manson, Q.F.; Balkenhol, M.; et al. Diagnostic Assessment of Deep Learning Algorithms for Detection of Lymph Node Metastases in Women with Breast Cancer. *JAMA* **2017**, *318*, 2199–2210. [CrossRef] [PubMed]
86. Liu, X.; Faes, L.; Kale, A.U.; Wagner, S.K.; Fu, D.J.; Bruynseels, A.; Mahendiran, T.; Moraes, G.; Shamdas, M.; Kern, C.; et al. A Comparison of Deep Learning Performance against Health-Care Professionals in Detecting Diseases from Medical Imaging: A Systematic Review and Meta-Analysis. *Lancet Digit. Health* **2019**, *1*, 271–297. [CrossRef] [PubMed]
87. Khan, A. Transforming Healthcare through AI: Unleashing the Power of Personalized Medicine. *Int. J. Multidiscip. Sci. Arts* **2023**, *2*, 67–77. [CrossRef]
88. Mehta, V. Artificial Intelligence in Medicine: Revolutionizing Healthcare for Improved Patient Outcomes. *J. Med. Res. Innov.* **2023**, *7*, e000292. [CrossRef]
89. Rastogi, M. The Growth and Potential of AI Applications in Medicine and Healthcare. *Indian. J. Appl. Res.* **2023**, *13*, 17–18. [CrossRef]
90. Majumdar, S. The Changing Landscape of AI-Driven System Optimization for Complex Combinatorial Optimization. In Proceedings of the 2022 ACM/IEEE 4th Workshop on Machine Learning for CAD (MLCAD), Snowbird, UT, USA, 12–13 September 2022; Institute of Electrical and Electronics Engineers (IEEE): New York, NY, USA, 2022; p. 49.
91. Cha, Y.-J.; Choi, W. Deep Learning-Based Crack Damage Detection Using Convolutional Neural Networks. *Comput.-Aided Civ. Infrastruct. Eng.* **2017**, *32*, 361–378. [CrossRef]
92. Mirhoseini, A.; Goldie, A.; Yazgan, M.; Jiang, J.W.; Songhori, E.; Wang, S.; Lee, Y.J.; Johnson, E.; Pathak, O.; Nazi, A.; et al. A Graph Placement Methodology for Fast Chip Design. *Nature* **2021**, *594*, 207–212. [CrossRef]
93. Shi, Y.; Han, L.; González-Moreno, P.; Dancey, D.; Huang, W.; Zhang, Z.; Liu, Y.; Huang, M.; Miao, H.; Dai, M. A Fast Fourier Convolutional Deep Neural Network for Accurate and Explainable Discrimination of Wheat Yellow Rust and Nitrogen Deficiency from Sentinel-2 Time Series Data. *Front. Plant Sci.* **2023**, *14*, 1250844. [CrossRef]
94. Brynjolfsson, E.; Mitchell, T. What Can Machine Learning Do? Workforce Implications: Profound Change Is Coming, but Roles for Humans Remain. *Science* **2017**, *358*, 1530–1534. [CrossRef]
95. Jiang, Z.; Xu, D.; Liang, J. A Deep Reinforcement Learning Framework for the Financial Portfolio Management Problem. *arXiv* **2017**, arXiv:1706.10059. [CrossRef]
96. Schmitt, M. Automated Machine Learning: AI-Driven Decision Making in Business Analytics. *Intell. Syst. Appl.* **2023**, *18*, 200188. [CrossRef]
97. Jhaveri, M.; Chirputkar, A.; Ashok, P. The Efficacy of Artificial Intelligence in Making Best Marketing Decisions. In Proceedings of the International Conference on Innovative Data Communication Technologies and Application, ICIDCA 2023—Proceedings, Dehradun, India, 14–15 March 2023; Institute of Electrical and Electronics Engineers Inc.: New York, NY, USA, 2023; pp. 225–229.
98. Van Der Pol, E.; Oliehoek, F.A. Coordinated Deep Reinforcement Learners for Traffic Light Control. In Proceedings of the 30th Conference on Neural Information Processing Systems (NIPS 2016), Barcelona, Spain, 5–10 December 2016.
99. Sun, P.; Kretzschmar, H.; Dotiwalla, X.; Chouard, A.; Patnaik, V.; Tsui, P.; Guo, J.; Zhou, Y.; Chai, Y.; Caine, B.; et al. Scalability in Perception for Autonomous Driving: Waymo Open Dataset. In Proceedings of the 2020 IEEE/CVF Conference on Computer Vision and Pattern Recognition (CVPR), Seattle, WA, USA, 13–19 June 2020; pp. 2443–2451.
100. Hawlader, F.; Robinet, F.; Frank, R. Vehicle-to-Infrastructure Communication for Real-Time Object Detection in Autonomous Driving. In Proceedings of the 2023 18th Wireless On-Demand Network Systems and Services Conference (WONS), Madonna di Campiglio, Italy, 30 January–1 February 2023; pp. 40–46. [CrossRef]
101. Hasanujjaman, M.; Chowdhury, M.Z.; Jang, Y.M. Sensor Fusion in Autonomous Vehicle with Traffic Surveillance Camera System: Detection, Localization, and AI Networking. *Sensors* **2023**, *23*, 3335. [CrossRef] [PubMed]
102. Self-Driving Car Technology for a Reliable Ride—Waymo Driver. Available online: https://waymo.com/intl/es/waymo-driver/ (accessed on 23 January 2024).
103. Yadav, S.K. The Development of AI & Self Driving Technology. *Interantional J. Sci. Res. Eng. Manag.* **2023**, *7*, 1–9. [CrossRef]
104. Tiwari, V.; Singh, N.K. A Study on Artificial Intelligence Based Automatic Vehicle Navigation. In Proceedings of the International Conferences on Contemporary Computing and Informatics, Uttar Pradesh, India, 14–16 December 2022; Institute of Electrical and Electronics Engineers Inc.: New York, NY, USA, 2022; pp. 1148–1152.
105. Pattnayak, P.; Patnaik, S. Space and Applications of Artificial Intelligence. In Proceedings of the 2022 OITS International Conference on Information Technology, OCIT 2022, Bhubaneswar, India, 14–16 December 2022; Institute of Electrical and Electronics Engineers Inc.: New York, NY, USA, 2022; pp. 158–163.
106. Al Sabt, I.; Farooqui, M.O. Navigating the Convergence of Artificial Intelligence and Space Law: Challenges and Opportunities. *HighTech Innov. J.* **2023**, *4*, 55–64. [CrossRef]
107. Slingerland, P.; Perry, L.; Kaufman, J.; Bycroft, B.; Linstead, E.; Mandrake, L.; Doran, G.; Goel, A.; Feather, M.S.; Fesq, L.; et al. Adapting a Trusted AI Framework to Space Mission Autonomy. In Proceedings of the 2022 IEEE Aerospace Conference (AERO), Big Sky, MT, USA, 5–12 March 2022. [CrossRef]

108. Duolingo—La Mejor Manera de Aprender Un Idioma a Nivel Mundial. Available online: https://www.duolingo.com/ (accessed on 23 January 2024).
109. Li, K.C.; Wong, B.T.M. Artificial Intelligence in Personalised Learning: A Bibliometric Analysis. *Interact. Technol. Smart Educ.* **2023**, *20*, 422–445. [CrossRef]
110. Risang Baskara, F.X. Personalised Learning With AI: Implications for Ignatian Pedagogy. *Int. J. Educ. Best Pract.* **2023**, *7*, 1–16. [CrossRef]
111. Essa, S.G.; Celik, T.; Human-Hendricks, N.E. Personalized Adaptive Learning Technologies Based on Machine Learning Techniques to Identify Learning Styles: A Systematic Literature Review. *IEEE Access* **2023**, *11*, 48392–48409. [CrossRef]
112. Fuentes, A.; Yoon, S.; Kim, S.C.; Park, D.S. A Robust Deep-Learning-Based Detector for Real-Time Tomato Plant Diseases and Pests Recognition. *Sensors* **2017**, *17*, 2022. [CrossRef] [PubMed]
113. Sladojevic, S.; Arsenovic, M.; Anderla, A.; Culibrk, D.; Stefanovic, D. Deep Neural Networks Based Recognition of Plant Diseases by Leaf Image Classification. *Comput. Intell. Neurosci.* **2016**, *2016*, 3289801. [CrossRef]
114. Arinichev, I.V. Using Digital Intelligent Technologies for the Diagnosis of Cereals Diseases in the Kuban. *Agrar. Sci. J.* **2022**, *5*, 70–73. [CrossRef]
115. Feng, J.; Hou, B.; Yu, C.; Yang, H.; Wang, C.; Shi, X.; Hu, Y.; Feng, J.; Hou, B.; Yu, C.; et al. Research and Validation of Potato Late Blight Detection Method Based on Deep Learning. *Agronomy* **2023**, *13*, 1659. [CrossRef]
116. Bracino, A.A.; Evangelista, D.G.D.; Concepcion, R.S.; Dadios, E.P.; Vicerra, R.R.P. Non-Destructive Classification of Paddy Rice Leaf Disease Infected by Bacterial and Fungal Species Using Vision-Based Deep Learning. *J. Adv. Comput. Intell. Intell. Inform.* **2023**, *27*, 333–339. [CrossRef]
117. Jouini, O.; Sethom, K.; Bouallegue, R. Wheat Leaf Disease Detection Using CNN in Smart Agriculture. In Proceedings of the 2023 International Wireless Communications and Mobile Computing, IWCMC 2023, Marrakesh, Morocco, 19–23 June 2023; Institute of Electrical and Electronics Engineers Inc.: New York, NY, USA, 2023; pp. 1660–1665.
118. Zhou, Y.; Pan, Z.; Liu, Z.; He, T.; Yan, J. SFPGRN: Spectral Detection Method of Plant Diseases Based on Deep Learning. *SPIE* **2023**, *12717*, 127171F. [CrossRef]
119. Hassan, S.M.; Maji, A.K. Plant Disease Identification Using a Novel Convolutional Neural Network. *IEEE Access* **2022**, *10*, 5390–5401. [CrossRef]
120. Narain, T.; Sahu, P.; Singh, A.P. Plant Disease Classification Using Siamese Convolutional Neural Network. *Lect. Notes Netw. Syst.* **2023**, *471*, 57–65. [CrossRef]
121. Tabbakh, A.; Barpanda, S.S. A Deep Features Extraction Model Based on the Transfer Learning Model and Vision Transformer "TLMViT" for Plant Disease Classification. *IEEE Access* **2023**, *11*, 45377–45392. [CrossRef]
122. Otero, M.; Velasquez, L.F.; Basile, B.; Onrubia, J.R.; Pujol, A.J.; Pijuan, J. Data Driven Predictive Models Based on Artificial Intelligence to Anticipate the Presence of Plasmopara Viticola and Uncinula Necator in Southern European Winegrowing Regions. *Front. Artif. Intell. Appl.* **2022**, *356*, 164–167. [CrossRef]
123. Lavanya, A.; Krishna, T.M. An AI and Cloud Based Collaborative Platform for PlantDisease Identification, Tracking and Forecasting for Farmers. *Int. J. Eng. Technol. Manag. Sci.* **2022**, *6*, 527–537. [CrossRef]
124. Zen, B.P.; Iqsyahiro Kresna, A.; Fransisca, D.C. Applications for Detecting Plant Diseases Based on Artificial Intelligence. *SinkrOn* **2022**, *7*, 2537–2546. [CrossRef]
125. Marco-Detchart, C.; Carrascosa, C.; Julian, V.; Rincon, J. Robust Multi-Sensor Consensus Plant Disease Detection Using the Choquet Integral. *Sensors* **2023**, *23*, 2382. [CrossRef]
126. Ojo, M.O.; Zahid, A. Improving Deep Learning Classifiers Performance via Preprocessing and Class Imbalance Approaches in a Plant Disease Detection Pipeline. *Agronomy* **2023**, *13*, 887. [CrossRef]
127. Vardhan, J.; Swetha, K.S. Detection of Healthy and Diseased Crops in Drone Captured Images Using Deep Learning. *arXiv* **2023**, arXiv:2305.13490. [CrossRef]
128. Dagwale, S.S.; Adakane, P. Prediction of Leaf Species & Disease Using Ai for Various Plants. *Int. J. Multidiscip. Res.* **2023**, *5*, 23034169.
129. He, K.; Zhang, X.; Ren, S.; Sun, J. Deep Residual Learning for Image Recognition. In Proceedings of the IEEE Conference on Computer Vision and Pattern Recognition, Las Vegas, NV, USA, 26 June–1 July 2016; pp. 770–778. [CrossRef]
130. Deng, J.; Dong, W.; Socher, R.; Li, L.-J.; Kai, L.; Li, F.-F. ImageNet: A Large-Scale Hierarchical Image Database. In Proceedings of the 2009 IEEE Conference on Computer Vision and Pattern Recognition, Miami, FL, USA, 20–25 June 2009; pp. 248–255. [CrossRef]
131. Bhati, R.; Mittal, S. The Role and Impact of Artificial Intelligence in Attaining Sustainability Goals. In Proceedings of the 2023 9th International Conference on Advanced Computing and Communication Systems (ICACCS), Coimbatore, India, 17–18 March 2023; Volume 1, pp. 2455–2458.
132. Tummapudi, S.; Sadhu, S.S.; Simhadri, S.N.; Damarla, S.N.T.; Bhukya, M. Deep Learning Based Weed Detection and Elimination in Agriculture. In Proceedings of the 6th International Conference on Inventive Computation Technologies, ICICT 2023—Proceedings, Lalitpur, Nepal, 26–28 April 2023; Institute of Electrical and Electronics Engineers Inc.: New York, NY, USA, 2023; pp. 147–151.
133. Kumar, P.; Nelson, A.; Kapetanovic, Z.; Chandra, R. Affordable Artificial Intelligence—Augmenting Farmer Knowledge with AI. *Digit. Agric. Action* **2023**. [CrossRef]

134. Buja, I.; Sabella, E.; Monteduro, A.G.; Chiriacò, M.S.; De Bellis, L.; Luvisi, A.; Maruccio, G. Advances in Plant Disease Detection and Monitoring: From Traditional Assays to In-Field Diagnostics. *Sensors* **2021**, *21*, 2129. [CrossRef]
135. Kumar, P.; Chandra, R.; Bansal, C.; Kalyanaraman, S.; Ganu, T.; Grant, M. Micro-Climate Prediction—Multi Scale Encoder-Decoder Based Deep Learning Framework. In Proceedings of the ACM SIGKDD International Conference on Knowledge Discovery and Data Mining, Virtual Event, 14 August 2021; Association for Computing Machinery: New York, NY, USA, 2021; pp. 3128–3138.
136. Sharmila, G.; Rajamohan, K. Image Processing and Artificial Intelligence for Precision Agriculture. In Proceedings of the 2022 International Conference on Innovative Computing, Intelligent Communication and Smart Electrical Systems (ICSES), Chennai, India, 15–16 July 2022. [CrossRef]
137. Sasikala, D.; Venkatesh Sharma, K. Future Intelligent Agriculture with Bootstrapped Meta-Learning Andϵ-Greedy Q-Learning. *J. Artif. Intell. Capsul. Netw.* **2022**, *4*, 149–159. [CrossRef]
138. Joseph, R.B.; Lakshmi, M.B.; Suresh, S.; Sunder, R. Innovative Analysis of Precision Farming Techniques with Artificial Intelligence. In Proceedings of the 2nd International Conference on Innovative Mechanisms for Industry Applications, ICIMIA 2020—Conference Proceedings, Bangalore, India, 5–7 March 2020; Institute of Electrical and Electronics Engineers Inc.: New York, NY, USA, 2020; pp. 353–358.
139. Raj, V.H.A.; De Carvalho, C.X. A Perspective on the Application of Artificial Intelligence in Sustainable Agriculture with Special Reference to Precision Agriculture. *SDMIMD J. Manag.* **2023**, *14*, 1–13. [CrossRef] [PubMed]
140. Williams, M.J.; Sikder, M.N.K.; Wang, P.; Gorentala, N.; Gurrapu, S.; Batarseh, F.A. The Application of Artificial Intelligence Assurance in Precision Farming and Agricultural Economics. In *AI Assurance: Towards Trustworthy, Explainable, Safe, and Ethical AI*; Elsevier: Amsterdam, The Netherlands, 2022; pp. 501–529. ISBN 9780323919197.
141. Benos, L.; Tagarakis, A.C.; Dolias, G.; Berruto, R.; Kateris, D.; Bochtis, D. Machine Learning in Agriculture: A Comprehensive Updated Review. *Sensors* **2021**, *21*, 3758. [CrossRef] [PubMed]
142. Singh, A.K.; Ganapathysubramanian, B.; Sarkar, S.; Singh, A. Deep Learning for Plant Stress Phenotyping: Trends and Future Perspectives. *Trends Plant Sci.* **2018**, *23*, 883–898. [CrossRef] [PubMed]
143. Gebbers, R.; Adamchuk, V.I. Precision Agriculture and Food Security. *Science* **2010**, *327*, 828–831. [CrossRef] [PubMed]
144. Mohanty, S.P.; Hughes, D.P.; Salathé, M. Using Deep Learning for Image-Based Plant Disease Detection. *Front. Plant Sci.* **2016**, *7*, 1419. [CrossRef]
145. Sparrow, R.; Howard, M.; Degeling, C. Managing the Risks of Artificial Intelligence in Agriculture. *NJAS Impact Agric. Life Sci.* **2021**, *93*, 172–196. [CrossRef]
146. Bronson, K. Smart Farming: Including Rights Holders for Responsible Agricultural Innovation. *Technol. Innov. Manag. Rev.* **2018**, *8*, 7–14. [CrossRef]
147. Dara, R.; Hazrati Fard, S.M.; Kaur, J. Recommendations for Ethical and Responsible Use of Artificial Intelligence in Digital Agriculture. *Front. Artif. Intell.* **2022**, *5*, 884192. [CrossRef]
148. Carolan, M. Automated Agrifood Futures: Robotics, Labor and the Distributive Politics of Digital Agriculture. *J. Peasant. Stud.* **2020**, *47*, 184–207. [CrossRef]
149. Carolan, M. 'Smart' Farming Techniques as Political Ontology: Access, Sovereignty and the Performance of Neoliberal and Not-So-Neoliberal Worlds. *Sociol. Rural.* **2018**, *58*, 745–764. [CrossRef]
150. Ribeiro, M.T.; Singh, S.; Guestrin, C. "Why Should i Trust You?" Explaining the Predictions of Any Classifier. In Proceedings of the ACM SIGKDD International Conference on Knowledge Discovery and Data Mining, San Francisco, CA, USA, 13–17 August 2016; Association for Computing Machinery: New York, NY, USA, 2016; pp. 1135–1144.
151. Okengwu, U.A.; Onyejegbu, L.N.; Oghenekaro, L.U.; Musa, M.O.; Ugbari, A.O. Environmental and Ethical Negative Implications of AI in Agriculture and Proposed Mitigation Measures. *Sci. Afr.* **2023**, *22*, 141–150. [CrossRef]
152. Plackett, B. The Rural Areas Missing out on AI Opportunities. *Nature* **2022**, *610*, S17. [CrossRef]
153. Goirand, M.; Austin, E.; Clay-Williams, R. Implementing Ethics in Healthcare AI-Based Applications: A Scoping Review. *Sci. Eng. Ethics* **2021**, *27*, 1–53. [CrossRef]
154. Fukuda-Parr, S.; Gibbons, E. Emerging Consensus on 'Ethical AI': Human Rights Critique of Stakeholder Guidelines. *Glob. Policy* **2021**, *12*, 32–44. [CrossRef]
155. Siegmann, C.; Anderljung, M. The Brussels Effect and Artificial Intelligence: How EU Regulation Will Impact the Global AI Market. *arXiv* **2022**, arXiv:2208.12645. [CrossRef]
156. Carbonell, I.M. The Ethics of Big Data in Big Agriculture. *Internet Policy Rev.* **2016**, *5*, 1–13. [CrossRef]
157. Stone, P.; Brooks, R.; Brynjolfsson, E.; Calo, R.; Etzioni, O.; Hager, G.; Hirschberg, J.; Kalyanakrishnan, S.; Kamar, E.; Kraus, S.; et al. Artificial Intelligence and Life in 2030: The One Hundred Year Study on Artificial Intelligence. *arXiv* **2016**, arXiv:2211.06318. [CrossRef]
158. Demilie, W.B. Plant Disease Detection and Classification Techniques: A Comparative Study of the Performances. *J. Big Data* **2024**, *11*, 1–43. [CrossRef]

horticulturae

MDPI

Article

Research on Real-Time Automatic Picking of Ground-Penetrating Radar Image Features by Using Machine Learning

Zhi Qiu [1,2], Junyuan Zeng [2], Wenhui Tang [1], Houcheng Yang [2], Junjun Lu [2] and Zuoxi Zhao [2,*]

1 School of Electrical and Mechanical Engineering, Lingnan Normal University, Zhanjiang 524048, China
2 College of Engineering, South China Agricultural University, Guangzhou 510642, China
* Correspondence: zhao_zuoxi@scau.edu.cn; Tel.: +86-136-0004-9101

Abstract: Hard foreign objects such as bricks, wood, metal materials, and plastics in orchard soil can affect the operational safety of garden machinery. Ground-Penetrating Radar (GPR) is widely used for the detection of hard foreign objects in soil due to its advantages of non-destructive detection (NDT), easy portability, and high efficiency. At present, the degree of automatic identification applied in soil-oriented foreign object detection based on GPR falls short of the industry's expectations. To further enhance the accuracy and efficiency of soil-oriented foreign object detection, we combined GPR and intelligent technology to conduct research on three aspects: acquiring real-time GPR images, using the YOLOv5 algorithm for real-time target detection and the coordinate positioning of GPR images, and the construction of a detection system based on ground-penetrating radar and the YOLOv5 algorithm that automatically detects target characteristic curves in ground-penetrating radar images. In addition, taking five groups of test results of detecting different diameters of rebar inside the soil as an example, the obtained average error of detecting the depth of rebar using the detection system is within 0.02 m, and the error of detecting rebar along the measuring line direction from the location of the starting point of GPR detection is within 0.08 m. The experimental results show that the detection system is important for identifying and positioning foreign objects inside the soil.

Keywords: ground-penetrating radar; YOLOv5 algorithm; intelligent technology; automatically picking; coordinate positioning

Citation: Qiu, Z.; Zeng, J.; Tang, W.; Yang, H.; Lu, J.; Zhao, Z. Research on Real-Time Automatic Picking of Ground-Penetrating Radar Image Features by Using Machine Learning. *Horticulturae* **2022**, *8*, 1116. https://doi.org/10.3390/horticulturae8121116

Academic Editors: Chenglin Wang and Lufeng Luo

Received: 4 November 2022
Accepted: 24 November 2022
Published: 28 November 2022

1. Introduction

GPR has the advantages of a fast and simple detection process and good detection performance, and it is widely used as a tool for underground soil foreign matter detection [1–3], road health detection [4,5], bridge quality inspection [6], and archaeological surveys [7,8], among other areas in detection experiments.

Since GPR is often disturbed by factors such as noise and reflected waves from other materials on the ground's surface during the detection process [9–11], and since raw GPR radar images rarely provide geometric information about buried target objects, these factors are not conducive to enabling researchers to judge and interpret the geometry and specific burial location of hard foreign objects in GPR images [12,13]; thus, a major component of interpreting GPR images relies on complex data processing and the professional research experience of researchers for interpretation [14]. However, when a large number of GPR data are involved and the GPR data need to be interpreted and recognized in real-time, human interpretation of GPR data may have reduced recognition efficiency and be prone to misclassification and omission [15], so it is important to explore a method to automate the detection of underground foreign objects. Therefore, it is necessary to explore a method with which to automate the detection of targets in practical engineering applications [16]. Some researchers have proposed the use of neural networks to automatically detect and identify the features of targets detected in GPR images, which are specifically parabolic in shape [17], and to discriminate the presence of foreign bodies within the soil shown in the images by means of deep-learning algorithms so as to identify the target features in GPR images [18]. Li et al. [19] demonstrated desirable detection results in a detection task involving 2D GPR image data using a deep-learning network framework.

In the last decade, some researchers have started to use deep-learning methods to automatically identify feature parts in GPR images [20–23], especially in the field of using machine learning methods to automatically identify the characteristic curves of reinforcements in ground-penetrating radar images [24–26]. Comparing several deep-learning neural network frameworks, the algorithm based on the YOLO series neural network framework is faster in terms of detection, and Li implemented the YOLOv3 algorithm for the real-time pattern recognition of GPR images using the TensorFlow framework developed by Google [27]. Compared with the YOLOv3 algorithm and YOLOv4 algorithm, the YOLOv5 algorithm has also made significant progress with respect to small data sets, and the models trained using the YOLOv5 algorithm have superior robustness to better distinguish the feature parts in GPR images [28].

Therefore, in order to further improve the accuracy of the GPR systems' detection of foreign matter in soil and the efficiency of real-time detection, in this study, a detection system based on ground-penetrating radar and the YOLOv5 algorithm that automatically detects the target characteristic curve in ground-penetrating radar images is built, and the YOLOv5 network framework is used to detect the feature curve of the GPR pictures and accurately locate the target after detection, which achieves real-time detection in GPR pictures and the accurate localization of soil-situated foreign objects in GPR pictures.

2. Materials and Methods

2.1. GPR Image Data Set Production

When the electromagnetic waves emitted from the GPR-transmitting antenna are propagated in a soil medium, the [10] electromagnetic waves will be reflected and refracted when they encounter foreign matter in soil with different electromagnetic characteristics from the soil medium. Therefore, in the GPR image, the foreign matter in soil is specifically shown as a parabolic feature. According to this target feature, we use the labelImg annotation tool to label the GPR images with the target.

Although different foreign objects are not exactly the same on the image due to their different material sizes, most of them are parabolic features with downward openings and some of them are non-parabolic features. Therefore, we use two types of labels to label parabolic features and non-parabolic features [29], and do not distinguish between them in terms of size and material used to label the target feature areas of interest in GPR images, as shown in Figure 1.

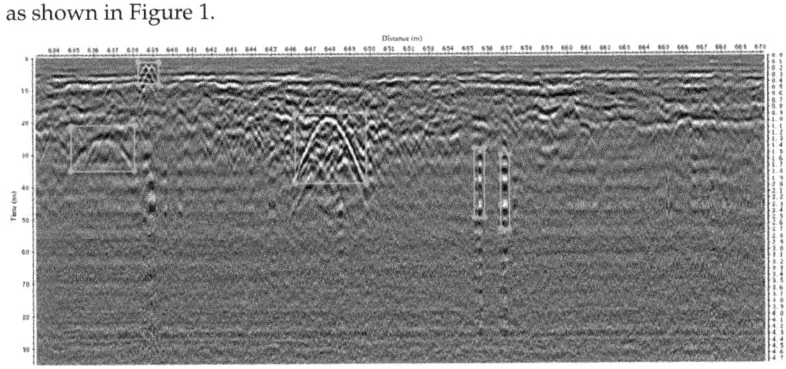

Figure 1. Parabolic target feature marker.

The dataset we used includes 295 GPR images with a total of 1679 tags, as shown in Figure 2. In addition, these tags will be divided into training set, validation set, and test set in the ratio of 7:2:1.

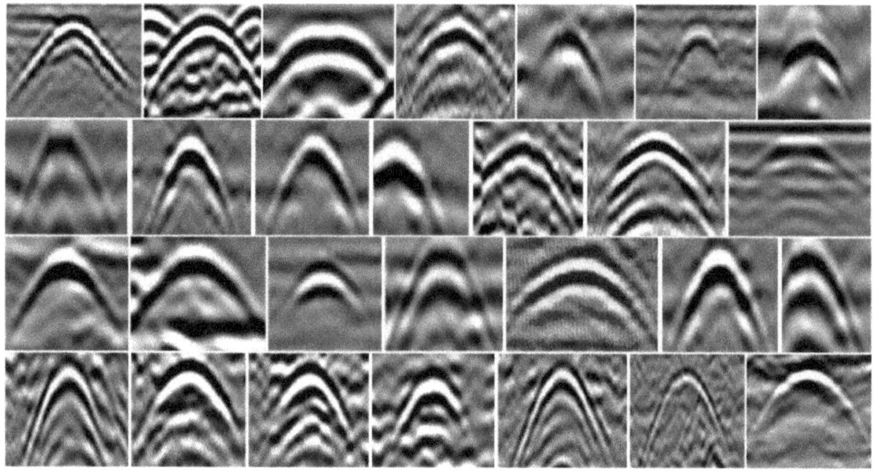

Figure 2. Partial GPR Dataset Diagram.

2.2. Real-Time Detection of GPR Image Targets Based on YOLOv5

Compared with other classic target detection algorithms, YOLOv5 is built on the PyTorch framework, with simpler support, easier deployment, and fewer model parameters, so it can be deployed on mobile devices, embedded devices, etc. It is the engineering version of the YOLO family of algorithms [30,31]. Since the experiment we conducted concerns fixing the GPR system on an all-terrain vehicle and performing background calculations through the upper computer, which should have good detection performance while ensuring real-time target detection, we decided to use the YOLOv5 network model. According to the size of this GPR image target detection dataset, we used the YOLOv5s network model in the sixth version of YOLOv5 release.

The basic framework of YOLOv5 version 6 can be divided into 4 parts: Input, Backbone, Neck, and Head. The Input part enriches the dataset by data augmentation to improve the robustness and generalization of the network model. The Backbone part mainly consists of Conv, C3, and SPPF modules for feature extraction. FPN + PANet is used in the Neck to aggregate the image features at that stage. The Head network performs target prediction and passes the predicted output, and its specific network structure is shown in Figure 3.

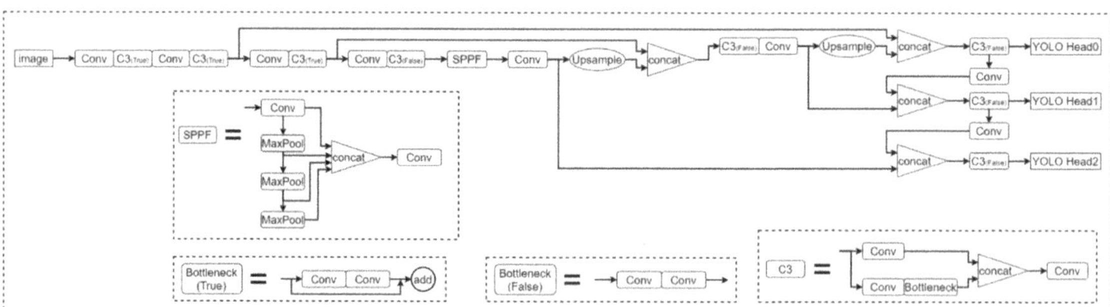

Figure 3. YOLOv5s network structure.

2.3. Evaluation of GPR Image Detection Model Based on YOLOv5 Algorithm

2.3.1. Evaluation Metrics of the Target Detection Algorithm

Judging the detection effectiveness of target detection algorithms usually requires evaluating some quantitative evaluation metrics so as to determine the algorithm's merit from an objective perspective. To quantify the effectiveness of the YOLOv5 algorithm for GPR image target detection, we chose some commonly used metrics to evaluate the model, which include accuracy (P), recall (R), average precision (AP), F1 score, inference time, and model size.

These metrics are defined as follows:

$$P = \frac{TP}{FP + TP} \tag{1}$$

$$R = \frac{TP}{FN + TP} \tag{2}$$

$$AP = \int_0^1 PRdR \tag{3}$$

$$F1 = \frac{2P \cdot R}{P + R} \tag{4}$$

where TP is the number of correctly detected targets, FP is the number of non-targets that the detector considers to be targets, and FN is the number of non-targets that the detector considers to be targets.

Generally, the precision rate (P) is the detection accuracy of the target detector at an IoU of 0.5, which is the proportion of correct targets detected by the target detector to all detected targets, the recall rate (R) is the recall rate of the target detector at an IoU of 0.5, which is the proportion of correct targets detected by the target detector to all targets in the data set. While precision and recall are mutually exclusive to some extent, i.e., high recall has lower precision and low recall has higher precision, in order to balance the indicators of recall rate and accuracy rate, we more often use average precision (AP). AP is the area of the PR curve in the range of 0–1, compared to single precision and recall, AP reflects the sum of precision at different recall rates and better expresses the detection performance of the detector.

2.3.2. Performance Comparison of Different Target Detection Algorithms in GPR Image Detection

We divided the data set processed according to Section 2.1, i.e., 295 GPR images, into training set, verification set, and test set corresponding to a ratio of 7:2:1, and trained the sets on a GPU server with a graphics card of Tesla V100 using the YOLOv5s network model in YOLOv5 version 6 mentioned in Section 2.2 with 300 epoch iterations.

We also compared other excellent target detectors such as Faster-RCNN and SSD in the same training scenario to verify that the YOLOv5 algorithm is more suitable for GPR real-time detection experiments. Due to the small number of samples of non-parabolic features, the detection results of parabolic features are quasi in this paper to compare the detection performance of each detector. The comparison results are shown in Table 1.

Table 1. Performance comparison of detection algorithms.

	Faster RCNN	**SSD**	**YOLOv5**
P	0.499	0.865	0.893
R	0.827	0.392	0.706
AP	0.715	0.604	0.805
F1	0.62	0.54	0.65
inference time	54.9 ms	9.6 ms	11 ms
weight	108 MB	91.1 MB	13.7 MB

As can be found in Table 1, the YOLOv5 network model has a detection accuracy of 89.3%, a recall of 70.6%, an average accuracy of 80.5%, an F1 score of 0.65, an inference time of 11 ms per frame, and a model size of 13.7 MB.

For the Faster RCNN algorithm, YOLOv5 algorithm has 12.1% less recall, but a 39.4% higher detection accuracy, 9% higher average accuracy, and 0.03 higher F1 score; notably, its inference time is 43.9 ms faster, and its model size is reduced by 94.3 MB. Through data comparison, it can be seen that YOLOv5 algorithm is more suitable for the real-time requirements of GPR than the Faster RCNN algorithm.

The YOLOv5 algorithm is also 1.4 ms slower than the SSD algorithm in terms of inference time, but it also meets the real-time requirements of the GPR detection system, and the YOLOv5 algorithm has a 2.8% higher detection accuracy, 31.4% higher recall, 20.1% higher average accuracy, 0.11 higher F1 score, and, most importantly, a 77.4 MB smaller model size. The data comparison shows that the YOLOv5 algorithm is more suitable for in-vehicle system deployment than the SSD algorithm, and it presents fewer false detections and omissions.

Overall, the detection speed of our YOLOv5 algorithm meets the requirement of real-time detection and has more balanced recall and accuracy compared with classical algorithms such as Faster RCNN and SSD; therefore, we obtain better detection performance in the target detection task of GPR images, and there are fewer errors and misses, which can better meet the needs of GPR image target detection in real-time. This means that we have better detection performance in GPR image target detection tasks, and fewer false and missed detections.

2.3.3. Comparison of the Effectiveness of Different Target Detection Algorithms in GPR Image Detection

We use YOLOv5 algorithm and Faster RCNN, SSD algorithm to detect GPR images and compare their effectiveness in detecting subsurface soil-situated foreign objects, including their correct recognition rate, missed detection rate, and false detection rate. The following Figure 4 comparison chart shows the effectiveness of YOLOv5 algorithm and the other two target detection algorithms in actual detection. In Figure 4, (a) is the original GPR image, (b) is the marker map of the GPR image feature curves, (c) is the detection effect of YOLOv5 algorithm, (d) is the detection effect of Faster RCNN algorithm, and (e) is the detection effect of SSD algorithm.

In Figure 4, there are several small parabolic features, two large parabolic features, and three non-parabolic features in the GPR image. Regarding the leakage phenomenon, the YOLOv5 algorithm detected the two large parabolic features, three non-parabolic features, and most of the small parabolic features. The Faster RCNN algorithm detected two large parabolic features and most of the small parabolic features but missed two non-parabolic features. The SSD algorithm detected two large parabolic features but missed three non-parabolic features and some small parabolic features were missed. In terms of misdetection, the YOLOv5 algorithm misidentifies a background feature as a small parabolic feature, the Faster RCNN algorithm misidentifies a non-parabolic feature as a parabolic feature, and the SSD misidentifies a background feature as a parabolic feature.

Overall, the YOLOv5 algorithm and the other two algorithms were able to detect most of the parabolic and non-parabolic features, and although the Faster RCNN algorithm had a higher confidence level for the target foreign matter features, the YOLOv5 algorithm had better regression accuracy for the GPR image-based target foreign matter compared to the other two target detection algorithms. The center of the upper border of the detection frame is roughly near the parabolic feature vertex, which means that the detected subsurface soil-situated foreign objects' position on the image coordinates can be roughly determined by the detection frame with better regression accuracy, which provides a certain theoretical basis for our research on the precise location of soil-situated foreign bodies.

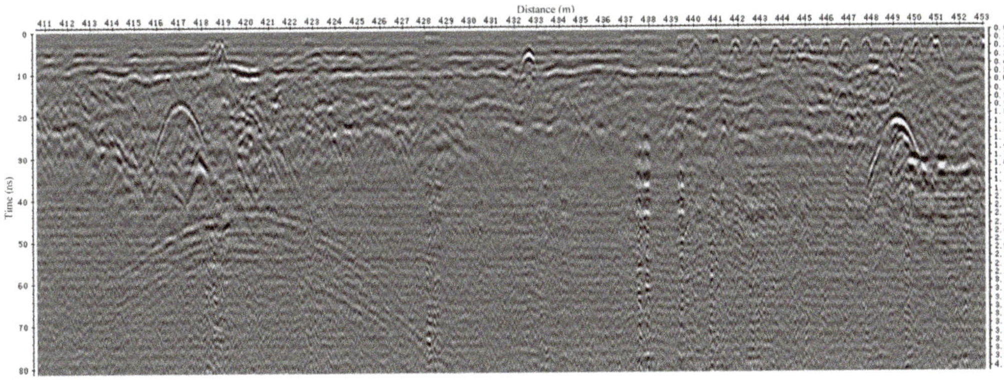

(**a**) Original GPR images

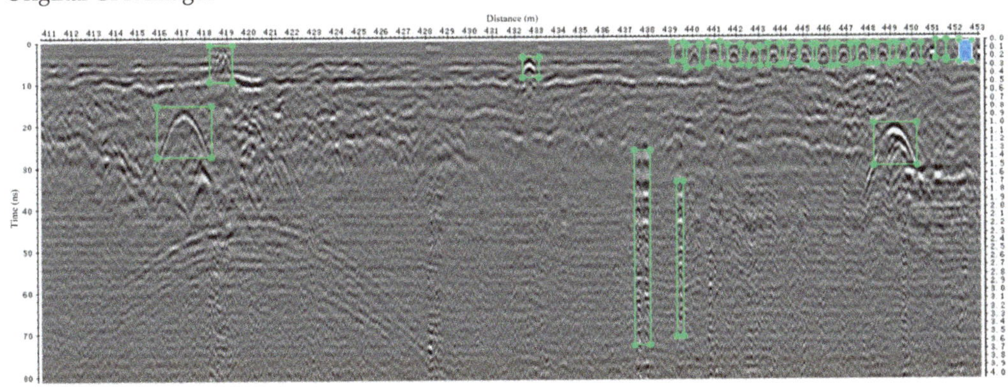

(**b**) Marker map of GPR image feature curves

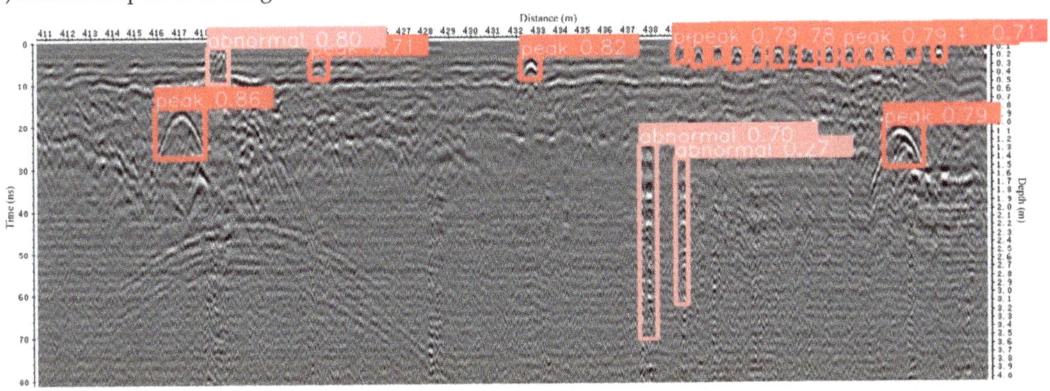

(**c**) Detection effect of the YOLOv5 algorithm

Figure 4. *Cont.*

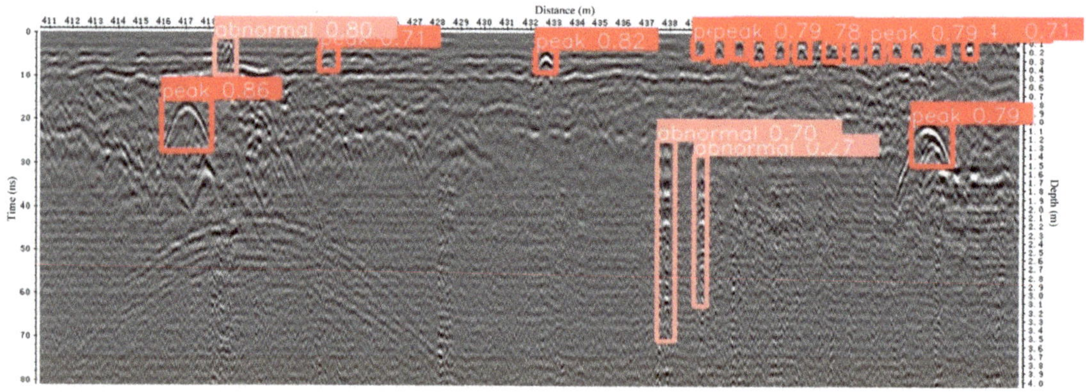

(**d**) Detection effect of Faster RCNN algorithm

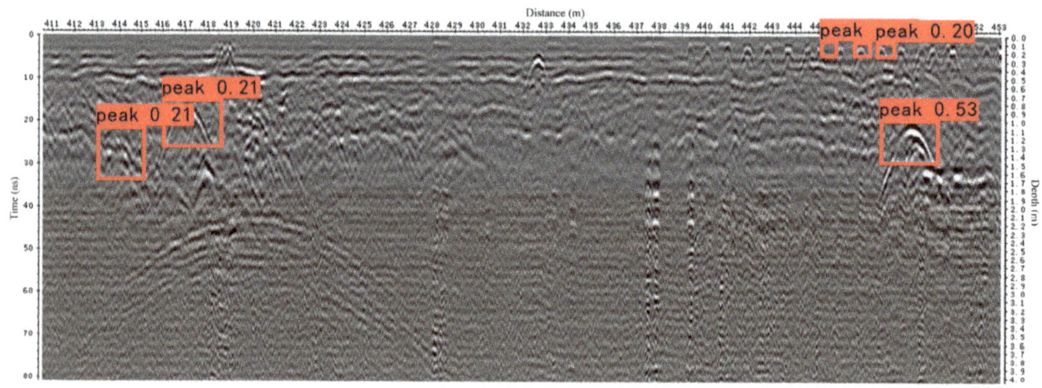

(**e**) Detection effect of SSD algorithm

Figure 4. Comparison of the effect of YOLOv5 algorithm and other two target detection algorithms in actual detection.

2.4. Building an Automatic Parabola Extraction and Detection System for Ground-Penetrating Radar Images Based on YOLOv5 Algorithm

In order to improve the accuracy and efficiency of identifying target objects in GPR images, we use a machine learning algorithm to automatically extract the foreign object feature curves of the ground-penetrating radar image in real time. The hardware components of the indoor real-time detection system built in this study include the GPR instrument, laptop, Android cell phone, and data cable; the software components include the GPR instrument equipped with MALA Controller software and YOLOv5 network structure. The GPR real-time detection system was built as shown in Figure 5.

According to the entire process of detecting the target by the ground-penetrating radar, we built a ground-penetrating radar image target feature curve automatic extraction detection system based on the ground-penetrating radar and YOLOv5 algorithms using artificial intelligence technology to achieve complex data-processing; reconstruct the GPR image coordinates; ascertain the target objects' buried location information, GPS information, and other parameters of the acquisition process; and perform real-time rapid detection and identification of target characteristic curves of GPR images obtained during the detection process. The detection and recognition process of the constructed automatic extraction and detection system of the target characteristic curves of ground-penetrating radar images is shown in Figure 6.

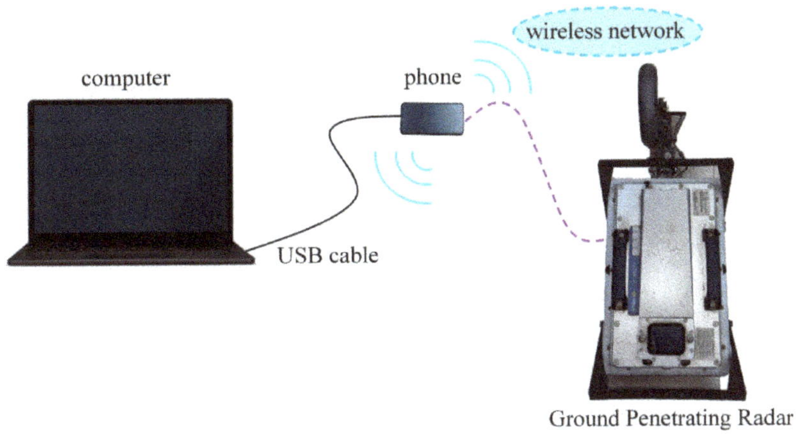

Figure 5. Components of the GPR real-time detection system.

In our experiments, we use MALA Controller software to image the GPR data, and the cell phone screen is read through the Android Debug Bridge (ADB) in order to obtain GPR images in real time. After acquiring the GPR images, the image pixel coordinates are re-established by finding the imaging section of the GPR image on the cell phone screen. The burial depth of the soil-situated foreign objects is obtained by using the linear interpolation method. Then, the value of the horizontal coordinate of the ground location in the GPR image is screenshotted, and the value of the horizontal coordinate in the GPR image is identified by using the OCR algorithm. Consequently, the position obtained using the linear interpolation method and the starting point position identified by OCR are summed. Finally, the positions obtained by the linear interpolation and the starting point positions identified by OCR are summed to obtain the distance positions of hard foreign objects in the GPR images along the measuring line in the detection process.

We acquire the GPR images by continuously reading the MALA Controller software interface in real time with the Android Debug Bridge (ADB), a development tool used for communication between computers and Android devices, which consists of three components: client (ADB client), server (ADB server), and service process (ADBD). The client (ADB client) runs on the computer and is used to send commands to the server (ADB server); the server (ADB server) runs on the computer as a client of the service process (ADBD) and is used to send commands to the Android device; the service process (ADBD) runs on the Android device and is used to execute commands on the Android device.

We first send the command os.system("adb shell screencap/sdcard/radar picture.png") to the computer client (ADB client) to communicate with the server (ADB server) through TCP port 5037; then, the server (ADB server) will communicate with the Android device through USB serial port or TCP. At this time, the service process (ADBD) will intercept the screen of the Android device in real time so as to obtain the GPR images recorded by the running MALA Controller software and store them in the memory card of the Android device. Similarly, the computer client (ADB client) sends the command os.system("adb pull/sdcard/radar image.png") to transfer the GPR image stored in the memory card of the Android device to the computer. The above operation realizes the real-time acquisition of GPR images. The principle of real-time acquisition of GPR images is shown in Figure 7.

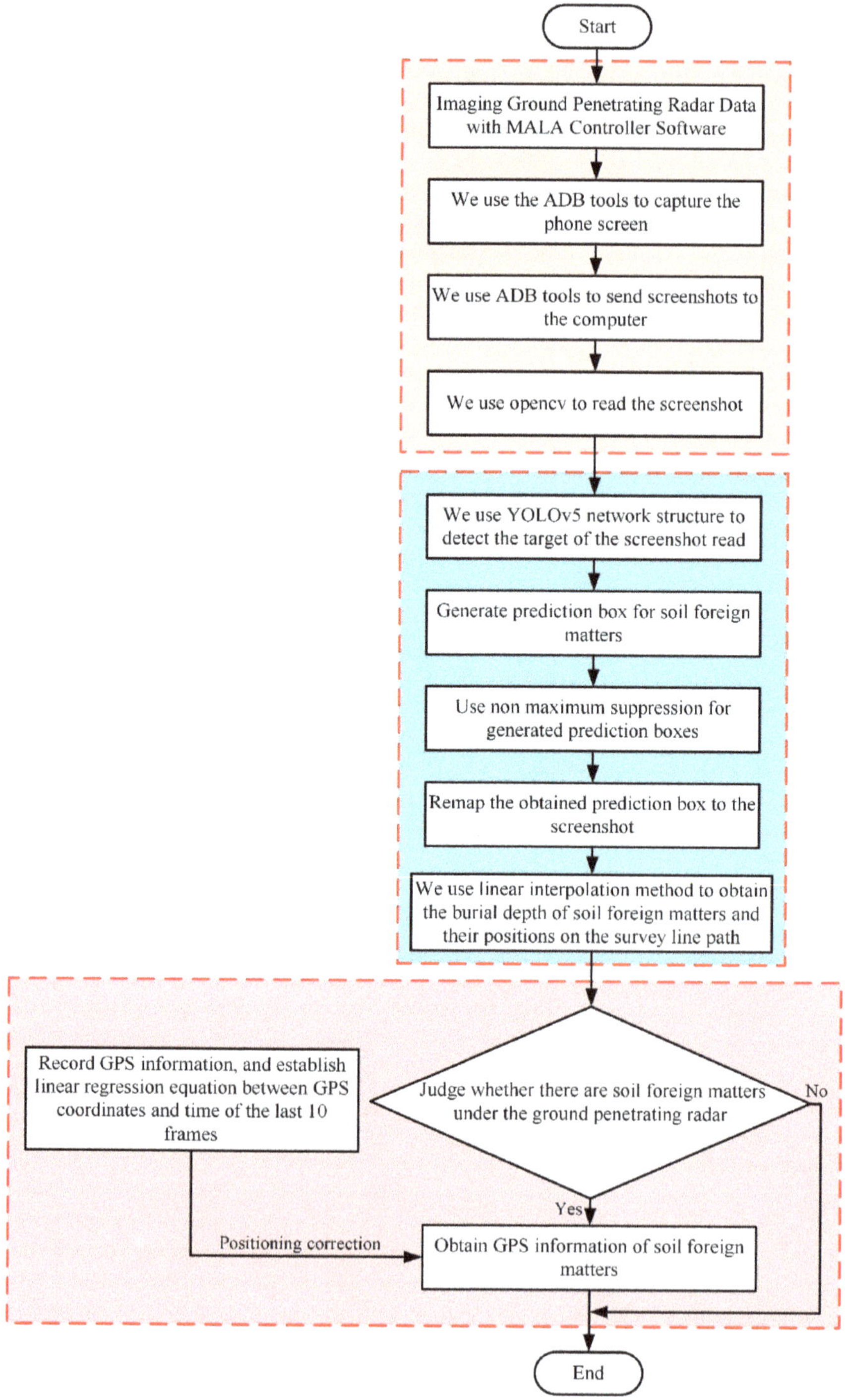

Figure 6. Flow chart for building a real-time detection and identification system.

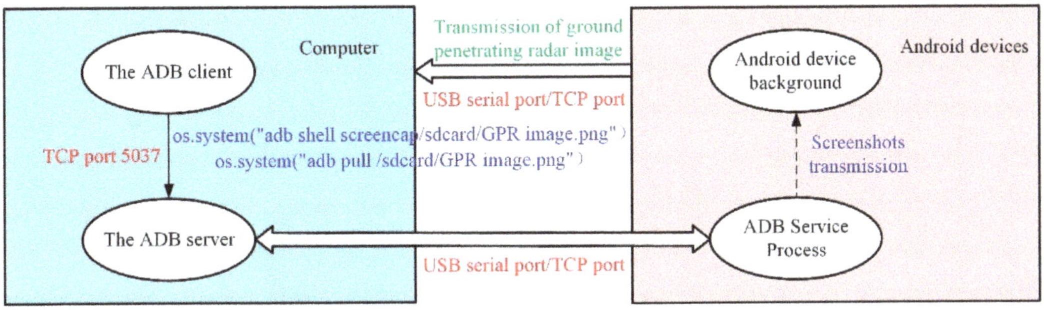

Figure 7. Schematic diagram of real-time acquisition of GPR image.

During the test, the GPR system uses a trigger wheel to control the triggering of the GPR pulse, transmitting and receiving an electromagnetic wave pulse every step. MALA Controller software communicates with the GPR, and the MALA Controller updates the image every time the GPR receives a new pulse, while the ABD tool takes continuous screenshots of the cell phone screen, thus realizing real-time data acquisition of the whole detection process of GPR.

2.5. Indoor Testing Based on the Detection System

We conducted indoor detection experiments in the Key Laboratory of Agricultural Machinery and Equipment of the Ministry of Education, which was built on common farmland soil platform in the south of China. The soil platform has dimensions of 6 m in length, 2 m in width, and 1.5 m in depth, and contains a batch of ferrous experimental materials, such as steel bars and iron plate materials. The high-dynamic GPR MALA GX750 HDR instrument was selected to detect the target objects buried inside the soil. In order to achieve real-time detection, we used MALA Controller, a cell phone software developed by MALA, to assist us to understand and interpret the detection results more intuitively, as well as obtain the GPR image data in real time.

In the soil platform in the Key Laboratory of Agricultural Machinery and Equipment of the Ministry of Education in the south, we used rebar as an example, whose diameters were 5 cm, 4 cm, and 3 cm, with lengths of 15 cm. The depth of the rebar was 15 cm, and the buried positions were 50 cm, 150 cm, and 250 cm from the starting point; the specific buried positions of the rebar are shown in Figure 8.

Figure 8. Test diagram of indoor real-time detection of steel bars.

We obtained the GPR image data recorded by MALA Controller software in real time by taking screenshots of the cell phone screen with the ABD tool, and part of this process is shown in Figure 9. Figure 9a shows a screenshot of the computer reading the interface of the MALA Controller software when the GPR instrument scans above the position of the first piece of rebar; Figure 9b shows a screenshot of the computer reading the interface of MALA Controller software when the GPR instrument has completely passed the buried position of the first piece of rebar, but before reaching the buried position of the second piece of rebar; Figure 9c is a screenshot of the computer reading the MALA Controller software interface when the GPR instrument reaches the top of the second piece of rebar; Figure 9d is a screenshot of the computer reading the MALA Controller software interface when the GPR instrument reaches the top of the third piece of rebar; and Figure 9e is a screenshot of the computer reading the MALA Controller software interface when the GPR instrument has completely crossed the buried position of the third piece of rebar. A screenshot of MALA Controller software interface is shown in Figure 9e.

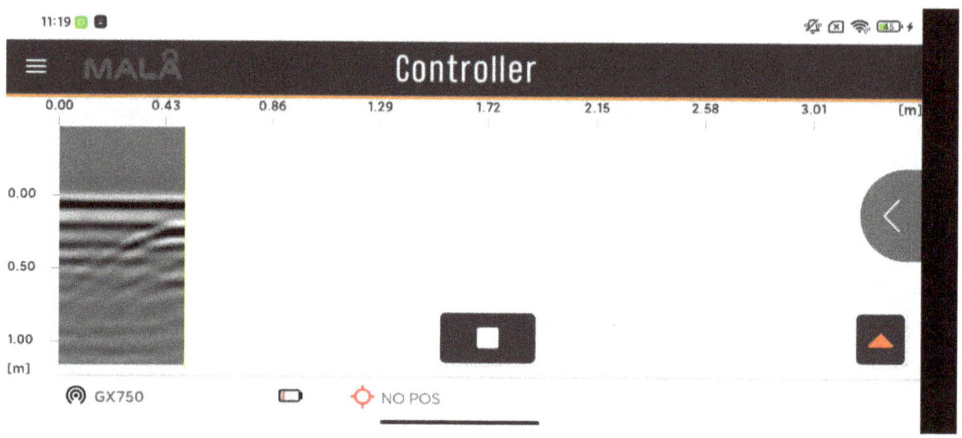

(**a**) The resulting image recorded by the ground-penetrating radar above the first steel bar.

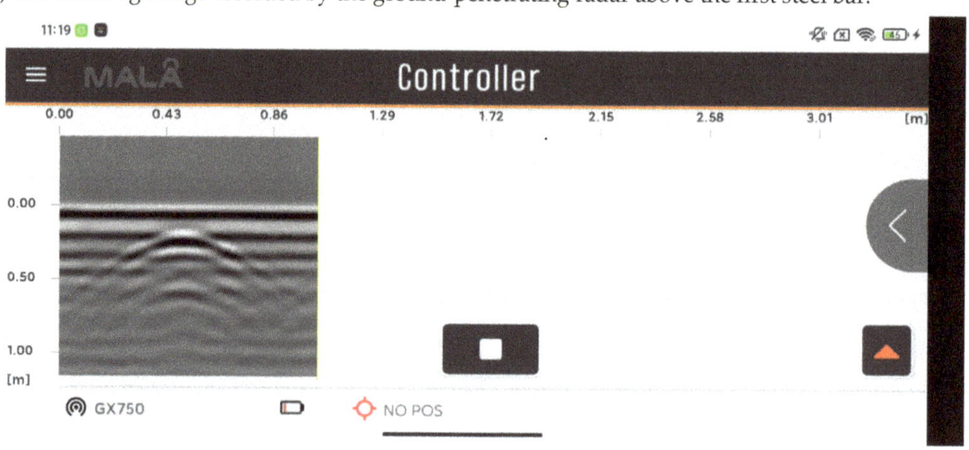

(**b**) The resulting image recorded by the ground-penetrating radar after passing over the first steel bar.

Figure 9. *Cont.*

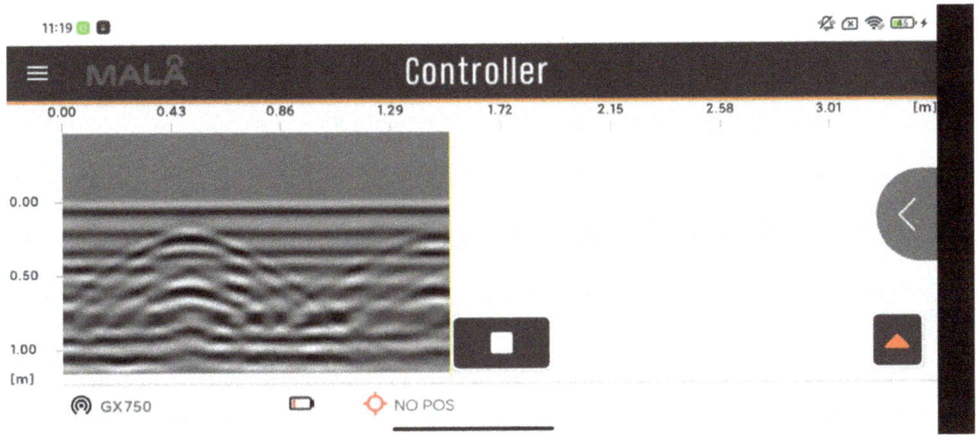

(**c**) The resulting image recorded by the ground-penetrating radar above the second steel bar.

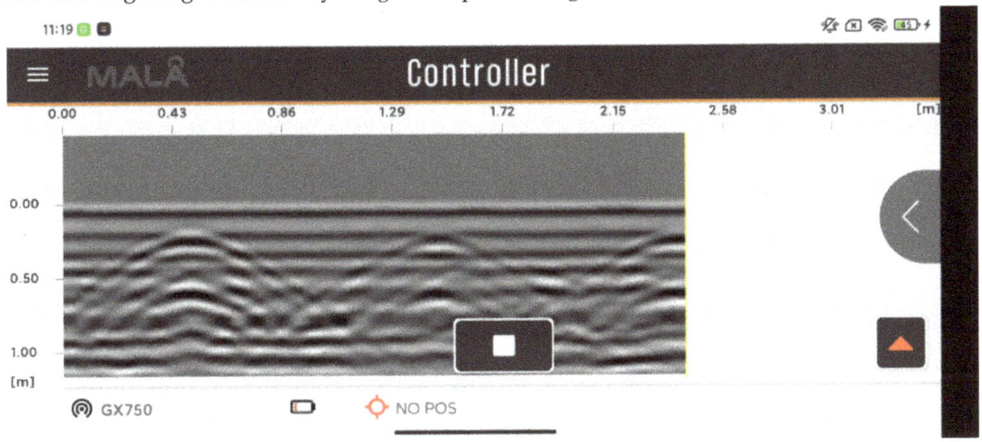

(**d**) The resulting image recorded by the ground-penetrating radar above the third steel bar.

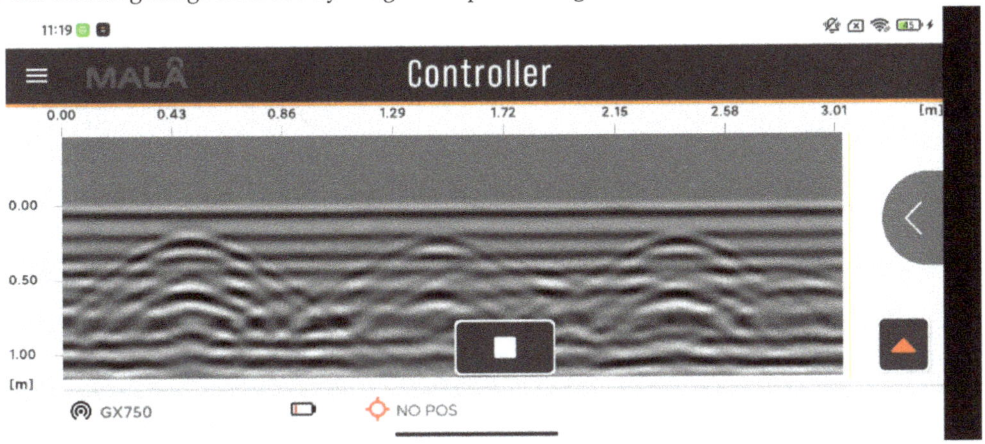

(**e**) Plot of the results recorded after GPR passed over all the steel bars.

Figure 9. The ABD tool captures part of the process of the mobile phone screen.

After we acquired the GPR images using MALA Controller software and intercepted the cell phone screen with ADB tool, according to the characteristics of the MALA Controller software interface, the image pixel coordinates were re-established after finding the imaging part of the GPR in the MALA Controller software interface. The image pixel coordinates before re-establishment are shown in Figure 10.

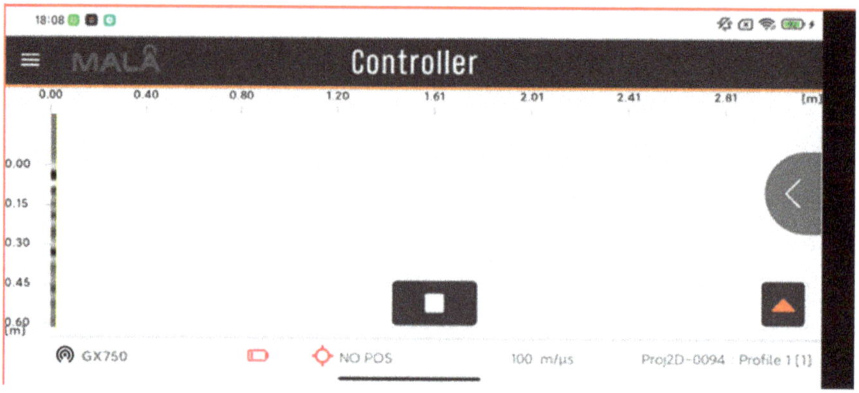

Figure 10. Image pixel coordinate diagram before reconstruction.

We found the GPR-imaging section in the MALA Controller software interface and established new image pixel coordinates based on this. The re-established image pixel coordinates are shown in Figure 11.

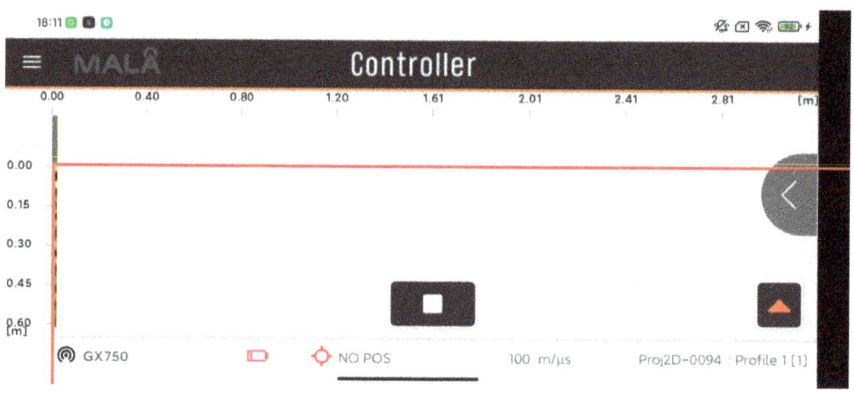

Figure 11. Reconstructed image pixel coordinate map.

Next, we choose a certain length of pixel vertical coordinates as the unit pixel depth; for example, note that the pixel vertical coordinate corresponding to 0.00 in the vertical coordinate is y_1, while the actual burial depth is d_1; note that the pixel vertical coordinate corresponding to 0.45 in the vertical coordinate is y_2, while the actual burial depth is d_2. Then, in the image pixel coordinate system, the difference between y_2 and y_1 describes the difference between d_2 and d_1 with respect to the local coordinate difference, which is the depth of 0.45 m.

We read the pixel coordinates of the YOLOv5 prediction frame on the image; find the pixel vertical coordinate Y of the foreign object, i.e., the peak of the hyperbola; and then use linear interpolation to solve for the burial depth of the foreign object in the local coordinate D. The solution formula is shown in Equation (5).

$$\frac{Y - y_1}{y_2 - y_1} = \frac{D - d_1}{d_2 - d_1} \tag{5}$$

Since, in the process of the GPR detection of underground foreign objects, the depth coordinate axis in the image does not change with the change in the GPR moving distance, then the burial depth obtained by selecting the 0.00 point under the new pixel coordinate system as the starting point is the burial depth in local coordinates.

However, when the GPR probes more than a certain distance along the survey line, i.e., when the GPR receives more than a certain number of columns of signal data, the MALA Controller software will only image the last N columns of signal data and refresh the horizontal coordinates of the MALA Controller software interface at the same time. Therefore, when the GPR moves more than a certain distance, the value of the horizontal axis of the MALA Controller software interface will change, i.e., the distance represented by the starting point in the new pixel coordinate system will change. Therefore, in order to obtain the distance of the target foreign object in the actual GPR moving along the measuring line's direction, in addition to the above-mentioned linear difference method, the distance of the GPR moving along the measuring line direction corresponding to the starting point under the new pixel coordinate system should be added. Take Figure 12 as an example: at this point the distance represented by the point 0.00 of the new pixel coordinates is 0.51 m.

Since the position of the measuring line distance axis of the ground-penetrating radar in the GPR image is fixed, we only need to identify the coordinates of the starting point and the ending point of the measuring line distance axis in the image and read out the pixel coordinates of the starting point and the ending point of the measuring line distance axis and the hyperbola vertex of the soil-situated hard foreign matter; then, we can obtain the measuring line distance of the soil-situated hard foreign matter through the linear interpolation method.

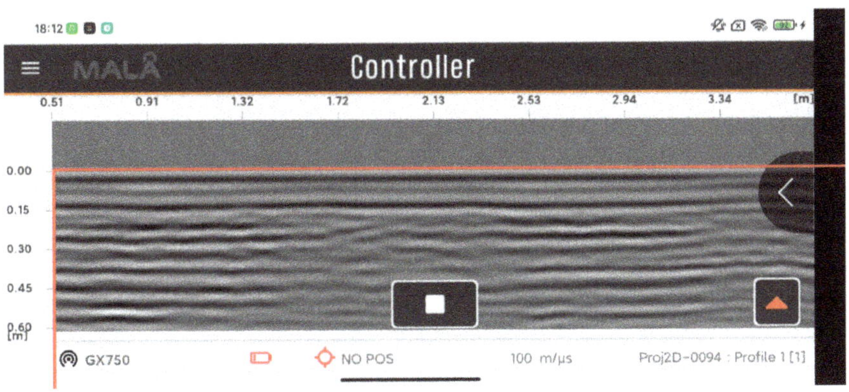

Figure 12. Reconstructed image pixel coordinate map.

In order to solve the above problem, we first obtain the digital images containing only the starting and ending points of the distance axis of the measurement line corresponding to 0.51 and 3.34 in Figure 12 by cutting the images, and then use the recognition engine of Tesseract-OCR to perform OCR text recognition on the cut digital picture, OCR text recognition is then performed on the clipped digital images using Tesseract OCR's recognition engine. Thus, the distance of the measured line from the start point of the axis and the end one are obtained.

Finally, the distance between the local coordinates of the starting point and the end point of the line of sight distance axis identified by OCR, as well as the starting point and the end point of the line of sight distance axis, and the pixel coordinates of the target foreign object in the ground-penetrating radar image are used to obtain the position of the soil hard foreign object along the measurement line by linear interpolation method. The position equation of the target foreign object in the GPR displacement direction is shown in Equation (6).

$$\frac{X - x_1}{x_2 - x_1} = \frac{S - s_1}{s_2 - s_1} \tag{6}$$

where X is the pixel horizontal coordinate of the target foreign object parabola in the GPR image, x_1 is the new pixel horizontal coordinate of the start point of the distance axis of the measurement line, x_2 is the new pixel horizontal coordinate of the end point of the distance axis of the measurement line, S is the distance of the target foreign object in local coordinates along the measurement line, s_1 is the distance of the start point of the distance axis of the measurement line in local coordinates along the measurement line, s_2 is the distance of the end point of the distance axis of the measurement line in local coordinates along the measurement line.

The abscissa in the GPR image is shown in Figure 13, we recognize that the starting point of the distance axis of the current survey line is 0.51 and the ending point is 3.34 through the OCR text recognition algorithm; then, we can read that the new pixel abscissa of the image corresponding to the starting point (0.51) is 0, the new pixel abscissa of the image corresponding to the end point (3.34) is 400, and the new pixel abscissa of the image corresponding to the hyperbola vertex of the soil-situated hard foreign object is 100. Consequently, the lateral distance (dist) of the soil-situated hard foreign object is shown in Equation (7).

$$\frac{\text{dist} - 0.51}{100 - 0} = \frac{3.34 - 0.51}{400 - 0} \tag{7}$$

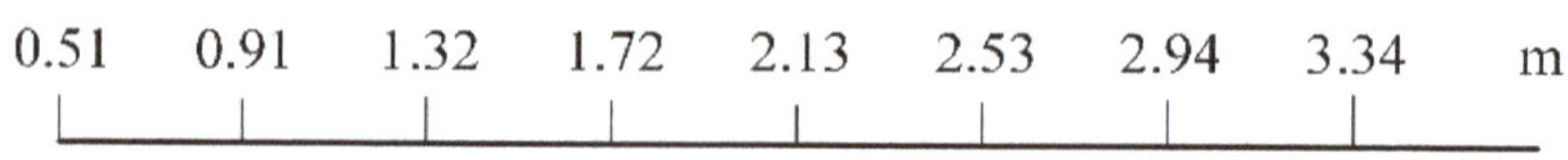

Figure 13. Abscissa in GPR image.

Similarly, through the observation of the depth coordinate axis and the reading of pixel coordinates, the burial depth of soil-situated hard foreign matter can be obtained with the help of linear interpolation method

The burial depth and displacement position obtained by linear interpolation method are shown in Figure 14.

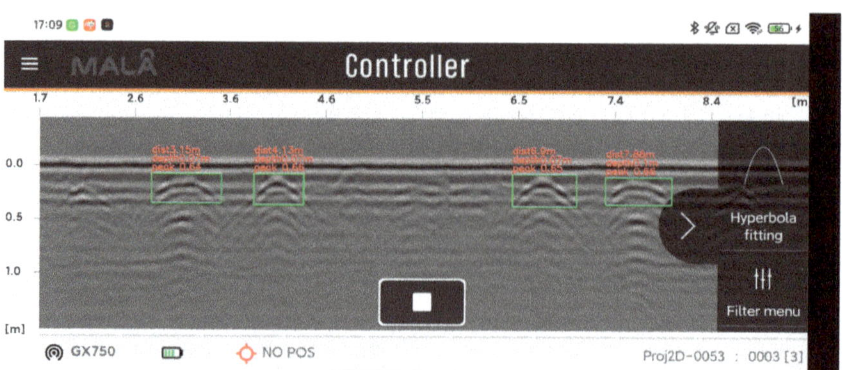

Figure 14. Buried depth and displacement position map obtained by linear interpolation.

3. Results and Discussion

We carried out detection experiments in the soil platform of the Key Laboratory of the Ministry of Education of Southern Agricultural Machinery and Equipment. We selected 15 cm long steel bars as foreign matter in the examined soil. The buried depth of the steel bars is 15 cm, and the buried positions are 50 cm, 150 cm, and 250 cm away from the starting point. The specific buried positions are shown in Figure 8.

We used the MALA GX 750 HDR instrument developed by the MALA Company to carry out detection experiments. The instrument collected 412 samples in each channel with a sampling spacing of 0.015 m; the diameter of the ranging wheel of the GPR instrument was 17 cm, and the propagation speed of the electromagnetic wave we selected in the soil medium was 100 m/μs.

In order to more clearly show the accuracy of the hyperbola position of the target in the GPR image detected by the detection system, the position where the GPR initiates detection each time is fixed during the GPR experiment. Therefore, we only need to read the pixel coordinates of the starting and ending points of the distance axis in the GPR image and the coordinates of the parabolic vertex in the GPR image, and then obtain the position of the rebar on the survey line through a linear interpolation.

Firstly, we used the MALA Controller mobile software to read the GPR data in real time and visualize it; then, the computer acquired the visualized GPR data in real time through the ADB tool, and the YOLOv5 algorithm automatically detected the hyperbola of the GPR image acquired by the ADB tool in real time. Consequently, the hyperbolic vertex pixel coordinates were obtained according to the position of the box, and the real-time location of the foreign matter in the soil was realized by linear interpolation. Finally, we compared and analyzed the error between the positioning results of our real-time detection system and the actual position.

We used the built real-time detection system for an indoor detection test; the real-time identification results output by the system in the detection process are shown in Figure 15; Figure 15a shows the real-time detection system used to identify the detection results of the GPR after the first piece of rebar was detected; Figure 15b shows the real-time detection system used to identify the detection results of GPR after detecting the second piece of rebar; and Figure 15c shows the real-time detection system used to identify the GPR results following the detection of the third piece of rebar. From Figure 15c, it can be seen that the three target bodies can be marked in real time, and the detection result map also marks the burial depth and the distance from the detection zero position of the three rebar pieces identified by the system.

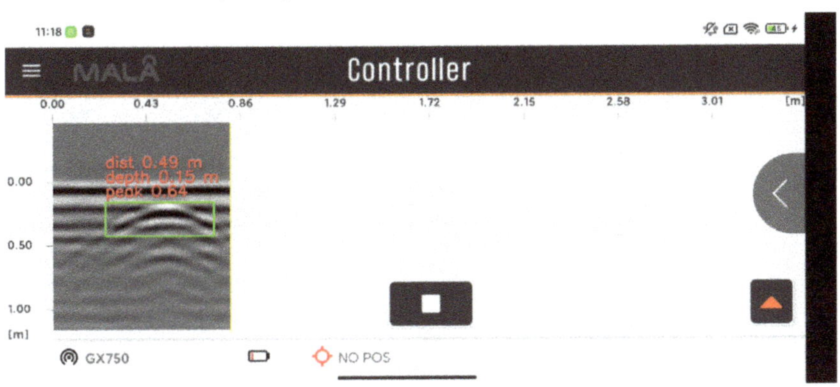

(**a**) The result of the real-time detection system used to identify the ground-penetrating radar results after passing over the first steel bar.

Figure 15. *Cont.*

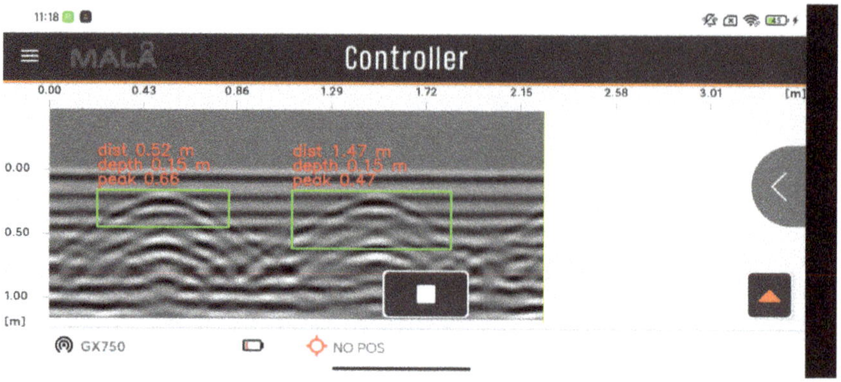

(**b**) The result of the real-time detection system used to identify the ground-penetrating radar results after passing over through the second steel bar.

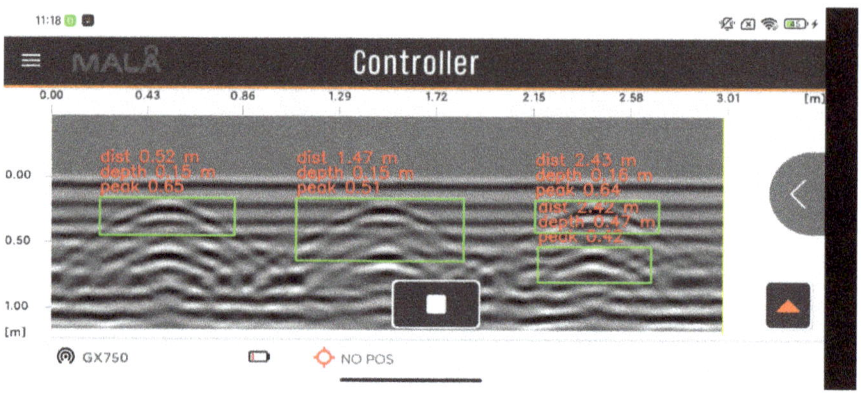

(**c**) The result of the real-time detection system used to identify the ground-penetrating radar results after passing over through the third steel bar.

Figure 15. Target location results of ground-penetrating radar images.

We used the real-time detection system along the detection direction to repeat the test five times. The comparison of the results regarding the detection of rebar along the detection line direction from the GPR detection starting point's location and the actual burial distance is shown in Table 2, while the comparison of the results obtained by the real-time detection system regarding the detection of the buried depth of the rebar and the actual buried location is shown in Table 3.

Table 2. Comparison of the distance between the steel bar detected by the ground-penetrating radar real-time detection system and the actual buried distance (unit/m).

	Rebar 1	**Rebar 2**	**Rebar 3**
The actual burial distance of the rebar	0.50	1.50	2.50
The value of the first measurement distance	0.52	1.47	2.42
The value of the second measurement distance	0.52	1.46	2.42
The value of the third measurement distance	0.52	1.47	2.43
The value of the fourth measurement distance	0.52	1.46	2.42
The value of the fifth measurement distance	0.52	1.46	2.42
The average error	0.020	0.036	0.078
Root mean square error	0.020	0.036	0.078

Table 3. Comparison of the depth between the steel bar detected by the ground-penetrating radar real-time detection system and the actual buried depth (unit/m).

	Rebar 1	Rebar 2	Rebar 3
The actual burial depth of rebar	0.15	0.15	0.15
The value of the first measurement depth	0.16	0.16	0.17
The value of the second measurement depth	0.15	0.15	0.16
The value of the third measurement depth	0.16	0.15	0.16
The value of the fourth measurement depth	0.16	0.17	0.17
The value of the fifth measurement depth	0.15	0.17	0.17
The average error	0.006	0.010	0.016
Root mean square error	0.008	0.013	0.017

From Tables 2 and 3, we can see that the average error of the depth of rebar detected by the constructed real-time detection system is within 0.02 m, and the error of detecting rebar along the measurement line from the location of the starting point of GPR detection is within 0.08 m, as seen from the test results regarding the detection of different diameters of rebar in five groups.

4. Conclusions

In this study, a detection system based on ground-penetrating radar and the YOLOv5 algorithm for automatically detecting target characteristic curves in ground-penetrating radar images was built, and the YOLOv5 network framework was used to detect the feature curves of the GPR pictures and accurately locate the targets after detection, which achieved the real-time detection of GPR pictures and the accurate localization of soil-situated foreign objects in GPR pictures. The contributions of this paper are as follows.

(1) We built a real-time detection system, used the ABD tool to take continuous screenshots of the employed cell phone screen when using the GPR instruments for detection, and used the YOLOv5 algorithm for the real-time target detection of GPR images, which successfully achieved ideal results regarding the detection of the target features of GPR images.

(2) Using soil-situated foreign object localization based on the linear interpolation method, we converted the pixel coordinates of soil-situated foreign object features in GPR images into local coordinates to determine the specific burial locations of soil-situated foreign objects. Then, we read the pixel coordinates of the YOLOv5 prediction frame on the images; ascertained the pixel longitudinal coordinates of foreign objects, i.e., the peak of the hyperbola; and then used the linear interpolation method to solve for the burial location of the soil-situated foreign objects in local coordinates.

Since the Faster RCNN, SSD, and YOLOv5 algorithms used herein are only image target detection algorithms, which are mainly used for the automatic detection of hyperbolas in GPR images, at this stage, the algorithms we used do not process the parameters of GPR signals. Therefore, in our future research work, we will use the algorithm to process the parameters of GPR signals and conduct in-depth research in this direction.

Author Contributions: Conceptualization, Z.Z. and Z.Q.; methodology, Z.Q.; software, J.Z.; validation, H.Y.; formal analysis, W.T.; investigation, J.L.; writing—original draft preparation, Z.Q.; writing—review and editing, J.Z. and H.Y.; funding acquisition, Z.Z. All authors have read and agreed to the published version of the manuscript.

Funding: The authors would like to acknowledge the support of this study provided by the Guangdong Provincial Department of Agriculture's Modern Agricultural Innovation Team Program for Animal Husbandry Robotics (Grant No. 2019KJ129), the State Key Research Program of China (Grant No. 2016YFD0700101), the Vehicle Soil Parameter Collection and Testing Project (Grant No. 4500-F21445), and the Special project of Guangdong Provincial Rural Revitalization Strategy in 2020 (YCN [2020] No. 39) (Fund No. 200-2018-XMZC-0001-107-0130).

Data Availability Statement: Not applicable.

Conflicts of Interest: The authors declare no conflict of interest.

References

1. Li, Y.; Zhao, Z.; Xu, W.; Liu, Z.; Wang, X. An effective FDTD model for GPR to detect the material of hard objects buried in tillage soil layer. *Soil Tillage Res.* **2019**, *195*, 104353. [CrossRef]
2. Meschino, S.; Pajewski, L.; Pastorino, M.; Randazzo, A.; Schettini, G. Detection of subsurface metallic utilities by means of a SAP technique: Comparing MUSIC- and SVM-based approaches. *J. Appl. Geophys.* **2013**, *97*, 60–68. [CrossRef]
3. Zhang, X.; Derival, M.; Albrecht, U.; Ampatzidis, Y. Evaluation of a Ground Penetrating Radar to Map the Root Architecture of HLB-infected Citrus Trees. *Agronomy* **2019**, *9*, 354. [CrossRef]
4. Hong, W.; Kang, S.; Lee, S.J.; Lee, J. Analyses of GPR signals for characterization of ground conditions in urban areas. *J. Appl. Geophys.* **2018**, *152*, 65–76. [CrossRef]
5. Krysiński, L.; Sudyka, J. GPR abilities in investigation of the pavement transversal cracks. *J. Appl. Geophys.* **2013**, *97*, 27–36. [CrossRef]
6. Abouhamad, M.; Dawood, T.; Jabri, A.; Alsharqawi, M.; Zayed, T. Corrosiveness mapping of bridge decks using image-based analysis of GPR data. *Autom. Constr.* **2017**, *80*, 104–117. [CrossRef]
7. Cuenca-García, C.; Risbøl, O.; Bates, C.R.; Stamnes, A.A.; Skoglund, F.; Ødegård, Ø.; Viberg, A.; Koivisto, S.; Fuglsang, M.; Gabler, M.; et al. Sensing Archaeology in the North: The Use of Non-Destructive Geophysical and Remote Sensing Methods in Archaeology in Scandinavian and North Atlantic Territories. *Remote Sens.* **2020**, *12*, 3102. [CrossRef]
8. Papadopoulos, N.; Sarris, A.; Yi, M.; Kim, J. Urban archaeological investigations using surface 3D Ground Penetrating Radar and Electrical Resistivity Tomography methods. *Explor. Geophys.* **2018**, *40*, 56–68. [CrossRef]
9. Li, W.; Cui, X.; Guo, L.; Chen, J.; Chen, X.; Cao, X. Tree Root Automatic Recognition in Ground Penetrating Radar Profiles Based on Randomized Hough Transform. *Remote Sens.* **2016**, *8*, 430. [CrossRef]
10. Jin, Y.; Duan, Y. A new method for abnormal underground rocks identification using ground penetrating radar. *Measurement* **2020**, *149*, 106988. [CrossRef]
11. Shuang-Fei, L.I.; Jia-Cun, L.I.; Zhang, D. Topographic correction of GPR profiles based on differential GPS. *J. Geomech.* **2022**, *22*, 771–777.
12. Chae, J.; Ko, H.Y.; Lee, B.G.; Kim, N. A Study on the Pipe Position Estimation in GPR Images Using Deep Learning Based Convolutional Neural Network. *J. Internet Comput. Serv.* **2019**, *20*, 39–46.
13. Wang, S.; Al-Qadi, I.L.; Cao, Q. Factors Impacting Monitoring Asphalt Pavement Density by Ground Penetrating Radar. *NDT E Int.* **2020**, *115*, 102296. [CrossRef]
14. Jin, Y.; Duan, Y. Wavelet Scattering Network-Based Machine Learning for Ground Penetrating Radar Imaging: Application in Pipeline Identification. *Remote Sens.* **2020**, *12*, 3655. [CrossRef]
15. Jiao, L.; Ye, Q.; Cao, X.; Huston, D.; Xia, T. Identifying concrete structure defects in GPR image. *Measurement* **2020**, *160*, 107839. [CrossRef]
16. Yuan, C.; Cai, H. Spatial reasoning mechanism to enable automated adaptive trajectory planning in ground penetrating radar survey. *Automat. Constr.* **2020**, *114*, 103157. [CrossRef]
17. Rajiv, K. Multi-Feature Based Multiple Pipelines Detection Using Ground Penetration Radar. *Int. J. Comput. Intell. Res.* **2017**, *13*, 1123–1138.
18. Shaw, M.R.; Millard, S.G.; Molyneaux, T.C.K.; Taylor, M.J.; Bungey, J.H. Location of steel reinforcement in concrete using ground penetrating radar and neural networks. *NDT E Int.* **2005**, *38*, 203–212. [CrossRef]
19. Li, Y.; Liu, C.; Yue, G.; Gao, Q.; Du, Y. Deep learning-based pavement subsurface distress detection via ground penetrating radar data. *Automat. Constr.* **2022**, *142*, 104516. [CrossRef]
20. Wang, Y.; Qin, H.; Tang, Y.; Zhang, D.; Wang, Z.; Pan, S. A deep learning network to improve tunnel lining defect identification using ground penetrating radar. *IOP Conf. Ser. Earth Environ. Sci.* **2021**, *861*, 42057–42058. [CrossRef]
21. Dinh, K.; Gucunski, N.; Duong, T.H. An algorithm for automatic localization and detection of rebars from GPR data of concrete bridge decks. *Automat. Constr.* **2018**, *89*, 292–298. [CrossRef]
22. Qin, H.; Zhang, D.; Tang, Y.; Wang, Y. Automatic recognition of tunnel lining elements from GPR images using deep convolutional networks with data augmentation. *Automat. Constr.* **2021**, *130*, 103830. [CrossRef]
23. Cui, F.; Ning, M.; Shen, J.; Shu, X. Automatic recognition and tracking of highway layer-interface using Faster R-CNN. *J. Appl. Geophys.* **2022**, *196*, 104477. [CrossRef]
24. Park, S.; Kim, J.; Jeon, K.; Kim, J.; Park, S. Improvement of GPR-Based Rebar Diameter Estimation Using YOLO-v3. *Remote Sens.* **2021**, *13*, 2011. [CrossRef]
25. Liu, H.; Lin, C.; Cui, J.; Fan, L.; Xie, X.; Spencer, B.F. Detection and localization of rebar in concrete by deep learning using ground penetrating radar. *Automat. Constr.* **2020**, *118*, 103279. [CrossRef]
26. Tešić, K.; Baričević, A.; Serdar, M. Non-Destructive Corrosion Inspection of Reinforced Concrete Using Ground-Penetrating Radar: A Review. *Materials* **2021**, *14*, 975. [CrossRef]
27. Li, Y.; Zhao, Z.; Luo, Y.; Qiu, Z. Real-Time Pattern-Recognition of GPR Images with YOLO v3 Implemented by Tensorflow. *Sensors* **2020**, *20*, 6476. [CrossRef] [PubMed]

28. Li, S.; Gu, X.; Xu, X.; Xu, D.; Zhang, T.; Liu, Z.; Dong, Q. Detection of concealed cracks from ground penetrating radar images based on deep learning algorithm. *Constr. Build. Mater.* **2021**, *273*, 121949. [CrossRef]
29. Ishitsuka, K.; Iso, S.; Onishi, K.; Matsuoka, T. Object Detection in Ground-Penetrating Radar Images Using a Deep Convolutional Neural Network and Image Set Preparation by Migration. *Int. J. Geophys.* **2018**, *2018*, 9365184. [CrossRef]
30. Redmon, J.; Divvala, S.; Girshick, R.; Farhadi, A. You only look once: Unified, real-time object detection. In Proceedings of the IEEE Conference on Computer Vision and PATTERN Recognition, Las Vegas, NV, USA, 27–30 June 2016; pp. 779–788.
31. Redmon, J.; Farhadi, A. YOLOv3: An Incremental Improvement. *arXiv* **2018**, arXiv:1804.02767.

horticulturae

Article

HeLoDL: Hedgerow Localization Based on Deep Learning

Yanmei Meng [1], Xulei Zhai [1], Jinlai Zhang [1], Jin Wei [1,*], Jihong Zhu [2] and Tingting Zhang [1]

1 College of Mechanical Engineering, Guangxi University, Nanning 530004, China
2 Department of Precision Instrument, Tsinghua University, Beijing 100190, China
* Correspondence: jixiegong@163.com

Abstract: Accurate localization of hedges in 3D space is a key step in automatic pruning. However, due to the irregularity of the hedge shape, the localization accuracy based on traditional algorithms is poor. In this paper, we propose a deep learning approach based on a bird's-eye view to overcoming this problem, which we call *HeLoDL*. Specifically, we first project the hedge point cloud top-down as a single image and, then, augment the image with morphological operations and rotation. Finally, we trained a convolutional neural network, *HeLoDL*, based on transfer learning, to regress the center axis and radius of the hedge. In addition, we propose an evaluation metric OIoU that can respond to the radius error, as well as the circle center error in an integrated way. In our test set, *HeLoDL* achieved an accuracy of 90.44% within the error tolerance, which greatly exceeds the 61.74% of the state-of-the-art algorithm. The average OIoU of *HeLoDL* is 92.65%; however, the average OIoU of the best conventional algorithm is 83.69%. Extensive experiments demonstrated that *HeLoDL* shows considerable accuracy in the 3D spatial localization of irregular models.

Keywords: green hedge; point cloud; model fitting; deep learning; CNN

1. Introduction

With the rise of the concept of "carbon peaking and carbon neutral", the task of achieving green development has become the mission of the times. Regular maintenance of hedges is an important task in the green development process. However, due to the increasing number of hedges, it has become increasingly difficult to rely on manual maintenance. Recently, automated hedge trimmers have been widely used, which can greatly reduce labor costs and avoid some potential safety hazards caused by manual garden maintenance. However, for hedges, it is difficult to locate them automatically because of their extremely irregular shape. As mentioned in the AdaHC [1] proposed by Li et al., the key to automating hedge maintenance is the precise localization of its center axis and radius. The precondition for the implementation of the AdaHC method is that the camera must be above the hedge. However, at this point, it is not certain where the hedge is, so the solution is not fully automated. Therefore, the 3D localization of hedges is the key to automated trimming.

The shape of the hedges studied in this paper was mainly sphere-like hedges. The hedge point cloud is used as the initial data. Thus, the problem was transformed into a localization problem for irregular spherical shapes. The target was to obtain the central axis of the hedge, as well as the height information. Previous methods for 3D object localization [2–4] were based on traditional algorithms, which can show relatively good results for regular models. However, because of the irregularity of the hedge, the fitting results have more uncertainties. Therefore, the location methods based on these traditional algorithms still have a big defect.

However, Krizhevsky et al. proposed the object classification algorithm AlexNet [5] based on deep learning in the ImageNet [6] contest. As a result, the deep learning algorithm [7–10] emerged, and its detection accuracy and speed far exceeded the effect based on the traditional algorithm at that time. With the development of autonomous

Citation: Meng, Y.; Zhai, X.; Zhang, J.; Wei, J.; Zhu, J.; Zhang, T. *HeLoDL*: Hedgerow Localization Based on Deep Learning. *Horticulturae* **2023**, *9*, 227. https://doi.org/10.3390/horticulturae9020227

Academic Editors: Chenglin Wang and Lufeng Luo

Received: 13 November 2022
Revised: 2 February 2023
Accepted: 3 February 2023
Published: 8 February 2023

driving technology in recent years, some excellent point-cloud-based perception solutions have been proposed [11–14]. These solutions have inspired us greatly, although they cannot locate hedges directly.

In this paper, we propose a novel 3D hedge localization method based on deep learning. Firstly, the height H of the hedge point cloud is quickly obtained by heap sorting. Secondly, the hedgerow's point cloud is projected onto a single-channel image with a resolution of 900×900. Each pixel width corresponds to the actual distance of 0.5 cm, and the center of the image is the centroid of the point cloud in the XOY plane. Thirdly, the two-dimensional convolution neural network was used to extract the image features. Finally, the detection head module was used to regress the hedge center axis (X, Y) and radius R. Moreover, inspired by the intersection over union (IoU), we propose the intersection and union ratio between circles as one of the evaluation indexes of the prediction results, which we call the OIoU. The main contributions of this paper are as follows:

a. A method of extracting green hedge point cloud information from images is proposed. This method uses pixel information to extract point cloud information by associating pixels with point clouds.

b. A method based on deep learning is proposed for the 3D localization of hedges. Its accuracy outperforms the state-of-the-art methods by a significant margin.

c. An evaluation index, the OIoU, is proposed. Combining center distance error and radius error, this index can measure the coincidence degree between circles more effectively.

d. A data augmentation method of rotating the image around the center of the mark is proposed. This method not only greatly improves the accuracy of the model, but also accelerates the convergence of the model.

The rest of this paper is organized as follows. First, the related works are introduced in Section 2. Section 3 introduces the dataset, as well as the details of the proposed method. The experimental results are given and discussed in Section 4. Finally, we give the conclusions of this study and the future work that needs to be carried out in Section 5.

2. Related Work

To solve the problem of hedge localization in a more intuitive way, we need 3D localization of hedges. According to the recent literature, there are two main schemes that can be used to achieve 3D hedge localization: (1) a hedge localization algorithm based on conventional algorithms for circle fitting; (2) a hedge localization algorithm based on conventional algorithms for sphere fitting.

The hedge localization method based on conventional algorithms for circle fitting first uses min-heap sorting [15] to obtain the height information of the hedge. Then, the point cloud is projected top-down, after which the circle fitting [3,16,17] is performed on the planar point cloud to obtain the center axis and radius of the hedge. The *Hough* transform was first proposed by Hough [16] and later popularized by [18] Duda and Hart. It is one of the basic methods to detect geometric shapes from images in image processing. The *Hough* transform uses the transformation between two coordinate spaces to map a curve or line with the same shape in one space to a point in another coordinate space to form a peak, thus transforming the problem of detecting arbitrary shapes into a statistical peak problem. A *Hough* circle detection algorithm of a cumulative array based on the voting strategy [19] was proposed by Kimme et al.. When the radius of the circle to be detected is known, the algorithm can accurately detect the circle of the corresponding size. However, if the radius is unknown, the algorithm needs to occupy more parameter space; not only is the calculation speed slow, but also the stability of the algorithm is poor, and it very easily results in false detection. Fischler and Bolles proposed random consistent sampling (*Ransac*) [3] for the first time. They used *Ransac* to solve the location determination problem. Grbić et al. proposed an ellipse detection algorithm [20] based on *Ransac* and the least-squares method, which has good practicability in medical image analysis and ultrasonic image segmentation. The idea of the *Ransac* algorithm is to estimate the parameters of the mathematical model iteratively from a group of observation datasets containing "external points". It has a certain probability to obtain a reasonable result, and the number of iterations must be increased to improve the probability. However, the hedge is not regular and spherical, as it may be raised in some places and depressed in others. Therefore, the *Ransac*-based method can be somewhat accurate, but there will always be some error. Therefore, it is an uncertain algorithm. For the same reason, the *minEnclosingCircle* method in OpenCV [17] has a rather good fit for regular circles, but the fit for the green hedge circle contour is not as good as it could be.

The hedge localization method based on conventional algorithms for sphere fitting directly fits the input hedge point cloud to the sphere [2,3,21] to obtain the sphere center and radius of the hedge. The X and Y coordinates of the sphere center are then used to determine the central axis of the hedge. The height of the hedge is still obtained by the min-heap sorting method [15]. Schnabel et al. presented an automatic algorithm [22] to detect elementary shapes in unorganized point clouds, which was used in PCL [21]. The method is based on random sampling to detect planes, spheres, cylinders, cones, and tori. The algorithm is robust even in the presence of many outliers and a high degree of noise. In addition, the algorithm is conceptually simple and easy to implement. However, the algorithm is tuned according to the a priori knowledge of different primitive shapes and parameters. Cao et al. proposed a spherical parameter detection method [2] based on the hierarchical HT (HHT). The HHT can reduce the storage space and calculation time significantly and is robust to noises and artifacts. However, the method fails completely when the gradient of the object is weaker than its background gradient. Ogundana et al. implemented the accumulator of the sparse 3D matrix model [4]. This method has the advantages of memory saving, high computational efficiency, and good 3D feature detection ability in sphere experimental shape data fitting. The method has proven to be fast and accurate. However, the method only detects the sphere center coordinates because it requires that the radius of the sphere is known. In fact, the hedge sphere to be fit is unknown, so the method is not applicable to hedge sphere fitting. Wang et al. [23] proposed an energy-based method for the automatic identification of multi-spheres, which is similar to the processing flow of *Ransac*. The main difference is that an energy-based judgment indicator is applied instead of the threshold-based indicator in *Ransac*. The results showed that the method outperforms the *Ransac*- and *Hough*-based methods in terms of accuracy and robustness. However, the experimental object of the method is its regular point cloud model, while such a regular shape hardly exists in the point cloud of hedgerows, so it is not applicable to the fitting of hedgerow spheres. Furthermore, the detection speed of the method is slow and cannot achieve a real-time processing effect.

3. Materials and Methods

3.1. Materials

The hedge point cloud was collected with a Livox-Horizon LIDAR. Some parameters of the solid-state LIDAR Livox-Horizon are as follows. The laser wavelength is 905 nm; the distance random error is (1σ@20 m) < 2 cm; the angular random error is $1\sigma < 0.05^\circ$; the beam divergence is 0.28° (vertical) $\times$ 0.03° (horizontal); the field of view angle is 81.7° (horizontal) $\times$ 25.1° (vertical); the measuring range (@10 klx) is $90m$@10% reflectivity; the data rate is 240,000 points per second. We collected 564 green hedge point cloud data. First, we fixed the Livox-Horizon LIDAR to the side of our unmanned car, as shown in Figure 1.

In the process of collecting data, we also noticed that the reflection intensity of the point cloud was higher when the Livox-Horizon collected at a close distance, which could not normally show the shape of the hedge at this time. Therefore, in the process of data collection, we avoided this phenomenon. The closest distance between the LIDAR and hedge was about 1 m, and the farthest distance was about 5 m. We walked along a straight line at different distances from the hedgerow to the car and recorded the point cloud bag based on the Robot Operating System [24]. It is necessary to be aware that the hedge is on the side of the car and not in front of it. In addition, due to the nature of the near-dense and far-sparse hedge point clouds, different acquisition distances were used to enrich the diversity of the samples and, thus, improve the robustness of the network model. We sampled the data at intervals of 10 frames. The traveling speed of the acquisition equipment was about 1 m/s, and the overlap rate of point clouds between two adjacent sampling points was between 27% and 79%. In addition, the distance between each hedge was about 1 m. Because the collection equipment was moving forward, we could extract the point cloud of the same hedge at different angles, which can also enrich the diversity of the data. Then, using the point cloud processing software CloudCompare [25], a total of 564 hedge point clouds were extracted from all frames of the data. In addition, we performed the dilation and erosion operations on the data, which were used to increase the amount of points in the dataset. We also rotated the data to expand the dataset. The workflow followed in the process of the morphological operation and image rotation will be described in detail in Sections 3.3.3 and 3.3.4. The dataset is divided as shown in Table 1.

Figure 1. Data acquisition equipment and scene. In order to ensure that the LIDAR can acquire the height of the hedge from bottom to top, the installation height of the LIDAR in this study was about 0.8 m. In other working scenarios, the height of the LIDAR needs to be adjusted according to the height of the hedge.

Table 1. The details of the dataset.

Dataset	Hedge's Point Cloud	Projected Image	Augmented Data
Training set	334	334	12,024
Validation set	115	115	4140
Testing set	115	115	\

Then, we circled the maximum outer circle of the point cloud contour in the original image without morphological operations based on the Labelme [26] software as the annotation information of the image. We invited three skilled and experienced gardeners to participate in our image annotation. Skilled gardeners were asked because it is more informative to use their experience as a criterion to determine the circles to be trimmed. Then, two rules were established that need to be followed during the annotation process. First, the annotated circle should coincide with the point cloud outline of the hedge as much as possible. Second, if it was difficult to identify the circles represented by the point cloud wheel outline, the labeling should strictly follow the actual boundary of the hedge. After the three gardeners were labeled separately, the average of the three labeling results was obtained and used as the final labeling result. Thus, the mean value of the measurements was the final labeling result for each sample. Finally, each labeled circle consisted of two key points, one of which was the center of the circle and the other a point on the boundary of the circle. The distance between these two key points was the radius. Additionally, we specifically counted the circle center coordinates X, circle center coordinates Y, and radius R for each sample and converted these coordinates to values in the LIDAR coordinate system to find their standard deviations. The respective maximum values of the three standard deviations were 0.0208, 0.0221, and 0.0242, respectively. On the other hand, the height information of the hedge was measured by three experienced gardeners using the CloudCompare software, based on the original point cloud of the hedge. As before, we averaged the measurements of the three people and used them as height annotation information for the hedge point cloud.

3.2. Problem Formulation

In the operation of the automatic trimmer, the trimmer needs to be placed directly above the hedge, so the height information at the top of the hedge and the center axis of the hedge need to be known. In addition, the hedge radius was used to determine whether the hedge needs to be trimmed because there are always some hedges that are not well grown and are too small to be trimmed. It is worth mentioning that if the LIDAR is mounted low, there may be a lack of information on the top of the hedge. This problem can be improved in two ways. Firstly, the hedge was localized at different measurement distances l. The localization results were converted to the world coordinate system, and their average value was taken as the final localization result. The other is to adjust the height h of the LIDAR mounting position to ensure that the LIDAR can detect the top of the hedge. Both l and h in the above two options need to be calibrated to the actual scene. In our experiments, the LIDAR was installed at a height of 0.8 m, and the data collected ranged from 1 m to 5 m.

In this paper, given a point cloud $\mathcal{X} \in \mathbb{R}^{N\times3}$ of a hedgerow, we aimed to obtain its center axis (X, Y), radius R, and height H. The center axis (X, Y) and height H were used to adjust the position of the trimming tool, and the radius R was used to determine the working range of the trimming tool. The overall framework of our approach is shown in Figure 2.

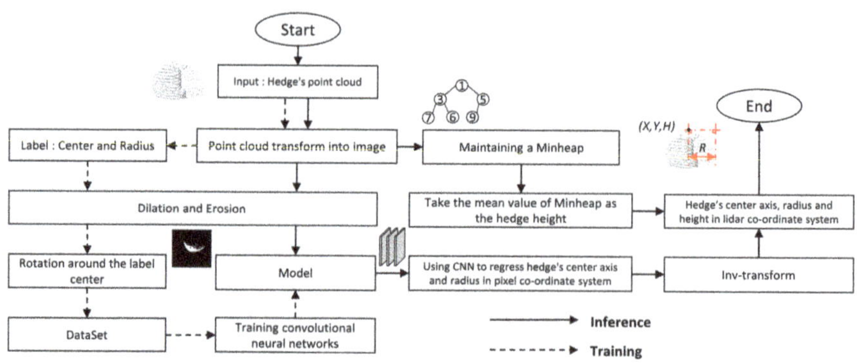

Figure 2. The pipeline of our *HeLoDL*. The solid arrows represent the inference process, and the dashed arrows represent the training process.

3.3. Pipeline of HeLoDL

The overall framework of our approach is shown in Figure 2, and the main steps are shown as follows:

(0) Input: Hedge's point cloud.

(1) Extract hedge's height: We used min-heap sorting to extract the hedge's height according to the Z coordinate of the point cloud.

(2) Transform the point cloud information into image information: The X coordinate and Y coordinate of all point clouds were averaged, respectively, to obtain $\bar{X}$ and $\bar{Y}$. Then, the center pixel of the image corresponds to $(\bar{X}, \bar{Y})$. At the same time, the actual distance 0.5 cm is represented by each pixel width. Then, all point clouds within 2.25 m from front to back and left to right were projected onto the image relative to $(\bar{X}, \bar{Y})$. Finally, if the point in the point cloud was projected onto a pixel, the pixel value was set to 255;

otherwise, the pixel value was set to 0. Then, a binary image with a resolution of 900 × 900 was obtained.

(3) Morphological operation: The aerial view obtained by Step 2 was dilated and eroded by the image morphology to fill the gap in the image and filter out some noise points on the edge.

(4) Rotation operation: All the images obtained from Step 3 were rotated around the center of the marked circle and then used as the dataset of the experiment.

(5) Using the CNN to regress center axis (u, v) and radius r: The image obtained in the fourth step was directly used as the input to the convolutional neural network. The features of the image were first extracted by the mainstream backbone, and then, the size and channels of the feature layers were further compressed by the convolution neural network. The final output feature sizes of the network were $2 \times 1 \times 1$ and $1 \times 1 \times 1$, which represent the center axis (u, v) and radius r of the hedge, respectively.

(6) Coordinate transformation: Transform u, v, and r in the pixel coordinate system to the corresponding point cloud coordinate system to obtain X, Y, and R.

(7) Output: Center axis (X, Y), height H, and radius R.

In the later subsection, we provided a detailed description of the main content of the above process.

3.3.1. Extract the Height Information of Hedge

This paper focused on how to obtain localization information for a given hedge point cloud. The point cloud data preprocessing, however, involves our earlier work. The earlier work mainly included joint LIDAR and camera calibration, hedge detection, frustum generation, ground filtering, direct-pass filtering, and point cloud clustering. Therefore, the work on point cloud data pre-processing was performed in the early stage and is not shown in the paper. Therefore, we can assume that the hedge point cloud objects in this paper are already the point cloud after filtering out the outliers. For the extraction of hedge height, we sorted the point cloud Z value from large to small, and the average value of the point cloud with the first 5% of the Z value represents the actual height of the hedge. We used the small root heap sorting algorithm to set the data capacity of the heap to 5% of the number of all point clouds, pressed the Z value of 5% of the point cloud into the heap, traversed the remaining 95% of the point cloud, and compared the Z value with the value of the top element of the heap. If it is greater than the top element of the heap, remove the top element of the heap, and put it into the heap according to the sorting rule of the small root heap; otherwise, skip the point. It is important to note that the value of 5% is based on the data we currently used for testing. This parameter was not fixed. It was obtained by calibrating the number of point clouds for different sizes of hedges and LIDAR according to the actual situation.

$$\begin{cases} H = \frac{1}{n} \sum_{i=k}^{n} Z_i \\ k = 0.95n \end{cases} \tag{1}$$

where H represents the height of the top of the hedge in the LIDAR coordinate system, n represents the number of points in the point cloud, and k represents 95% times the total number of points. The time complexity of the algorithm is $O(n \log m)$, and m represents 5% of the number of point clouds.

3.3.2. Transform Point Cloud Information into Image Information

After the height of the hedge was obtained, we transformed the point cloud information into image information. The process of transformation is shown in Figure 3. Firstly, these point clouds were projected on the plane to obtain point clouds in the XOY plane in the LIDAR coordinate system. Secondly, the average values of the X axis and Y axis of the plane point cloud in the LIDAR coordinate system $\bar{X}$ and $\bar{Y}$ were obtained, respectively, and then, a pixel matrix with a 900×900 resolution was created. This corresponds to the $(\bar{X}, \bar{Y})$ coordinates of the plane point cloud to the pixel at $(450, 450)$ in the pixel coordinates.

The positive direction of the X axis in the LIDAR coordinate system is inthe negative direction of the v axis in the pixel coordinate system, and the positive direction of the Y axis in the LIDAR coordinate system is inthe negative direction of the u axis in the pixel coordinate system. Then, a binary image containing the actual location information of the point cloud was generated. The correspondence between the coordinates of the planar point cloud in the LIDAR coordinate system and the pixel coordinates is shown as follows:

$$\begin{cases} u_0 = \frac{\bar{Y}-Y}{0.5} * 100 + 450 & \bar{Y} - 2.25 \leq Y \leq \bar{Y} + 2.25 \\ v_0 = \frac{\bar{X}-X}{0.5} * 100 + 450 & \bar{X} - 2.25 \leq X \leq \bar{X} + 2.25 \end{cases} \quad (2)$$

where (u_0,v_0) represents the coordinates of each pixel on the generated image. It should be noted that the range of 2.25 m from the center of mass was chosen based on the actual situation. The first thing was to make sure that the point cloud of the hedge did not exceed this range. Secondly, it was considered that no hedge point cloud would be outside this boundary when we rotate the image around the center of the label.

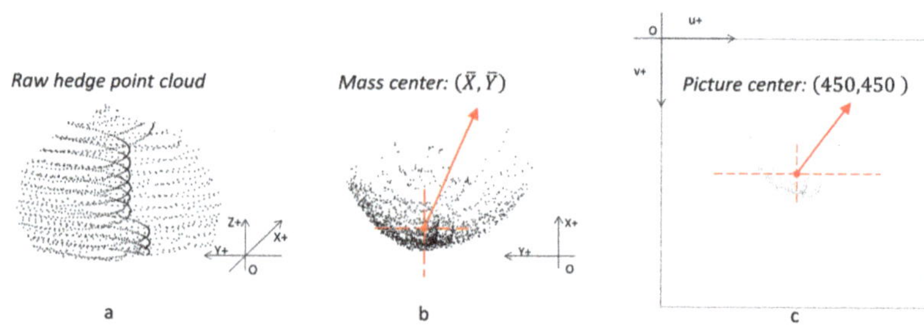

Figure 3. The transformation from spatial point cloud to planar image. To clearly show the position relationship between (**b**) and (**c**), we swapped the pixel values of 0 and 255 in (**c**).

3.3.3. Morphological Operation

There is a problem with the images obtained from the above steps. Due to the sparsity of the point cloud, the percentage of pixels of the point cloud on the image was very small, and the distribution was relatively discrete. Intuitively, the point clouds occupied a small number of pixels after projection onto the image. From the statistics in our dataset, the largest cluster of point clouds had 8000 points, and the resolution of the image was 900×900. The number of pixels in the point cloud after projection was less than 1% of the pixels on the image, so we believe that it is a relatively small percentage. Therefore, we performed morphological operations on the image to make its features more obvious and to filter out some noise at the same time. There are corrosion and dilation operations in image morphological processing. The corrosion operation takes the minimum value in a neighborhood of each location as the output pixel value of that location, while the dilation operation is on the contrary. Therefore, for the sparse point cloud projection map, we used the operation of dilating and then eroding to increase its features and filter out the excess noise, as shown in Figure 4.

The principle to dilate the original image is as follows:

$$A \oplus B = \{(x,y)|(B)_{xy} \cap A \neq \phi\} \quad (3)$$

where $A \oplus B$ means to dilate Image A with Structural Element B. In the above formula, $(B)_{x,y}$ means that the origin of the structural element B is translated to (x,y). If the element in B intersecting with A is not empty, then the pixel value at (x,y) is 0; otherwise, it is 255.

The principle to erode the original image is as follows:

$$A \ominus B = \{(x,y)|(B)_{xy} \subseteq A\} \quad (4)$$

where $A \ominus B$ means to erode Image A with Structural Element B. If all the elements in B belong to A, then the pixel value at (x, y) is 255; otherwise, it is 0.

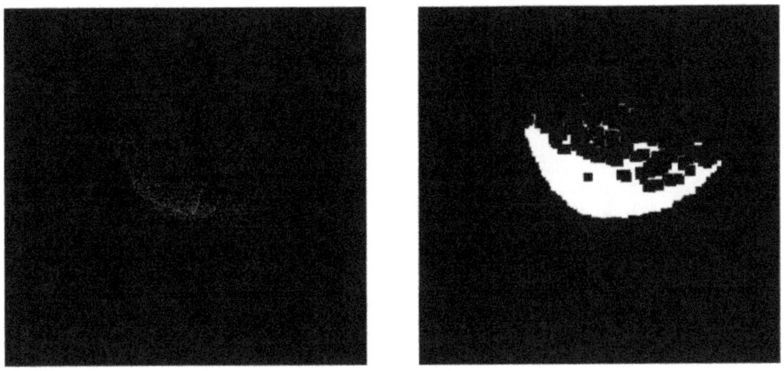

Figure 4. Image morphological operation.

3.3.4. Rotation Operation

After the previous steps, we obtained the image from point cloud projection and processed by dilation and erosion, but the images we input were only irregular figures, similar to a crescent moon. Because the point cloud of the hedge scanned by LIDAR is only single-sided and the orientation of the point cloud shape on the image is also single after conversion to a picture, we conjectured that, by the rotation of the image, the neural network can learn more features of the circle to achieve better prediction. Although the input data are only one-sided in practice, this does not impact enhancing the learning of circular features by rotating the initial dataset. Additionally, since the center of rotation is the center of the labeled circle, it does not change the central axis of the hedge point cloud, nor the radius, even after rotation. Therefore, we rotated the point cloud image around the center point of our label according to a series of angles. The corresponding relationship of pixel coordinates before and after image rotation is shown in the following formula:

$$\begin{bmatrix} u_1 \\ v_1 \end{bmatrix} = \begin{bmatrix} \cos\theta & -\sin\theta \\ \sin\theta & \cos\theta \end{bmatrix} \begin{bmatrix} u_0 - Center[0] \\ v_0 - Center[1] \end{bmatrix} + \begin{bmatrix} Center[0] \\ Center[1] \end{bmatrix} \quad (5)$$

where *Center* represents the center of the labeled circle, θ represents the angle of counter-clockwise rotation, and (u_1, v_1) represents the coordinates of each pixel after rotation.

The details of the rotation and the working criteria can be described as follows. First, the center of rotation is the center of the labeled circle. Second, the object of rotation is the pixel with the pixel value of 255 in the original image. Third, the rotation angle interval is 10°. The purpose of the third step is to train the neural network to learn richer features, so the rotation angle interval was set relatively small. The image data after rotation is shown in Figure 5.

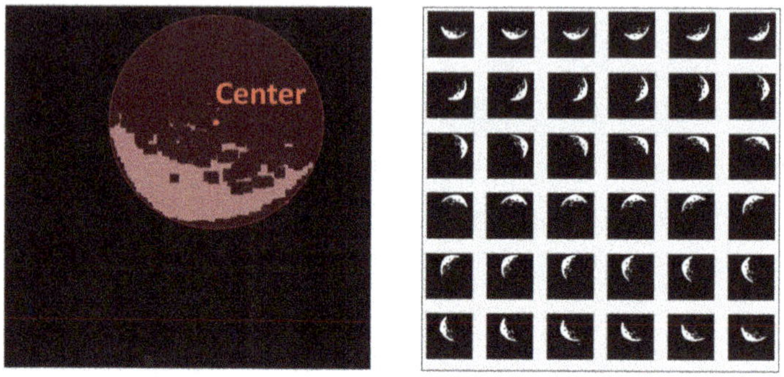

Figure 5. Image rotation operation. It should be noted that all of our annotations were made in the original image, as shown in Figure 4. However, due to the poor visualization of the original images, we put the annotation results on the morphologically processed images.

3.3.5. Using CNN to Regress Center Axis (u, v) and Radius r

As shown in Figure 6, the input of the neural network is a single-channel image with a resolution of 900 × 900, while the input images of the more popular backbone are three-channel RGB images. To be able to use transfer learning to accelerate the convergence of the network model, we first expanded the image to three channels by a layer of 3 × 3 convolution. At this point, the backbone and its pre-trained model, which are the mainstream in academia, were used to extract the features of the images. The output of the backbone is the feature layer of 512 × 29 × 29, and the visualization of some of its channels is shown in Figure 7. The maximum pool and convolution layer were used to further compress the feature layers, and the number and size of channels were compressed to the size of 128 × 7 × 7.

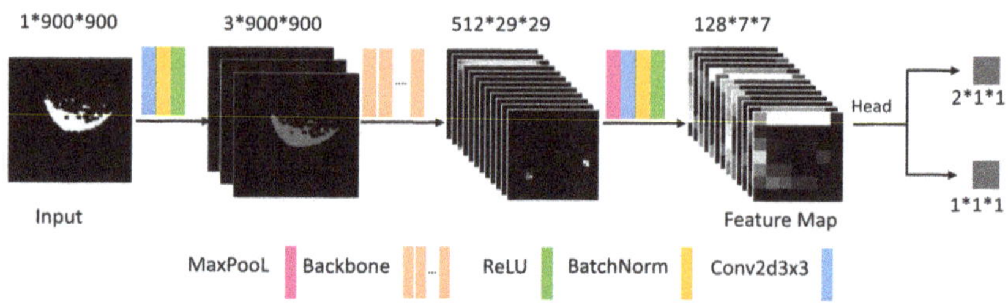

Figure 6. CNN was used to extract key information from the hedges.

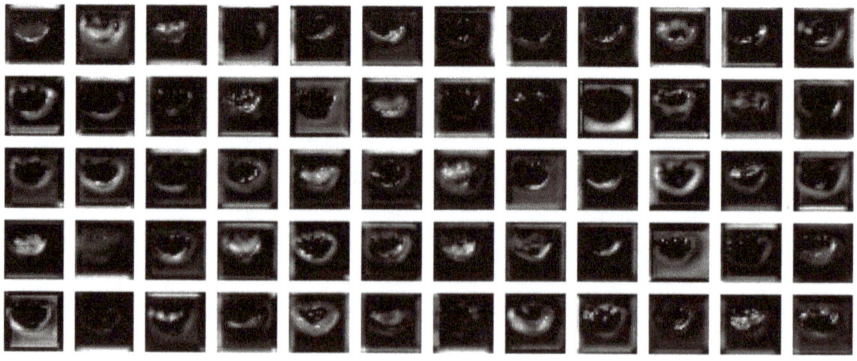

Figure 7. Partial feature maps extracted from backbone network.

The head part of the model is mainly divided into two branches, namely CT_{Head} and R_{Head}. In order to improve the overall performance of the network, we added batch normalization [27] and ReLU activation functions [28] in each layer. Among them, the output of CT_{Head} is a 1×1-sized feature layer of two channels, and the two channels represent the X value and Y value of the center axis of the hedge, respectively. The output of R_{Head} is a 1×1-sized feature layer of a single channel, which was used to represent the radius R of the hedge. The structure of the head module is shown in Figure 8.

Since there is only one hedge to be fit in each image, no classification operation was performed in our scheme. Instead, the radius R and the center axis (X, Y) of the hedge to be fitted were regressed directly, which is a typical regression problem. It is inevitable that there will be some noises after the point cloud is projected to the image, and the L1 loss function is robust against the outliers in the data, so we used the L1 loss function to measure the prediction of the model.

$$Loss = \lambda_{CT}\frac{1}{n}\sum_{i=1}^{n}(|u_i - \widehat{u_i}| + |v - \widehat{v_i}|) + \lambda_R\frac{1}{n}\sum_{i=1}^{n}(|r_i - \widehat{r_i}|) \tag{6}$$

where $(\widehat{u_i}, \widehat{v_i})$ denotes the coordinates of the circle center of the i-th labeled circle in the pixel coordinate system and $\widehat{r_i}$ is the radius of the labeled circle; (u_i, v_i) denotes the coordinates of the circle center of the i-th predicted circle in the pixel coordinate system, and r_i is the radius of the predicted circle. λ_{CT} is the weight of the center axis error, and λ_R is the weight of the radius prediction error. As the trimming radius of our trimming tool was already determined, the accuracy of the central axis of the hedge was more important than the accuracy of the radius. Therefore, we set λ_{CT} to 1 and λ_R to 0.5 as a way to make the training process more focused on the task for central axis prediction.

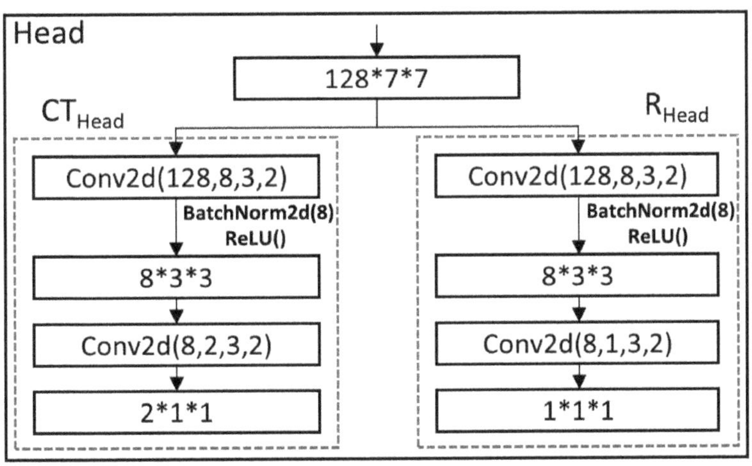

Figure 8. The structure of the prediction head.

3.3.6. Training and Inference

After we transformed the point cloud into a picture with actual distance information, we obtained a single-channel image with a resolution of 900×900. We used the Adam optimizer to adjust the parameters of the model. The initial learning rate of the optimizer was set to 0.01, and the learning rate was adjusted dynamically according to the change of the loss function during the training process. In addition, we used the backbone network of *ResNet* [29] and *ShuffleNet* [30], respectively; meanwhile, we used transfer learning in the training process. At the initial stage of training, we first froze the weight parameters of the backbone network to prevent them from being modified prematurely. Later in the training process, we unfroze the weight parameters of the backbone network and trained them together with the whole convolutional neural network.

The inference process is different from the training process, and the rotation operation is dropped in inference. The neural network directly regresses the pixel coordinates of the center axis of the hedge point cloud (u, v) and the pixel radius r. Using the formula 7, we can transform the pixel coordinate center point (u, v) to the center axis (X, Y) in the LIDAR coordinate system; at the same time, the pixel radius r is transformed to the radius R in the LIDAR coordinate system.

$$\begin{cases} R = \frac{r*0.5}{100} \\ X_{CT} = \bar{X} - 0.5 * \frac{v-450}{100} \\ Y_{CT} = \bar{Y} - 0.5 * \frac{u-450}{100} \end{cases} \tag{7}$$

4. Results

The convolutional neural network was trained on a server equipped with an NVIDIA TITAN RTX GPU with 24GB of memory. The inference process of the *HeLoDL* model and the various comparison methods were performed on a Lenovo laptop with an AMD R7 5800H CPU, 16GB RAM, and an Nvidia RTX 3060 GPU with 6GB of memory.

4.1. Evaluation Index

4.1.1. Evaluation Index Related to Center Axis and Radius

To fairly evaluate the performance of different algorithms, we considered the accuracy of the position of the center axis of the hedge (X, Y) and the radius R, respectively. In our evaluation index, we stipulated that the offset of the center axis of the hedge cannot exceed 5 cm, which is a rigid index. Since each pixel grid represents the actual distance of 0.5 cm, the distance between the location of our predicted center axis and the labeled center axis cannot exceed 10 pixels. Then, the evaluation index of circle center accuracy CT_{ACC} is defined as follows:

$$f(i) = \begin{cases} 1 & if \sqrt{(u_i - \widehat{u_i})^2 + (v_i - \widehat{v_i})^2} \le 10 \\ 0 & otherwise \end{cases} \tag{8}$$

$$CT_{ACC} = \frac{1}{n} \sum_{i=1}^{n} f(i) \tag{9}$$

CT_{ACC} cannot specifically express the accuracy of the algorithm. Therefore, we designed two other indicators, the average error of the radius R_{ERR} and the center distance error CT_{ERR}, to represent the error of the algorithm in the center point and radius, respectively, which are defined as follows:

$$R_{ERR} = \frac{1}{n} \sum_{i=1}^{n} (|r_i - \widehat{r_i}|) \tag{10}$$

$$CT_{ERR} = \frac{1}{n} \sum_{i=1}^{n} \sqrt{(u_i - \widehat{u_i})^2 + (v_i - \widehat{v_i})^2} \tag{11}$$

4.1.2. OIoU

For a single prediction circle, the distance error CT_{ERR} of the center of the circle can effectively evaluate the accuracy of the center prediction, while the radius error R_{EER} of the circle can effectively evaluate the accuracy of the radius prediction. However, the center error CT_{ERR} and radius error R_{EER} cannot directly reflect the degree of overlap of the two circles. Inspired by the classical IoU, we propose the OIoU. As shown in Figure 9, the OIoU is the ratio of the area of the intersection of two circles to the union of the areas of two circles. The OIoU can intuitively reflect the coincidence of the predicted circle with the ground truth circle, which is defined as follows:

$$OIoU = \frac{Intersection}{Union} \tag{12}$$

if $R_1 + R_2 < L_{12}$,

$$Intersection = 0, Union = \pi\left(R_1^2 + R_2^2\right) \tag{13}$$

if $R_1 \geq R_2 + L_{12}$,

$$Intersection = \pi R_2^2, Union = \pi R_1^2 \tag{14}$$

if $R_2 \geq R_1 + L_{12}$,

$$Intersection = \pi R_1^2, Union = \pi R_2^2 \tag{15}$$

other cases,

$$\begin{aligned} Intersection &= \theta_{O1}{R_1}^2 + \theta_{O2}{R_2}^2 - R_1 L_{12} \sin\angle AO_1O_2 - R_2 L_{12} \sin\angle AO_2O_1 \\ Union &= \pi\left({R_1}^2 + {R_2}^2\right) - Intersection \end{aligned} \tag{16}$$

where $\theta_{O1} = \arccos\left(\frac{{L_{12}}^2 + {R_1}^2 - {R_2}^2}{2L_{12}R_1}\right)$, $\theta_{O2} = \arccos\left(\frac{{L_{12}}^2 + {R_2}^2 - {R_1}^2}{2L_{12}R_2}\right)$.

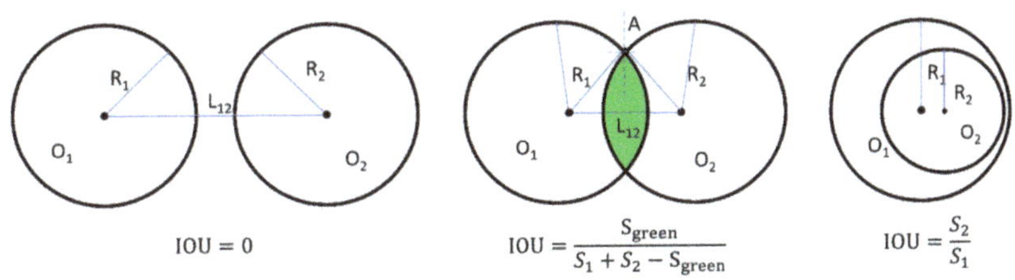

Figure 9. Several situations of the OIoU.

4.2. Experimental Results

We performed experiments with other baseline methods in the test set. The baseline methods to obtain the three-dimensional information of hedges through point clouds are presented in Figure 10. In the first method, the *Ransac* sphere fitting algorithm was directly used to fit the original point cloud of the hedge. In the second method, the original green hedge point cloud was projected onto the XOY plane to obtain the plane point cloud, then the boundary points of the plane point cloud were obtained based on the normal estimation, and the circle was estimated by the *Ransac* circle fitting algorithm based on the boundary points. In the third method, firstly, the original point cloud was projected in the top view and transformed into an image. Next, the image was dilated and eroded. Then, the contour with the most boundary points was extracted. Finally, the outer circle of the contour was extracted using the *minEnclosingCircle* function in OpenCV.

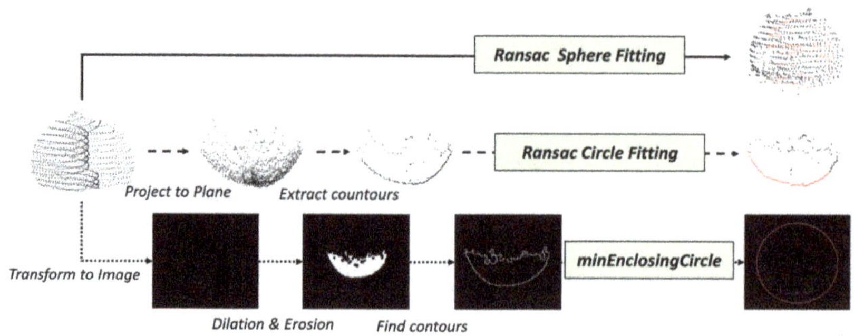

Figure 10. The pipeline of several baseline methods.

In our method, considering the accuracy and complexity of the model, we chose a *ResNet*34-based network model and compared it with the other three solutions. It can be seen from Figure 11 that our method is obviously better than the other solutions in predicting the center and radius of the circle. As shown in Table 2, our method is significantly better than the two traditional methods in terms of center accuracy. The average center distance error in our method was 2.67 cm, which fully meets the accuracy requirements of 5 cm. The importance of the prediction radius task is second only to the central point prediction task, and the radius error of our method was 1.61 cm, also showing strong stability and accuracy. Of course, we can see that the time consumption of our method is very short, and we can achieve real-time processing. The OIoU metric we designed also showed a high level at the same time. The results of the predicted radius and circle center tasks can also prove the reasonableness and comprehensiveness of the OIoU.

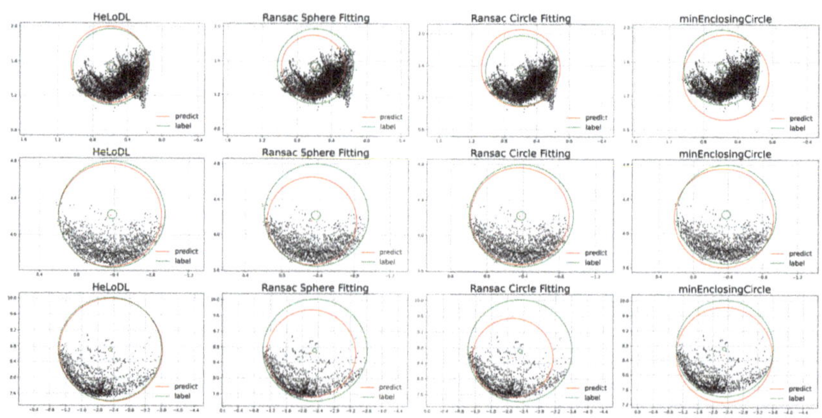

Figure 11. Visualization of the prediction of different methods. The red circle represents the predicted circle, and the green circle represents the ground truth. It is worth noting that the small green circle indicates the allowable error range, and if the center of the red circle can fall inside that circle, the prediction result meets the error requirement.

Table 2. Comparison with the state-of-the-art algorithm.

Method	CT_{ACC} (%)	R_{ERR} (cm)	CT_{ERR} (cm)	$OIoU_{mean}$ (%)	$Time_{mean}$ (ms)
Ransac Sphere Fitting	36.522	7.091	6.935	73.775	16.603
Ransac Circle Fitting	61.739	4.398	5.458	83.689	13.132
minEnclosingCircle	28.696	2.430	9.265	80.914	13.509
HeLoDL	90.435	1.635	2.712	92.654	12.727

It is also necessary to note that since the same height information acquisition scheme is used in several algorithms. Therefore, we will not compare the height information in the experimental results. Of course, the process of height information extraction is included in the inference process of each algorithm, so the time consumption of several algorithms regarding the inference process is fair.

4.3. Comparative Experiment

4.3.1. The Effect of Different Backbones

In this section, we tried different backbones, such as $ResNet$ [29] and $ShuffleNetV2$ [30]. $ResNet$ adopts the structure of the skip connection, which greatly increases the depth of the network and can extract more deep features. In addition, the $ResNet$ backbone network suppresses the problems of gradient explosion and gradient disappearance. $ShuffleNetV2$ is a lightweight backbone network in terms of speed and accuracy. In addition, in order to fit the lightweight network on embedded devices, we also tested the model based on $ShuffleNetV2$. The results are shown in Table 3. It can be seen that the lightweight network model has lower complexity, less computational time, and faster training speed, but the detection accuracy is not as good as that of the backbone network with a greater depth.

Table 3. The effect of applying different backbone networks to the model.

Backbone	CT_{ACC} (%)	R_{ERR} (cm)	CT_{ERR} (cm)	OIoU (%)	$Infer_{Time}$ (ms)	$Train_{Time}$ (hours)	Gflops	Parameters (millions)	Memory (M)
ResNet18	83.482	1.865	3.314	91.045	11.384	16.255	30.115	11.794	445.205
ResNet34	90.435	1.635	2.712	92.654	12.727	26.352	60.732	21.893	641.014
ResNet50	90.435	1.565	2.694	92.465	16.493	43.164	68.115	25.893	1818.297
ShuffleNet_v2_x0_5	65.224	2.573	5.559	87.364	15.103	8.261	0.792	1.543	212.841
ShuffleNet_v2_x1_0	79.132	2.114	3.691	90.062	14.844	11.943	2.583	2.455	370.992
ShuffleNet_v2_x1_5	76.527	2.452	3.687	89.674	15.213	15.697	5.132	3.684	510.53

4.3.2. Ablation Study

In this section, we validated the performance of different combinations of transfer learning, the rotation operation, and the morphological operation. The pre-training weight of the model is generally based on the rich dataset, but the color of the image dataset in this paper is relatively simple. Therefore, we hoped to study whether the transfer learning works well for the model proposed in this paper. At the same time, we also wanted to know the effect of image rotation and the morphological operation on the accuracy of the model. Based on this, we designed eight groups of experiments, as shown in Table 4. We trained the model according to the requirements of the eight groups of experiments. This experiment was based on $ResNet34$.

Table 4. The design of the ablation experiment.

Experimental Serial Number	0	1	2	3	4	5	6	7
Transfer Learning	✘	✔	✘	✘	✔	✘	✔	✔
Rotation operation	✘	✘	✔	✘	✔	✔	✘	✔
Morphological operation	✘	✘	✘	✔	✘	✔	✔	✔

To analyze the influence of several operations on the model training process and model accuracy, we visualize the accuracy and loss of the model on the validation set during the training in Figure 12. In addition, in order to directly show the performance of center prediction on the validation set in each group of experiments, we used the area S of the val_ct_acc curve and the X axis of the coordinate system, which is defined as:

$$S = \sum_{i=1}^{200} val\_ct\_acc(Epoch_i) \tag{17}$$

where val_ct_acc denotes the CT_{ACC} of the model on the validation set at the time of training in each epoch. $Epoch_i$ denotes the i-th time the neural network completes the forward computation and backward propagation. Moreover, in order to verify the real effects of the models in each experiment, we tested the models of each experiment using 115 sets of experimental data in the test set, respectively, and the experimental data are shown in Table 5.

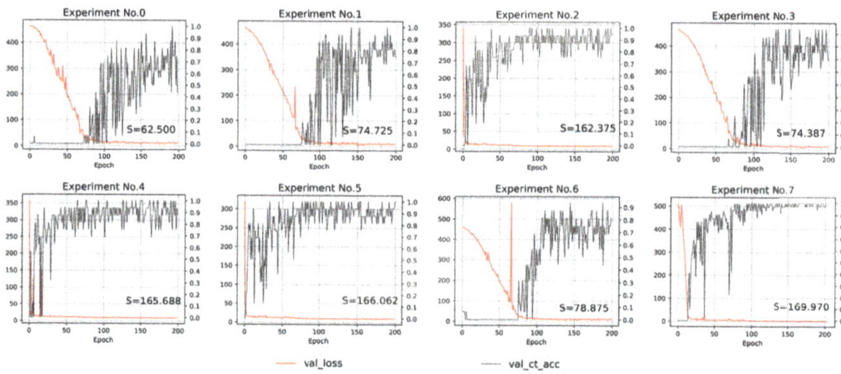

Figure 12. Ablation study of loss and CT_{ACC} on the validation set during model training, where S denotes the area enclosed by the curve *val_ct_acc* and the X-axis.

Comparing the results of Experiments 0 and 1, 2 and 4, 3 and 6, and 5 and 7, we can observe from Table 5 that transfer learning plays a critical role in improving CT_{ACC}. Comparing the results of Experiments 0 and 2, 1 and 4, 3 and 5, and 6 and 7, in Table 5, we can estimate that the rotation operation improves the CT_{ACC} of the model on the test set by an average of 14%. Furthermore, it can also be clearly seen from Figure 12 that the rotation enhancement operation of the training set not only accelerates the convergence speed of the model training, but also greatly improves the accuracy of the model. Comparing the results of Experiments 0 and 3, 1 and 6, 2 and 5, and 4 and 7, as we can see in Table 5, the morphological operation of the image also improves the CT_{ACC}.

Table 5. Ablation study results for the test set.

Experimental Serial Number	0	1	2	3	4	5	6	7
R_{ERR} (cm)	2.980	2.494	1.849	2.688	1.816	1.720	2.196	1.635
CT_{ERR} (cm)	4.696	5.060	2.959	4.941	2.815	2.785	4.237	2.712
OIoU (%)	88.022	88.084	91.767	88.311	91.699	92.200	89.815	92.654
CT_{ACC} (%)	71.304	72.174	86.957	74.783	87.826	88.696	76.522	90.435

Comparing the results of Experiments 0 and 1, 2 and 4, 3 and 6, and 5 and 7, we can observe from Table 5 that transfer learning plays a critical role in improving CT_{ACC}. Comparing the results of Experiments 0 and 2, 1 and 4, 3 and 5, and 6 and 7, in Table 5, we can estimate that the rotation operation improves the CT_{ACC} of the model on the test set by an average of 14%. Furthermore, it can also be clearly seen from Figure 12 that the rotation enhancement operation of the training set not only accelerates the convergence speed of the model training, but also greatly improves the accuracy of the model. Comparing the results of Experiments 0 and 3, 1 and 6, 2 and 5, and 4 and 7, as we can see in Table 5, the morphological operation of the image also improves CT_{ACC}.

As shown in Figure 12 and Table 5, Experiment No. 7 showed the best results compared with the other groups due to the simultaneous use of transfer learning, the image rotation, and the morphological operation of the image. This strongly proves the necessity of the method of our data pre-processing and transfer learning.

4.4. Visualization

In order to demonstrate the working principle of the method in this paper more clearly, we selected a set of experiments from the actual test set data and visualize the process in Figure 13. The first three subfigures of Figure 13 show the scene acquisition, hedge point cloud extraction, and center axis localization, respectively. Finally, we used the localization results as the input for the robot arm planning control, and the last subfigure shows the final localization effect of the robot arm. It is necessary to note that the point cloud data in Figure 13 were intercepted by setting the point cloud range, because the focus of this paper was on how to obtain the accurate position information from the hedge point cloud. The experimental data in the previous section and the final robot localization position can prove that the localization accuracy of the proposed method can fully meet the practical requirements.

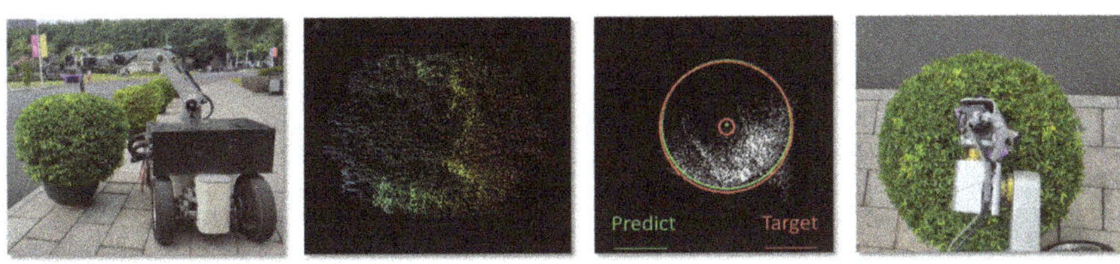

Figure 13. Visualization of the process of an experimental case in the testing set.

5. Conclusions

We investigated the 3D localization problem of hedges in a garden environment. In order to solve this problem, a novel algorithm *HeLoDL* for the 3D localization of hedges via deep learning was proposed for the first time. We also proposed an evaluation metric, the OIoU, to evaluate the effectiveness of the localization algorithm. Additionally, we provide two networks with different model sizes so that the model can run on different devices. Extensive experiments validated that the *HeLoDL* outperformed strong baselines by a large margin. There were some limitations in this study. First, the ground for the work should be flat. Additionally, the mounting height of the LIDAR needs to be adjusted according to the height of the hedge in the working scenario. Therefore, in order for the system to show high working performance in practical applications, the following requirements can be made for the environment. Firstly, the ground on which the device runs should be as flat as possible. Secondly, the trimmed hedges should be near the flat road surface and the hedges of different large and low heights should be arranged separately as much as possible. Additionally, the data acquisition solution of *HeLoDL* we currently provide is still not automated enough. In the future, we will consider using the point cloud segmentation scheme [31,32] to simplify the process of generating hedge point cloud datasets to some extent.

Author Contributions: Conceptualization, Y.M. and J.W.; methodology, X.Z. and T.Z.; software, X.Z.; validation, J.Z. (Jinlai Zhang) and J.W.; formal analysis, J.W. and T.Z.; investigation, X.Z.; resources, Y.M.; data curation, X.Z.; writing—original draft preparation, X.Z.; writing—review and editing, Y.M., J.Z. (Jinlai Zhang) and T.Z.; visualization, X.Z.; supervision, Y.M., J.W. and J.Z. (Jihong Zhu); project administration, Y.M.; funding acquisition, Y.M. All authors have read and agreed to the published version of the manuscript.

Funding: This research was funded by Guangxi Innovation Driven Development Project grant number GuiKeAA19254019.

Data Availability Statement: Not applicable.

Conflicts of Interest: The authors declare no conflict of interest.

References

1. Li, Z.; Xu, E.; Zhang, J.; Meng, Y.; Wei, J.; Dong, Z.; Wei, H. AdaHC: Adaptive Hedge Horizontal Cross-Section Center Detection Algorithm. *Comput. Electron. Agric.* **2022**, *192*, 106582.
2. Cao, M.; Ye, C.; Doessel, O.; Liu, C. Spherical parameter detection based on hierarchical Hough transform. *Pattern Recognit. Lett.* **2006**, *27*, 980–986. [CrossRef]
3. Fischler, M.A.; Bolles, R.C. Random sample consensus: A paradigm for model fitting with applications to image analysis and automated cartography. *Commun. ACM* **1981**, *24*, 381–395.
4. Ogundana, T.; Coggrave, C.R.; Burguete, R.; Huntley, J.M. Fast Hough transform for automated detection of spheres in three-dimensional point clouds. *Opt. Eng.* **2007**, *46*, 051002.
5. Krizhevsky, A.; Sutskever, I.; Hinton, G.E. Imagenet classification with deep convolutional neural networks. *Commun. ACM* **2017**, *60*, 84–90. [CrossRef]
6. Deng, J.; Dong, W.; Socher, R.; Li, L.J.; Li, K.; Li, F.F. ImageNet: A Large-Scale Hierarchical Image Database. In Proceedings of the 2009 IEEE Conference on Computer Vision and Pattern Recognition, Miami, FL, USA, 20–25 June 2009; pp. 248–255. [CrossRef]
7. Girshick, R.; Donahue, J.; Darrell, T.; Malik, J. Rich feature hierarchies for accurate object detection and semantic segmentation. In Proceedings of the IEEE Conference on Computer Vision and Pattern Recognition, Columbus, OH, USA, 23–28 June 2014; pp. 580–587.
8. Ren, S.; He, K.; Girshick, R.; Sun, J. Faster r-cnn: Towards real-time object detection with region proposal networks. *Adv. Neural Inf. Process. Syst.* **2015**, *28*. [CrossRef]
9. Wei, L.; Dragomir, A.; Dumitru, E.; Christian, S.; Scott, R.; Cheng-Yang, F.; Berg, A.C. *SSD: Single Shot MultiBox Detector*; Springer: Cham, Switzerland, 2016.
10. Redmon, J.; Farhadi, A. Yolov3: An incremental improvement. *arXiv* **2018**, arXiv:1804.02767.
11. Zhou, Y.; Tuzel, O. Voxelnet: End-to-end learning for point cloud based 3d object detection. In Proceedings of the IEEE Conference on Computer Vision and Pattern Recognition, Salt Lake City, UT, USA, 18–23 June 2018; pp. 4490–4499.
12. Lang, A.H.; Vora, S.; Caesar, H.; Zhou, L.; Yang, J.; Beijbom, O. Pointpillars: Fast encoders for object detection from point clouds. In Proceedings of the IEEE/CVF Conference on Computer Vision and Pattern Recognition, Long Beach, CA, USA, 15–20 June 2019; pp. 12697–12705.
13. Ge, R.; Ding, Z.; Hu, Y.; Wang, Y.; Chen, S.; Huang, L.; Li, Y. Afdet: Anchor free one stage 3d object detection. *arXiv* **2020**, arXiv:2006.12671.
14. Zhang, F.; Guan, C.; Fang, J.; Bai, S.; Yang, R.; Torr, P.H.; Prisacariu, V. Instance segmentation of LIDAR point clouds. In Proceedings of the 2020 IEEE International Conference on Robotics and Automation (ICRA), Paris, France, 31 May–31 August 2020; pp. 9448–9455.
15. Williams, J.W.J. Algorithm 232: Heapsort. *Commun. ACM* **1964**, *7*, 347–348.
16. Hough, P.V.C. Machine Analysis of Bubble Chamber Pictures. *Conf. Proc. C* **1959**, *590914*, 554–558.
17. Bradski, G. The OpenCV Library. *Dr. Dobb's J. Softw. Tools* **2000**, *25*, 120–123.
18. Duda, R.O.; Hart, P.E. Use of the Hough transformation to detect lines and curves in pictures. *Commun. ACM* **1972**, *15*, 11–15.
19. Kimme, C.; Balard, D.; Sklansky, J. Finding circles by an array of accumulators. *Commun. ACM* **1975**, *18*, 120–122. [CrossRef]
20. Grbić, R.; Grahovac, D.; Scitovski, R. A method for solving the multiple ellipses detection problem. *Pattern Recognit.* **2016**, *60*, 824–834.
21. Rusu, R.B.; Cousins, S. 3D is here: Point Cloud Library (PCL). In Proceedings of the IEEE International Conference on Robotics & Automation, Shanghai, China, 9–13 May 2011.
22. Schnabel, R.; Wahl, R.; Klein, R. Efficient RANSAC for point-cloud shape detection. In *Computer Graphics Forum*; Blackwell Publishing Ltd.: Oxford, UK, 2007; Volume 26, pp. 214–226.
23. Wang, L.; Shen, C.; Duan, F.; Lu, K. Energy-based automatic recognition of multiple spheres in three-dimensional point cloud. *Pattern Recognit. Lett.* **2016**, *83*, 287–293.
24. Quigley, M.; Conley, K.; Gerkey, B.; Faust, J.; Foote, T.; Leibs, J.; Wheeler, R.; Ng, A.Y. ROS: An open-source Robot Operating System. In Proceedings of the ICRA Workshop on Open Source Software, Kobe, Japan, 12–17 May 2009; Volume 3, p. 5.
25. Dewez, T.J.B.; Girardeau-Montaut, D.; Allanic, C.; Rohmer, J. FACETS: A Cloudcompare Plugin To Extract Geological Planes From Unstructured 3D Point Clouds. In Proceedings of the 23rd Congress of the International-Society-for-Photogrammetry-and-Remote-Sensing (ISPRS), Prague, Czech Republic, 12–19 July 2016; Volume 41, pp. 799–804. [CrossRef]
26. Russell, B.C.; Torralba, A.; Murphy, K.P.; Freeman, W.T. LabelMe: A Database and Web-Based Tool for Image Annotation. *Int. J. Comput. Vis.* **2008**, *77*, 157–173. [CrossRef]
27. Ioffe, S.; Szegedy, C. Batch normalization: Accelerating deep network training by reducing internal covariate shift. In Proceedings of the International Conference on Machine Learning. PMLR, Lille, France, 6–11 July 2015; pp. 448–456.

28. Glorot, X.; Bordes, A.; Bengio, Y. Deep sparse rectifier neural networks. In Proceedings of the Fourteenth International Conference on Artificial Intelligence and Statistics, JMLR Workshop and Conference Proceedings, Fort Lauderdale, FL, USA, 11–13 April 2011; pp. 315–323.
29. He, K.; Zhang, X.; Ren, S.; Sun, J. Deep Residual Learning for Image Recognition. In Proceedings of the 2016 IEEE Conference on Computer Vision and Pattern Recognition (CVPR), Las Vegas, NV, USA, 27–30 June 2016.
30. Ma, N.; Zhang, X.; Zheng, H.T.; Sun, J. ShuffleNet V2: Practical Guidelines for Efficient CNN Architecture Design. In Proceedings of the European Conference on Computer Vision, Munich, Germany, 8–14 September 2018.
31. Wang, W.; Yu, R.; Huang, Q.; Neumann, U. Sgpn: Similarity group proposal network for 3d point cloud instance segmentation. In Proceedings of the IEEE Conference on Computer Vision and Pattern Recognition, Salt Lake City, UT, USA, 18–23 June 2018; pp. 2569–2578.
32. Wei, H.; Xu, E.; Zhang, J.; Meng, Y.; Wei, J.; Dong, Z.; Li, Z. BushNet: Effective semantic segmentation of bush in large-scale point clouds. *Comput. Electron. Agric.* **2022**, *193*, 106653.

horticulturae

MDPI

Article

Reasearch on Kiwi Fruit Flower Recognition for Efficient Pollination Based on an Improved YOLOv5 Algorithm

Haili Zhou [1,2], Junlang Ou [1], Penghao Meng [1], Junhua Tong [1,2,*], Hongbao Ye [3] and Zhen Li [1]

1 School of Mechanical Engineering, Zhejiang Sci-Tech University, Hangzhou 310018, China; zy077520@163.com (H.Z.)
2 Key Laboratory of Transplanting Equipment and Technology of Zhejiang Province, Hangzhou 310018, China
3 Institute of Agricultural Equipment, Zhejiang Academy of Agricultural Sciences, Hangzhou 310021, China
* Correspondence: jhtong@zstu.edu.cn

Abstract: A close relationship has been observed between the growth and development of kiwi fruit and the pollination of the kiwi flower. Flower overlap, flower tilt, and other problems will affect this plant's pollination success rate. A pollination model based on YOLOv5 was developed to improve the pollination of kiwi flowers. The K-means++ clustering method was used to cluster the anchors closer to the target size, which improved the speed of the algorithm. A convolutional block module attention mechanism was incorporated to improve the extraction accuracy with respect to kiwi flower features and effectively reduce the missed detection and error rates. The optimization of the detection function improves the recognition of flower overlap and the accuracy of flower tilt angle calculation and accurately determines flower coordinates, pollination point coordinates, and pollination angles. The experimental results show that the predicted value of the YOLOv5s model is 96.7% and that its recognition accuracy is the highest. Its mean average precision value is up to 89.1%, its F1 score ratio is 90.12%, and its memory requirements are the smallest (only 20 MB). The YOLOv5s model achieved the highest recognition accuracy as determined through a comparison experiment of the four sets of analysed models, thereby demonstrating its ability to facilitate the efficient target pollination of kiwi flowers.

Keywords: machine vision; target recognition; YOLOv5; kiwi fruit pollination

Citation: Zhou, H.; Ou, J.; Meng, P.; Tong, J.; Ye, H.; Li, Z. Reasearch on Kiwi Fruit Flower Recognition for Efficient Pollination Based on an Improved YOLOv5 Algorithm. *Horticulturae* **2023**, *9*, 400. https://doi.org/10.3390/horticulturae9030400

Academic Editors: Chenglin Wang and Lufeng Luo

Received: 12 March 2023
Accepted: 16 March 2023
Published: 20 March 2023

1. Introduction

Kiwi has a short flowering period, usually only 3 to 5 days, which increases the difficulty of pollination. Under natural conditions, kiwi pollination mainly relies on wind and insect pollination, but insect pollination is easily affected by weather factors. Female flowers cannot be fully pollinated, resulting in a low fruit-setting rate, small fruit, an increased deformed fruit rate, reduced economic benefits, and other problems, and a single kiwifruit needs at least 1000 seeds to produce more than 100 g of fruit [1].

Automatic targeting technology is rapidly developing. The automatic target-spraying machine studied by Jiang et al. [2] can selectively spray-pollinate a target in accordance with the changes in the target's position and characteristics, effectively improve the adhesion rate of the division on the crop, remarkably reduce the settlement of the division in the nontarget area, and obtain superior pollination performance, which can reduce production costs and effectively reduce farmers' burdens.

The key to automatic target pollination technology is target recognition and detection. Object recognition and detection mainly employ machine vision; accordingly, deep learning (DL) convolutional neural networks have developed rapidly in the field of machine vision in recent years and play pivotal roles in different fields. With the emergence of DL, more complex, in-depth models can now be constructed through continuous training with large sums of data so as to achieve automatic feature extraction and continuous optimization, thereby improving the accuracy of recognition. In 2012, the AlexNet network designed by

Krizhevsky et al. [3] won the ImageNet Contest by an absolute margin. Subsequently, more and better convolutional neural networks, such as VGG, ResNet, GoogleNet, YOLO and other networks, have been proposed.

Liu et al. [4] fine-tuned and modified YOLOv3 in terms of model feature weights and proposed the YOLOv3-SE model, which achieved better identification accuracy for Dongjujube in a dense environment where branches and leaves interweave and obscure targets. Zhang et al. [5] replaced the feature extraction network of YOLOv3 with the SEResNet50 network and proposed a new and improved model, ISDYOLOv3, which solved the problem of important information loss in the convolution process and achieved an average accuracy of 94.91% in the detection of mature mangos. Li et al. [6] realised the real-time monitoring of weeds in a cotton field by optimising the YOLOv3 model in the feature extraction network and building and testing the embedded platform. Based on the VGG network model, Yue et al. [7] added high-order residuals and parameter-sharing feedback subnetworks into the VGG network to identify crop diseases under complex natural conditions, resulting in higher identification accuracy (90.98%) and better robustness in actual environments.

Li et al. [8] proposed an efficient grape detection model, YOLO-GRAPE, which considered the complex growing environment of grapes. A subsampling fusion structure was added to the network to improve the accuracy of network recognition. Wang et al. [9] used the ResNeXt model with modified structural weight parameters to build a new feature enhancement module to address the lack of features in images with fog and used the attention mechanism to help the detection network focus on more useful features in such images. Lv et al. [10] proposed a BIFPN-S structure imitating the BiFPN model, which enhanced feature diffusion and feature reuse, and replaced the network SILU activation function with the ACON-C activation function to improve its network performance.

Chaschatzis et al. [11] used the DL method to detect the characteristic compressed tissue of sweet cherries. ERICA data sets, including an entire sweet cherry tree and single leaf images, were built to provide better recognition results over unbalanced data sets. Ficzere et al. [12] used the YOLOv5 algorithm to identify and classify the capsule defects of tablets and measure the capsule thickness, obtained a classification accuracy of 98.2%. Jin et al. [13] proposed a defect identification method based on DCGAN and YOLOv5 and solved problems regarding an insufficient number of samples and the uneven distribution of defect types in the defect detection of bonded structural parts by fine-tuning the structure and loss function of DCGAN. Li et al. [14] tested the food recognition algorithm of YOLOv5 on CFNet-34, a self-extended dataset based on the Chinese food dataset ChineseFood-Net. The average recognition accuracy of this algorithm was 89.7%, demonstrating good accuracy and robustness. Xue et al. [15] increased the attention mechanism and output layer to enhance feature extraction and feature fusion in order to improve recognition accuracy when analysing a complex background. The accuracy of target recognition was remarkably improved, and the mean average precision (mAP) value reached 80.5%. To detect postnatal defects of kiwifruit, Yao et al. [16] added a small target detection layer, improved the detection ability of the model for small defects, and introduced the loss function Ciou to improve the accuracy of regression. The CosineAnnealing algorithm was used to improve the effect of training. Zhao et al. [17] proposed a Wheat Grain Detection Network (WGNet) based on the training of the benchmark and used sparse network pruning and a mixed attention module to solve the problem of degradation. Zhang et al. [18] proposed an improved CoordConv feature extraction method to address the characteristics of weak echo intensity and small target area common in forward-facing sonar images, thereby endowing high-level features with corresponding coordinate information and improving the accuracy of network detection regression. Zhang et al. [19] proposed a method based on DL: YOLOv5-CA. A coordinated attention (CA) mechanism was integrated into YOLOv5 to highlight downy-mildew-related visual features and thus improve detection performance. Dai et al. [20] replaced Conv in the C3 module with CrossConv, which mitigated the problem of feature similarity loss in the fusion process and enhanced the module's feature representation capacity. The SPP algorithm was improved by using the fast space pyramid pool algorithm, which reduces the number of feature fusion parameters and accelerates the feature fusion speed.

In the process of feature recognition, many researchers have proposed solutions to the problem of object overlap and occlusion. Gao et al. [21] integrated data enhancement to boost network generalizability. Moreover, non-maximum suppression (NMS) was optimized to improve the accuracy of the network, which improved its ability to recognize fruit overlap. Ye et al. [22] used methods such as data augmentation, Test Time Augmentation (TTA), and the Weighted Boxes Fusion (WBF) to improve the robustness and generalization of a model for the identification of terminal overlapping buds in different growth states. Li D H et al. ([23–25]) used the mean shift algorithm to pre-segment images and set the radius parameter range in the detection process to further accelerate the speed of the algorithm. Additionally, the centre coordinates and radius of the target were obtained by detection; then, the overlapping targets were identified.

In this study, we present the results of the use of an improved YOLOv5 neural network model to address certain problems, such as the inability to accurately identify the 3D position of kiwi flower buds in the natural environment due to the complex background images and lighting conditions present therein, which results in a large range of coarse pollination during pollination and the inability to accurately achieve the full pollination of flower buds. The target detection of kiwifruit flowers was conducted, the parameters were optimized, and the training parameters were adjusted in accordance with the shooting distance, the effect of capturing the pictures, and the intensity of the flowers so as to achieve the highest accuracy in terms of target detection. In this study, the method of single target and double frame was adopted. The angles of the flowers were accurately identified through the position relationship of the flower and the stamen selection box, combined with the actual angles of the flowers, which provided angle data to support the accurate detection of the target. The pollination strategy of overlapping flowers was studied. Based on the pollination strategy, the detection function of YOLOv5 was modified such that it would meet the ideal requirements. A double-flow nozzle was used to pollinate the flowers, the spray parameters were determined, and the pollination area relationship of the inclined flowers was analysed. The droplet escape rate was measured, the spray compensation time was calculated, and a verification test was performed to prove the feasibility and practicability of this study and provide reliable support for the precise selection of kiwifruit using target spray technology.

2. Materials and Methods

2.1. Image Acquisition

A PyTorch DL framework was built, and PyCharm platform was used to achieve the training and testing of the model with respect to kiwi flower and bud recognition on a desktop computer (Intel core i5 10400fCPU, 2.6 GHz, 16 G memory, NVIDIA GeForce RTX 2060GPU, 6 GB video memory, Windows10 system, and CUDA and Cudnn libraries) made by the Intel Corporation, which is located in Arizona, the United States. The initial learning rate was set to 0.01, the threshold of IOU was set to 0.01, 8 samples were used as a batch-processing unit during model training, and the number of training rounds was 200. The optimal training weight parameter file was obtained after model was trained. The optimal weight file was used to distinguish kiwi flower pictures, and the performance of the recognition model was evaluated in accordance with the identification scenario.

A red kiwi plantation in Qiaosi, Hangzhou, was selected as the base of this experiment, as shown in Figure 1. The size of the kiwifruit plantation greenhouse was 6 m $\times$ 50 m and 150–170 cm off the ground, with melon rack spacing of 1.5 m. The light conditions in the greenhouse were sufficient, the kiwifruits were planted orderly, and the flowers were oriented in different directions, wherein about 70% of them were oriented downward, while the rest were scattered. The angle data must be identified and obtained to provide data support for the precise pollination process of the target.

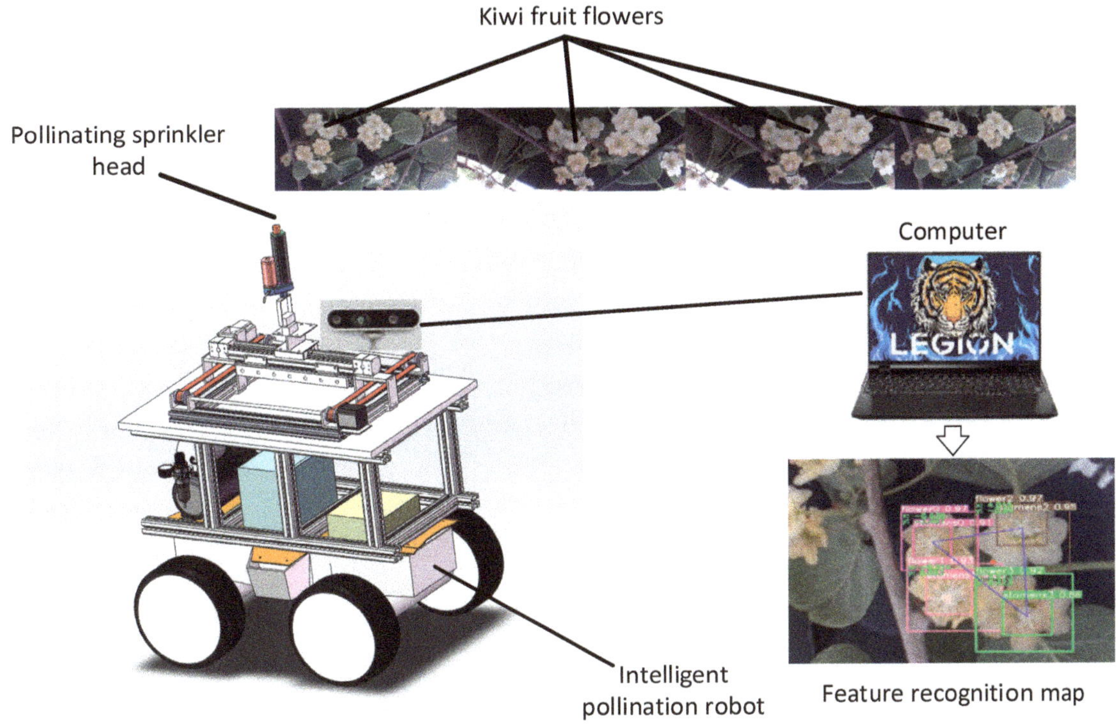

Figure 1. Kiwi flower feature recognition process.

An Intel RealSenseD415 camera made by the Intel Corporation, which is located in Arizona, the United States, was selected for this study. It is equipped with D400 series depth module, and its field-of-view angle is about 70°. It can effectively capture 10 m area within sight and supports the output of 1280 × 720 resolution depth. D415 has high pixel density and can achieve better, more accurate scanning for users in any given area of the same point. It has a special antiglare-processing function and can be used outdoors. The shooting height range of the camera in this study was 30–50 cm, and the camera was placed horizontally so that it faced upward when shooting.

2.2. Based on YOLOv5 Algorithm

There are a variety of object detection algorithms, such as YOLOv5, MobileNet, SSD, faster R-CNN, VGG, ResNet, etc. Compared with YOLOv5, MobilleNet, as a lightweight deep neural network, has fewer parameters and higher precision and can generate required feature maps with less computing time. However, due to its small number of convolutional layers, its feature extraction ability is very insufficient. The SSD model is fast but has poor recognition performance for small objects. Faster-RCNN uses RPN to generate candidate regions, and then extracts features through RoI pooling, so its accuracy is relatively high, and its processing speed is fast. However, its time complexity is high, it is difficult to use in practical applications, and its training and reasoning times are extensive. The VGG model has fewer parameters and operates through the convolutional series method, and it is stable and easy to transplant. However, due to its large number of full connection points and deep network structure, its training speed is slow. The residuals block is added to the ResNet network, which is helpful for the back propagation of the gradient in the training process, but there is a great deal of redundancy in the deep residuals network. In conclusion, the yolov5 model can significantly separate features, enhance mesh feature fusion, and ensure faster training speed.

YOLOv5 is a further improvement of the YOLOv4 algorithm, offering superior detection performance and rendering the results of testing via the coco test set more remarkable than before. The main innovation of YOLOv5 is its integration of multiple shortcuts. Accordingly, this algorithm can be used to develop a set of fast training and deployment schemes.

The network structure of YOLOv5 algorithm is mainly composed of input terminal, backbone network, neck network, and prediction component, and its network structure is shown in Figure 2. YOLOv5 can be divided into four models, namely, YOLOv5s, YOLOv5m, YOLOv5l, and YOLOv5x, in accordance with different depths and feature widths of the models. The larger the model, the greater its detection accuracy. However, despite its longer detection time, the YOLOv5s model is the smallest and most suitable for rapid detection tasks, and it offers superior detection accuracy. It can meet the requirements of daily agricultural and industrial applications.

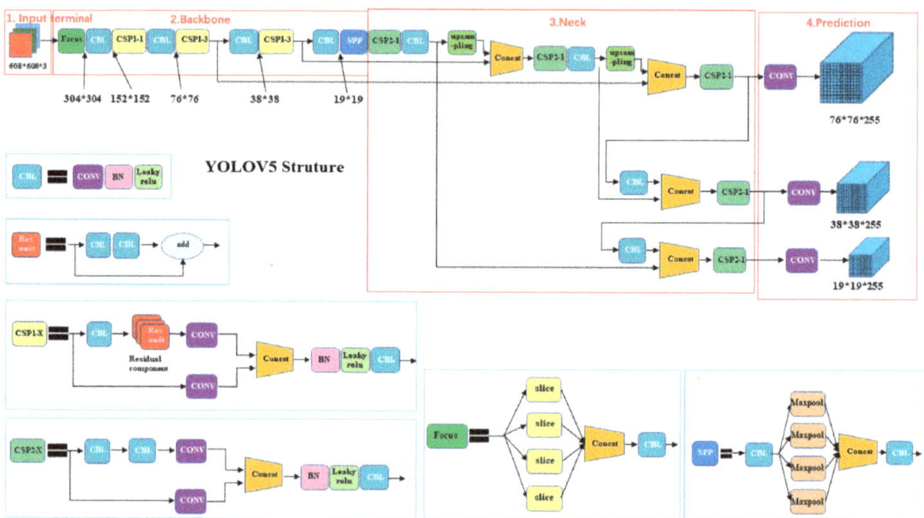

Figure 2. YOLOv5 frame diagram.

First, YOLOv5 was used to enhance the mosaic data of the input image at the input end and then randomly selected four photos in the data set to be pieced together into one by means of random scaling, cropping, and arrangement for training. This step enriched the data set, improved the training speed of the network, and reduced the memory requirements of the model. Adaptive anchor frame calculation and adaptive picture scaling were conducted together with stitching. In the network-training stage, the prediction box corresponds to the initial anchor frame output, and the size of the input picture is not limited. The pictures of different sizes are fitted to a fixed, common size; then, the pictures are input into the detection network. In this way, the image input process is simplified.

The Focus structure and SPP structure are added to the BackBone network. The focus structure can slice the input image. The original $608 \times 608 \times 3$ image outputs a feature mapping of $302 \times 302 \times 32$ after slicing, Concat, and Conv operations. The SSP module improves the receptive field, remarkably separates the features, and strengthens the network feature fusion ability so that images of any feature size can be inputted, and the size of the input images can be diversified.

The feature pyramid (FPN) + path aggregation structure (PAN) is added to the Neck network, which can allow for image feature fusion from multiple angles. Its complete structure is composed of an FPN layer that conveys semantic features from the top down and a PAN layer that conveys strong positioning features from the bottom up. Under the synergistic action of the two layers, the feature extraction effect is enhanced, and the precision of network training is improved.

In the prediction phase, the object sizes of the extracted and spliced feature maps are predicted and classified, and the predicted results are compared with the actual measured results and optimised. As shown in the Figure 2, the model outputs three sizes of the prediction models, and the models of different sizes have remarkable differences in terms of the field of sensitivity and resolution. An image output with a small multiple is suitable for the detection of small objects, whereas an image output with a large multiple is suitable for the detection of large objects. The output multiple of the image should be selected in accordance with the size of the detected object.

2.3. Methods Used in This Study

The main working methods used in this study are as follows: The K-means++ clustering method was used to cluster out the prediction anchor frame closer to the target size, and convolutional block attention module (CBAM) mechanism was added. *CIOU* was used to replace *GIOU*. The modules for calculating the flower inclination angle, judging the overlap, determining the pollination point coordinates, and searching for the pollination angles were added into the detection function.

2.3.1. Use of K-Means++ to Cluster out New Anchor Boxes

By default, YOLOv5 uses the K-means algorithm to cluster data sets for generating anchor boxes and uses genetic algorithm to adjust the size of anchor boxes during training. The k-means clustering algorithm initialises K points randomly as clustering centres, calculates the distance between samples and each clustering centre, divides the samples to the nearest clustering centre point, calculates the mean value of all sample features divided into each category, and uses the mean value as the new clustering centre. However, the initial random selection of clustering centre is uncertain, which affects the convergence speed of the algorithm or leads to false classification. Therefore, this study uses the K-means++ clustering algorithm to obtain a prior anchor box that is more consistent with the data sample.

The essential difference between k-means++ and k-means algorithms depends on the initial selection of K clustering centres. The basic principle of K-means++ in the initialisation process of the cluster centre is to ensure that the distance between the initial cluster centres is as far as possible. First, a sample point is randomly selected as the initial cluster centre; then, the sample point with the largest distance is selected as the next cluster centre through the roulette wheel method until K cluster centre points are selected. Finally, the K-means algorithm is used to determine the final clustering centre. Although the K-means++ algorithm takes a certain amount of time for the initial determination of cluster centre points, it reduces the influence of the random selection of cluster centre points on the size of anchor frame selection, selects better initial cluster centres, and improves the speed of the algorithm.

2.3.2. Adding CBAM Mechanism

When observing objects, the human visual system will transfer more attention to areas whose objects' attributes can be quickly judged. This type of attention selection mechanism can accelerate our recognition of an object's attributes and use attention resources more effectively. In this study, the attention mechanism is generated by referring to this feature of the human eye. In essence, the model allocates a certain weight to different features of an object and draws attention to the most useful features of the object so as to improve the object recognition rate.

Attention is divided into two main types, namely, spatial attention (CAM) and channelled attention (SAM), as shown in the figure. Compared with other attention mechanisms, CBAM attention mechanism is a simple and effective attention module for feedforward convolutional neural networks, which consists of two modules, namely, spatial attention module (CAM) and channel attention module (SAM). For the input feature map, the CBAM module inferences the attention diagram in two separate dimensions (channel and space). The CAM module first applies global maximisation and global average pooling to the input feature graph, transmits the processed feature graph to a shared neural network, performs elementwise addition operation on the output feature, and outputs the attention feature through calculation. The attention features generated by the CAM are input into SAM after multiplication. The spatial attention features are obtained through sigmoid nonlinear operation in SAM. Finally, the attention map is multiplied by the input feature map to perform adaptive feature optimisation, save parameters and computational power, and ensure that the function can be realised immediately in the existing network.

2.3.3. CIOU Replaces GIOU

By default, YOLOv5 adopts *GIOU*-loss as the bounding box's loss function and uses binary cross entropy and Logits loss functions to calculate the loss probability and target score. However, the loss function is the same as *IOU*-loss when the values of *IOU* and *GIOU* obtained during calculation are the same, which makes it impossible to measure the real position relationship. Therefore, in this study, *CIOU*-loss is chosen to replace *GIOU*-loss, and the formula is expressed as

$$CIOU = IOU - \frac{\rho^2(b, b^{gt})}{c^2} - \alpha v \quad (1)$$

CIOU provided two penalty terms, and box regression considered the coverage area, centre distance, and aspect ratio, which more effectively solved the three problems and allowed the model to obtain a better regression effect. As shown in Figure 3.

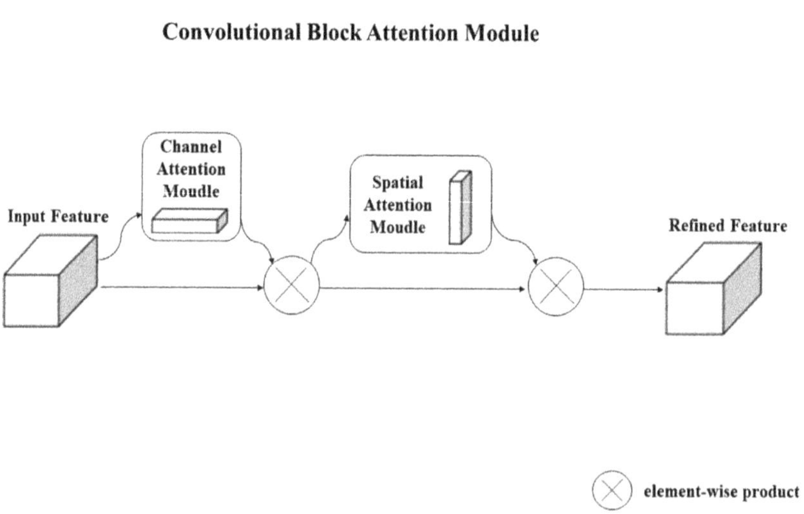

Figure 3. Attention model.

2.3.4. Detection Function for Flower Angle Calculation Module

The research shows that different locations of the recognition box will have a distinct influence on the detection of the tilt angle of the flower. When the distance between the recognition box changes, the detected tilt angle of the flower will also change. When the distance between the centres of two flower frames is minimal, the flower tilt can't be detected, but when the distance between the centres of two flower frames is large, the detected flower tilt will show a certain error.

In accordance with the above research and analysis, a certain correlation is observed between the tilt angle of the flower and the position of the recognition box. Therefore, we needed to establish a model including the flower and the flower core. We can further explore the position relationship between the flower tilt angle and different flower's centre distance through the calculation and analysis of this model. As shown in Figure 4, the original length of the flower is L, the horizontal length of the flower is b, the distance between the centre point in the horizontal direction of the flower and stamen is Δx, and the thickness between the flower and the stamen is a when the tilt angle of the flower is α.

Figure 4. Flower tilt model.

As shown in Figure 4, the tilt angle of flowers has the following functional relationship with the horizontal length of flowers and the thickness of flowers and stamens

$$\cos\alpha = \frac{b}{L} \quad (2)$$

$$\sin\alpha = \frac{\Delta x}{a} \quad (3)$$

The formula for calculating the tilt angle of the flower can be obtained by combining the above two formulas and simplifying the calculation, as follows:

$$\alpha = \arctan\left(\frac{L}{a} \times \frac{\Delta x}{b}\right) \quad (4)$$

We can find the expression of the tilt angle of the flower by using this formula. However, the flower's own gravity has an effect on the actual tilt in a real scenario. Considering that the flower stalk is long and not rigid, it is prone to deformation and bending under the influence of the gravity of the flower itself after the flower tilts at a certain angle, thereby reducing the tilting angle. We conducted field research in an orchard and collected relevant pictures and data of flowers tilting at a certain angle while considering the influence of gravity on the tilt angle of flowers, as shown in Figure 5. By combining the actual data and the calculated tilt angle, the calculation formula of the tilt angle of the flower was further optimised and improved, and the accuracy of the calculation formula of the tilt angle of the flower was improved.

Figure 5. Flower tilt photo.

The collected pictures and data were processed by using the regression function to reduce the adverse influence of gravity on the calculation of the actual tilt angle of flowers, and the final regression formula can be obtained as follows:

$$y_a = 8.663\sqrt{\frac{100\Delta x}{b}} \quad (5)$$

where y_a is the calculated angle. This formula can satisfy the calculation of the tilt angle of flowers, and the error is controlled within 5%. In accordance with the actual situation, the maximum tilt angle of flowers is 30°.

The error between the calculated tilt angle and the actual measured tilt angle is calculated as follows

$$Y = \frac{|y_b - y_a|}{y_a} \times 100\% \quad (6)$$

where Y is the error rate and y_b is the actual tilt angle by experimental measurement.

2.3.5. Search for Pollination Points Based on Flower Overlap and Flower Angle Identification

The analysed flowers have four distribution conditions: single flower, two overlapping flowers, three overlapping flowers, and four or more overlapping flowers. In accordance with the spray pollination method, the final pollination area is a circle with a certain area. Therefore, a pollination strategy based on polygons was developed to find the edge points of polygons and thus ensure that the flowers can be fully pollinated within a given range. The intersection of two flower frames is defined as an overlap point. The specific pollination strategies are as follows.

For a single flower, the centre point of the flower core is selected as the pollination point; thus, the nozzle moves toward the pollination point, and the nozzle angle is not tilted.

When two flowers overlap, the centre point of the two flowers is selected as the feature point, and the two feature points are connected. The centre point of the feature point is designated as the position of the nozzle. The nozzle tilts at a certain angle after it moves to this position and is then aimed at the two flower core feature points for pollination.

When three flowers overlap, the centre point of the three flowers is selected as the feature point, and the three feature points are connected to form a triangle. The centre of the triangle is selected as the position of the nozzle. The nozzle tilts at a certain angle after it moves to this position and is aimed at the three flower core feature points for pollination.

When four or more flowers overlap, the top, rightmost, bottom, and leftmost feature points are searched. If four feature points are found, then the quadrilateral centre is selected as the position of the nozzle. The nozzle tilts at a certain angle after it moves to this position and is aimed at the four flower core feature points for pollination, ensuring that the entire overlapping area is covered. Three feature points are regarded as three flowers overlapping, and two feature points are treated as two flowers overlapping. The specific pollination scenario is shown in Figure 6.

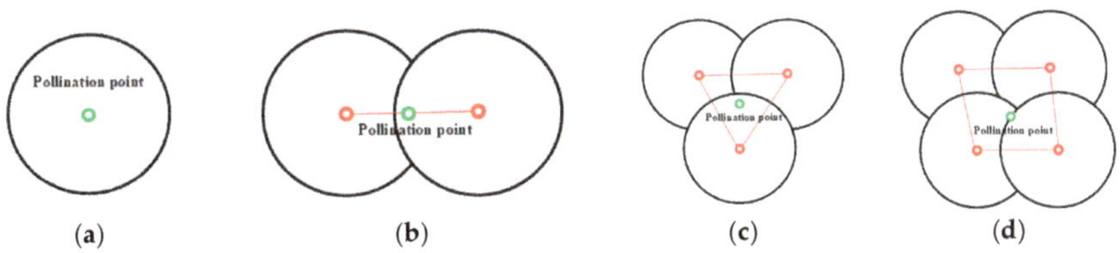

Figure 6. Schematic of pollination points under overlapping conditions: (**a**) single flower; (**b**) overlap of two; (**c**) overlap of three; (**d**) four or more overlapping.

In accordance with the overlap relationship, the recognization results of overlap is considered to be matched with the actual when the intersection area of two flower frames accounts for more than 60% of the flower frame area. In accordance with the Gauss area shoelace formula, the centroid of the polygon in the overlapping case is calculated. The area of the polygon is calculated in accordance with the shoelace formula.

$$A = \frac{1}{2}\left|\sum_{i=1}^{n} x_i(y_{i+1} - y_{i-1})\right| = \frac{1}{2}\left|\sum_{i=1}^{n} y_i(x_{i+1} - x_{i-1})\right| = \frac{1}{2}\left|\sum_{i=1}^{n} det\begin{pmatrix} x_i & x_{i+1} \\ y_i & y_{i+1} \end{pmatrix}\right| \tag{7}$$

where A is the polygon area of overlapping flowers, x_{i-1}, x_i, x_{i+1} is the abscissa of the overlapping flower centroid, y_{i-1}, y_i, y_{i+1} is the ordinate of the overlapping flower centroid. The position of the centroid can be calculated by using the root equation, as follows:

$$c_x = \frac{1}{6A}\sum_{i=0}^{n-1}(x_i + x_{i+1})(x_i y_{i+1} - x_{i+1} y_i) \tag{8}$$

$$c_y = \frac{1}{6A}\sum_{i=0}^{n-1}(y_i + y_{i+1})(x_i y_{i+1} - x_{i+1} y_i) \tag{9}$$

where c_x is the abscissa of the pollination point, c_y is the ordinate of pollination point.

The centroid's coordinates are outputted, and the tilt angle with respect to the target is calculated in accordance with the coordinates of the centre points of the stamens of each overlapping polygon, as shown in Figure 7.

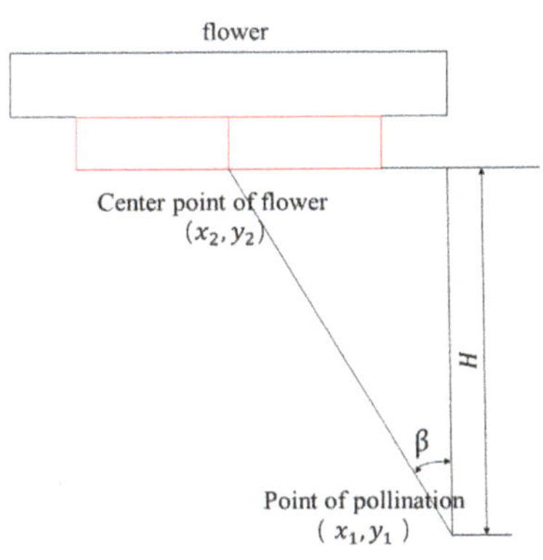

Figure 7. Pollination of the centre point of the flower.

The formula for calculating the pollination angle is as follows:

$$\beta = \frac{|x_1 - x_2|}{H} \quad (10)$$

where β is the pollination angle, x_1 and x_2 are the horizontal coordinates of the pollination point and the flower centre point, respectively, and H is the pollination distance.

In accordance with the above regression function of tilt angle and overlapping pollination strategy, the detection function can be modified, as shown in Figure 8, and the tilt angle calculation module, overlapping flower scenario identification module, and pollination point discovery module can be incorporated.

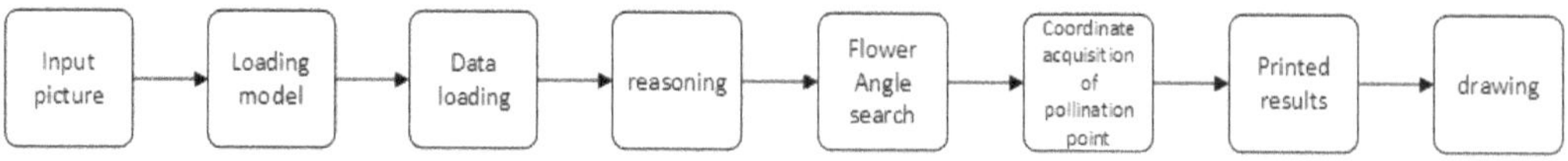

Figure 8. Image-reasoning process diagram.

The input image is identified in accordance with the weight obtained from the training data set. The setting of the double-target frame will identify the flowers and stamens and match them in accordance with their intersection relationship. The tilting angle of the flowers in the X and Y directions after matching will be determined by using the regression equation of the tilting angle of the flowers; then, the overlap situation of the flowers will be determined in accordance with the intersection ratio. Thus, the coordinates of pollination points and the angle at which the nozzle tilts with respect to the overlapping flowers in X and Y directions at the pollination points are determined in accordance with different overlapping conditions, and the results are printed and drawn.

2.4. Model Evaluation

The precision, recall rate, and mAP value of the model are used as the evaluation criteria for the feature recognition quality of YOLOv5 model, which can be calculated as follows:

$$Precision = TP/(TP + FP) \quad (11)$$

$$Recall = TP/(TP + FN) \quad (12)$$

$$F1 = [2/(1/Precision) + (1/Recall)] \quad (13)$$

$$AP = 1/11 \times \sum_{r\in(0,0.1,0.2,......1)} pinterp(r) \quad (14)$$

$$mAP = \frac{\sum AP}{N} \quad (15)$$

where *Precision*—the ratio of true position in the recognized image; *Recall*—the proportion of the left and right positive samples in the test set that are correctly identified as positive samples; *F*1—the weighted harmonic average of Precision and Recall; *AP*—the interpolated average precision of the detection algorithm; *TP*—the result wherein kiwi flowers are correctly identified; *FP*—the result wherein kiwi flowers are incorrectly identified; and *FN*—the result wherein kiwi flowers are not detected.

2.5. Data Set Construction and Model Parameter

In the test base shown in Figure 9, more than 2000 kiwi flower pictures with a resolution of 1920 × 1080 were collected. The shooting period was from 3:00 pm to 5:00 pm, and multiple shooting angles were selected During the shooting of kiwifruit orchard, the distribution of kiwifruit flowers in full bloom was complex, and many overlapping situations were observed. As shown in Figure 10, the main overlapping situations consisted of two flowers overlapping, three flowers overlapping, and four or more flowers overlapping.

Figure 9. Test base.

Figure 10. Distribution of flower overlap.

A total of 880 kiwifruit flower images with good image quality were selected as the original data set, and the image data were manually labelled, including the flower and the core area of the same flower. The smallest external rectangle of the flower and the core area were used as the marking frame to ensure that the subject occupied a higher proportion of the marking frame. Flowers were labelled as the 'flower' class and stamens as the 'stamens' class; then, the images were converted to xml format to obtain more complete coordinate information. Data augmentation was performed with respect to the original data set, mainly in the form of rotation and the reduction and enhancement of brightness and Gaussian noise. The purpose of data augmentation is to enhance the useful information in an image, improve the visual properties of the image, purposefully emphasise the overall or local characteristics of the image, clarify the original image, or emphasise some interesting features. This process was conducted to distinguish different object features more effectively and improve the robustness of the model. Following data enhancement, a total of 3344 images were used as training data set used to train the recognition model. As shown in Figure 11.

Figure 11. Images after data enhancement.

Data sets D1, D2, D3, and D4 were constructed to compare the influence of data sets participating in the modelling of flower overlap recognition. For each overlap, 800 pieces were selected for model training, 200 pieces for model verification, and 200 pieces for model testing. The data set structure is shown in Table 1. Before the training procedure, the initial parameters of the YOLOv5 model were set (as shown in Table 2).

Table 1. Data set structure.

Data Set	Place	Condition of Overlap	Set of Training	Set of Verification	Set of Tests
D1	Orchard greenhouses	Single flower	800	200	200
D2		Overlap of two	800	200	200
D3		Overlap of three	800	200	200
D4		Four or more overlapping	800	200	200

Table 2. Parameters for model training.

Parameters	Value
Input size/pixels	640×640
Initial learning rate	0.032
Momentum	0.843
Cyclical learning rate	0.12
Iteration	200

The size of the input image determines the amount of information calculated in the training process. An appropriate image size can achieve high training accuracy and reduce hardware loss in order to reduce computational work. According to preliminary studies, the size of the input image was selected as 640×640. Since VOC data set is used, the default parameter value obtained by optimization training applied to VOC data set is selected. According to the data set size and the difficulty of target feature recognition, the number of iterations is set to 200.

The distribution of specific label information obtained after the training of the recognition model is shown in Figure 12. The flower image is of a moderate size due to the selection of an appropriate shooting distance. In the image of the recognition model, the flower, as a medium-sized target, enables the system to accurately identify its features. The dark-blue colour denotes the distribution area of labels. The size of labels is moderate, and the size of targets identified is relatively uniform. This data set is suitable for the close pollination of targets.

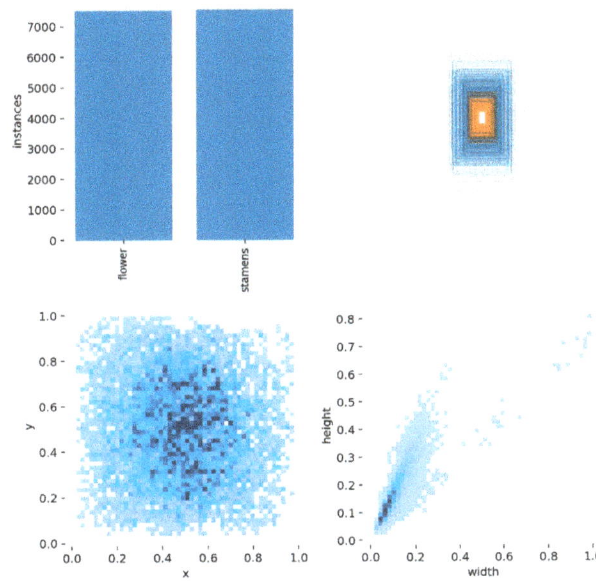

Figure 12. Label information distribution.

3. Results

3.1. Training Results

When collecting data on the kiwi flower distribution in a kiwifruit plantation, the number and overlap of flowers collected through the employed imaging system will be affected by changes in recognition distance. When the target pollination device is close to the kiwi flower, there are fewer flowers and flower overlap characteristics that can be recognised, but the overlap information of the kiwi flower can be identified more accurately. On the contrary, when the target pollination device is far from the kiwi flower, there will be more kiwi flowers in the visual system's recognition range, and there will also be more overlapping features of the flowers, which prevents the employed imaging device from effectively identifying the nature of the overlap of the kiwi flowers, thereby hampering the employed device's ability to identify the characteristics of the flowers. In the kiwi flower recognition training program relayed herein, the YOLOv5 with K-means++ and CBAM attention mechanism were used to identify kiwi flowers, and the optimized detect function was used to judge and calculate the flower's overlap and tilt angle.

A total of 800 training sets, 200 validation sets, and 200 test sets were used for each flower overlap. The test sets of flower overlap were divided into four categories: (1) single flower; (2) overlap of two flowers; (3) overlap of three flowers; and (4) overlap of four or more flowers. The training results are shown in Figure 13. As shown in Figure 13, the predicted value and recall rate tend to be stable at 100 rounds, with a maximum forecast confidence of 0.962, a maximum recall rate of 0.92, and a maximum mAP@0.5 value of 0.89.

From 100 rounds to 200 rounds, the above parameters are relatively stable and fluctuate only in a small range. Therefore, the model trained for 200 rounds was selected as the kiwi flower recognition model in the spray pollination process. The total number of kiwi flowers and the overlap number of the kiwi flowers can be obtained, and the accuracy, mAP value, maximum forecast confidence, and maximum recall rate can be calculated through this model.

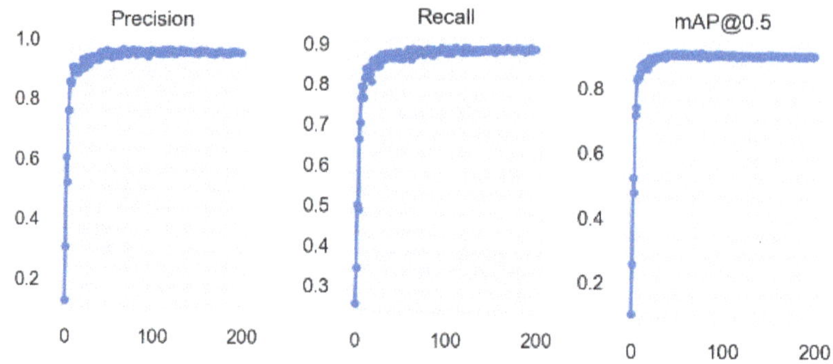

Figure 13. Prediction curve, Recall curve, and mAP value of the training results.

With regard to the visual system on the target device, its feature recognition accuracy can be maintained when the number of kiwifruit flowers is low. However, instances of false detection and failed detection will gradually increase when the number of flowers is increases and the degree of overlap becomes more complex. Therefore, the K-means++ algorithm and CBAM attention mechanism were incorporated into the YOLOv5 model, which can adjust the size of the anchor frame and select better clustering centres during training. The CBAM attention mechanism gives a higher weight to the overlapping features of kiwi flowers, so the visual system's ability to recognise the overlapping situation of flowers is enhanced, and a higher mAP value and maximum prediction confidence can be obtained compared with the results detected under the YOLOv5 model. These experimental results show that the optimised YOLOv5 model can more accurately identify the number of overlapping flowers and the tilt angle of flowers at higher quantities than those previously documented.

3.2. *Accuracy Rate of Flower Angle Recognition*

A certain error is observed between the actual tilt angle and the theoretical tilt angle of a kiwi flower due to the influence of the flower's own gravity. We calculated the regression function of the collected data and obtained a more accurate formula for calculating the tilt angle of the flower (see Formulae (4) and (5)). It was only necessary to identify the coordinates of the flower and the flower bud; then, the tilt angle of the flower could be calculated using Formula (5).

The flower tilt angle measured via image recognition was compared with the actual measured flower tilt angle, and the error rate (refer to Formula (6)) between the real tilt angle and the identified tilt angle was obtained through multiple groups of comparison experiments, as shown in Figure 14.

The tilt angles of 50 groups of flowers were measured using image recognition, and the data on the tilt angles of the two groups were compared. In accordance with the error rate chart of the tilt angle measurement of the flowers, the error between the tilt angle of the flowers measured by the calculation formula and the actual tilt angle was between 0.5% and 0.8%, which proved that the revised regression function tilt angle calculation formula was highly reliable. It can be used to measure the actual tilt angle of flowers, thereby providing accurate coordinate information that can be used to allow kiwi flowers to accurately pollinate the target and laying the foundations for effectively improving the precision rate of pollination. The accuracy of flower tilt angles measured using this method is high, and the established model can accurately obtain the flower tilt angle with high reliability.

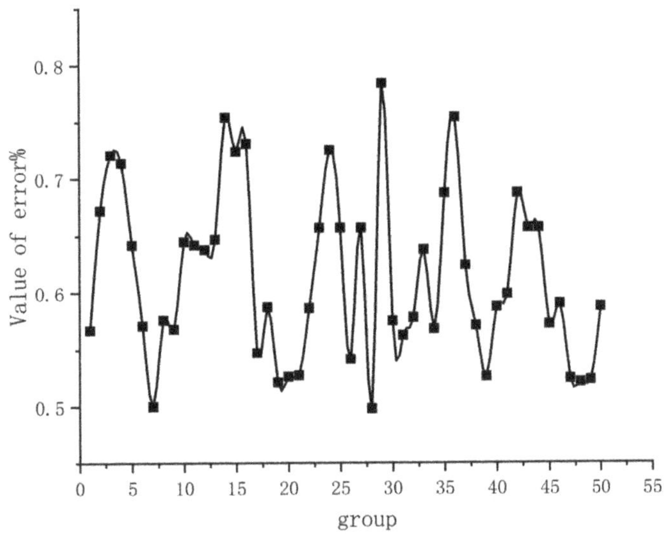

Figure 14. Error rate of flower tilt angle measurement.

3.3. Comparison Experiment Conducted to Identify Kiwi Flower Overlap

Kiwi flower overlap can be divided into four main conditions: single-flower overlap, two flowers overlapping, three flowers overlapping, and four and more flowers overlapping. The overlap of flowers is relatively complicated due to the complexity and variability of the shooting environments and angles. This condition leads to the failure of the employed visual system of the target device to accurately identify the features of the flowers and obtain the coordinates of the flowers and flower buds, which increases the probability of failed and false detection, reduces the accuracy of identification, and affects the success rate of pollination of flowers.

In this paper, YOLOv5 was used to identify the overlap of macaque peach blossoms. To compare the performance of different algorithms, such as Faster-RCNN, SSD, VGG, etc., in identifying flower overlap, the specific results of different target detection algorithm models are listed in Table 3.

Table 3. Test results of the five models regarding flower overlap.

Model	Predicted Value (Flowers)	Rate of Recall (Flowers)	Predicted Value (Stamens)	Rate of Recall (Stamens)	mAP@0.5
YOLOv5s	96.7%	89.1%	91.1%	78.3%	90.1%
Faster-RCNN-ResNet50	57.4%	98.9%	58.4%	97.9%	92.6%
Faster-RCNN-VGG	68.5%	98.9%	67.9%	98.0%	95.6%
SSD-VGG	76.6%	87.4%	82.8%	65.6%	82.3%
SSD-MobileNetv2	86.7%	70.2%	89.6%	55.0%	81.1%

Table 3 shows that the predicted accuracy of flowers and stamens obtained by YOLOv5s reached 96.7% and 91.1%, respectively, while maintaining high recall values. The predicted values of Fster-RCNN-ResNet50 and Faster-RCNN-VGG are both less than 70%, which is because the processes performed by the two algorithms are complex and their real-time performance is poor, which are both factors that are not conducive to the recognition of a large number of overlaps between flowers and stamens. Compared with YOLOv5, the

SSD-MobileNetv2 and SSD-VGG networks had deficiencies in terms of their small target identification features and extraction, and the predicted value decreased to less than 90%. In summary, in terms of overlapping target detection and small target detection, the overall ability of YOLOv5 is the best among the analysed models, and its predicted values, recall rate, and map@0.5 remain stable, indicating that the model can accurately identify kiwi flower overlap and obtain flower tilt angles.

The YOLOv5s model was used to conduct batch reasoning with the detection function for the recognition of flower overlap. The reasoning results are shown in Figure 15. As shown in Figure 15, the model we selected for kiwi flower overlap obtained good recognition results and can remarkably distinguish flowers from stamens. The model has a fast response speed, allowing it to identify the overlap of kiwi flowers and ensure a certain recognition speed. The YOLOv5s model can accurately calculate the centroid in accordance with the overlap situation and locate the pollination centre point after the modification of the detection function. In accordance with the calculation formula of the tilt angle, the tilt angle of the centre point of the edge of the overlapping polygon with respect to the target can be found. For multiple targets in the same picture, an overlap situation can be identified and judged simultaneously, which improves the efficiency of the subsequent targeting task; that is, it guarantees the accuracy of targeting.

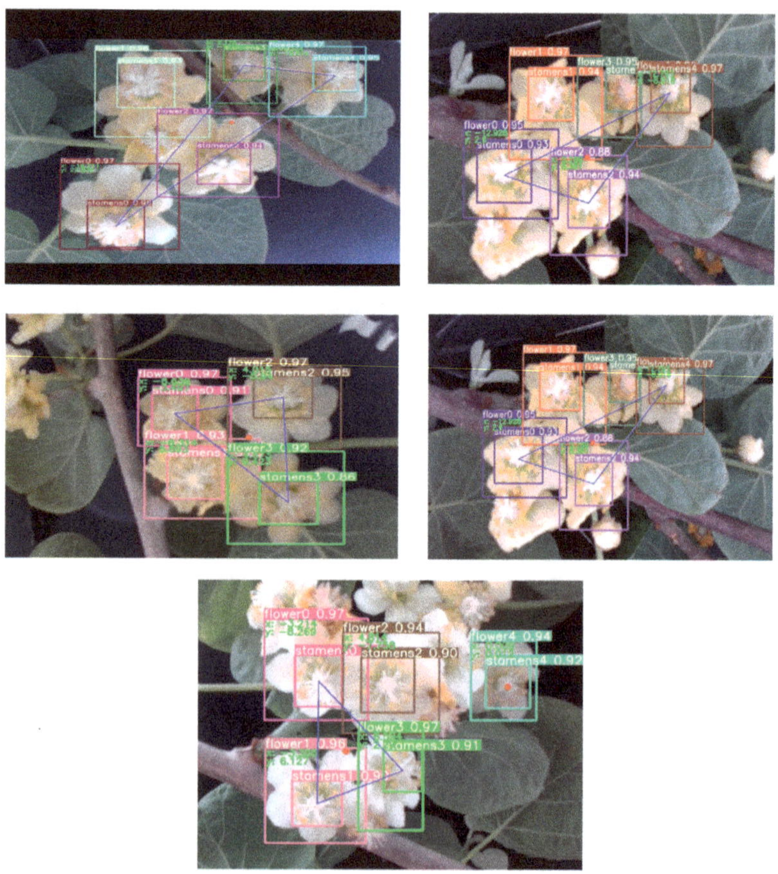

Figure 15. Inference result diagram of the detection function.

3.4. Comparative Test of Four YOLOv5 Models

An identification test was conducted for the four YOLOv5 series models to verify the rationality of the model selected in this study. The comprehensive performance of each model was determined to prove whether it was an optimal model. The test data of each model are shown in Table 4.

Table 4. Test results of the four models.

Model	F1	Average Time Per Frame (Milliseconds)	Memory (MB)
YOLOv5s	90.12%	8.64	20
YOLOv5m	93.97%	40.5	134
YOLOv5l	85.74%	425	278
YOLOv5x	91.24%	13.87	60.8

As shown in Table 4, all the models have high F1 rating values that surpassed 85%. The YOLOv5s model has the fastest average time per frame, reaching 8.64 milliseconds, which is higher than the other three. The YOLOv5s model only has 20 MB of memory, which is far smaller than the 100 MB or more of other models. The above data prove that YOLOv5s has high precision and a fast response speed, which is suitable for real-time detection used in small-scale agricultural operations.

Therefore, the proposed target identification technology of single-target and double-frame flower identification based on YOLOv5 can adequately realise the precise targeting of the pollination of kiwi flowers.

4. Discussion

In this study, a recognition model based on YOLOv5 was proposed for kiwi flower pollination. The experimental results show that the model has a high degree of recognition in terms of the overlap and tilt angle of kiwi flowers, can ensure extremely high accuracy, and can provide accurate coordinate information regarding kiwi flowers and flower buds, thus facilitating the development of an accurate pollination strategy.

In model's training process, the K-means++ clustering algorithm was adopted. Compared with the K-means algorithm, the optimised clustering algorithm can obtain a better initial clustering centre, select a more appropriate anchor frame size, and improve the calculation speed of the algorithm. The CBMA mechanism was incorporated into the model, which improved the model's extraction accuracy of kiwi flower features, effectively reduced the rate of missed detection and error detection, thus saving computational workload such that it was easier for the visual system to extract the features in the image. The optimization of the detection function improved the recognition of flower overlaps and the accuracy of flower tilt angle calculation and accurately determined the flower coordinates, pollination point coordinates, and pollination angles.

Different algorithms were used to identify the features of overlapping flowers. The effectiveness of YOLOv5 and other algorithms, such as Faster-RCNN, SSD, VGG, etc, was compared. Consequently, it was determined that the YOLOv5s model performs the best as the predicted values of flowers and stamens can reach 96.7% and 91.1%, respectively, while maintaining high recall values, which can meet the identification requirements regarding kiwi flower overlap and tilt angles. Among the four YOLOv5 series models, compared with YOLOv5m, YOLOv5l, and YOLOv5x, YOLOv5s has higher F1 values, the shortest average recognition time, and the smallest memory size, demonstrating its advantages of a fast response speed and high precision.

The algorithmic model of this study is suitable for kiwifruit plantations with flat ground but requires suitable lighting conditions. For kiwifruit plantations with rugged ground, it is very likely that smaller targets may appear, so a small target detection layer needs to be incorporated into the algorithm model to improve its recognition accuracy. Moreover, considering the night pollination to improve overall pollination efficiency, it is necessary to enhance the illumination and corresponding filtering treatment of the images selected at night to improve the accuracy of night target recognition.

Author Contributions: Conceptualization, H.Z. and J.T.; Data curation, P.M. and Z.L.; Formal analysis, P.M. and J.T.; Funding acquisition, H.Z. and J.T.; Investigation, H.Y.; Methodology, H.Z., J.O. and J.T.; Project administration, H.Y.; Supervision, H.Y.; Validation, H.Z., J.O. and P.M.; Visualization, J.O. and Z.L.; Writing—original draft, H.Z. and P.M.; Writing—review and editing, J.T. All authors have read and agreed to the published version of the manuscript.

Funding: This research was funded by the financial support provided by Zhejiang Key Research and Development Program (grant number 2022C02055), National Natural Science Foundation of China (grant number 32201676), and Fundamental Research Funds of Zhejiang Sci-Tech University (grant number 19022456-Y).

Data Availability Statement: The data that support the findings of this study are available on request from the corresponding author, Junhua Tong, upon reasonable request.

Conflicts of Interest: The authors declare no conflict of interest. The funders had no role in the design of the study; in the collection, analyses, or interpretation of data; in the writing of the manuscript; or in the decision to publish the results.

References

1. Guan, L. Japan's kiwi artificial pollination technology. *J. Deciduous Fruit Trees* **2002**, *5*, 60.
2. Jiang, Z.J. Optimization of Double-Flow Spray Parameters and Development of Kiwifruit Pollination Device. Master's Thesis, Northwest Agriculture and Forestry University of Science and Technology, Xi'an, China, 2020.
3. Krizhevsky, A.; Sutskever, I.; Hinton, G.E. ImageNet classification with deep convolutional neural networks. *Adv. Neural Inf. Process. Syst.* **2012**, *25*, 1097–1105. [CrossRef]
4. Liu, T.; Teng, G.; Yuan, Y.; Liu, B.; Liu, Z. Recognition method of winter jujube fruit in natural scene based on improved YOLO-v3. *Trans. Chin. Soc. Agric. Mach.* **2021**, *52*, 17–25.
5. Zhang, H.; Fu, Z.; Han, W.; Yang, G.; Niu, D.; Zhou, X. Maize seedling number acquisition method based on improved YOLO. *Trans. Chin. Soc. Agric. Mach.* **2021**, *52*, 221–229.
6. Li, K.J. Research and Application of Weed Detection Algorithm in Cotton Field based on YOLOv3. Master's Thesis, Xinjiang University, Ürümqi, China, 2021.
7. Yue, Y.; Li, X.; Zhao, H.; Wang, H. Crop disease image recognition based on improved VGG networks. *J. Agric. Mech. Res.* **2022**, *44*, 18–24.
8. Li, H.; Li, C.; Li, G.; Chen, L. A real-time table grape detection method based on improved YOLOv4-tiny network in complex background. *Biosyst. Eng.* **2021**, *212*, 347–359. [CrossRef]
9. Wang, H.; Xu, Y.; He, Y.; Cai, Y.; Chen, L.; Li, Y.; Angel Sotelo, M.; Li, Z. YOLOv5-Fog: A Multiobjective Visual Detection Algorithm for Fog Driving Scenes Based on Improved YOLOv5. *IEEE Trans. Instrum. Meas.* **2022**, *71*, 1–12. [CrossRef]
10. Lv, J.; Xu, H.; Han, Y.; Lu, W.; Xu, L.; Rong, H.; Yang, B.; Li, Z.; Ma, Z. A visual identification method for the apple growth forms in the orchard. *Comput. Electron. Agric.* **2022**, *197*, 106954. [CrossRef]
11. Chaschatzis, C.; Karaiskou, C.; Mouratidis, E.G.; Karagiannis, E.; Sarigiannidis, P.G. Detection and Characterization of Stressed Sweet Cherry Tissues Using Machine Learning. *Drones* **2021**, *6*, 3. [CrossRef]
12. Ficzere, M.; Mészáros, L.A.; Kállai-Szabó, N.; Kovács, A.; Antal, I.; Nagy, Z.K.; Galata, D.L. Real-time coating thickness measurement and defect recognition of film coated tablets with machine vision and deep learning. *Int. J. Pharm.* **2022**, *623*, 121957. [CrossRef]
13. Jin, Y.; Gao, H.; Fan, X.; Khan, H.; Chen, Y. Defect Identification of Adhesive Structure Based on DCGAN and YOLOv5. *IEEE Access* **2022**, *10*, 79913–79924. [CrossRef]
14. Li, H.; Yang, G. Dietary Nutritional Information Autonomous Perception Method Based on Machine Vision in Smart Homes. *Entropy* **2022**, *24*, 868. [CrossRef]
15. Xue, J.; Zheng, Y.; Dong-Ye, C.; Wang, P.; Yasir, M. Improved YOLOv5 network method for remote sensing image-based ground objects recognition. *Soft Comput.* **2022**, *26*, 10879–10889. [CrossRef]
16. Yao, J.; Qi, J.; Zhang, J.; Shao, H.; Yang, J.; Li, X. A Real-Time Detection Algorithm for Kiwifruit Defects Based on YOLOv5. *Electronics* **2021**, *10*, 1711. [CrossRef]
17. Zhao, W.; Liu, S.; Li, X.; Han, X.; Yang, H. Fast and accurate wheat grain quality detection based on improved YOLOv5. *Comput. Electron. Agric.* **2022**, *202*, 107426. [CrossRef]
18. Zhang, H.; Tian, M.; Shao, G.; Cheng, J.; Liu, J. Target Detection of Forward-Looking Sonar Image Based on Improved YOLOv5. *IEEE Access* **2022**, *10*, 18023–18034. [CrossRef]
19. Zhang, Z.; Qiao, Y.; Guo, Y.; He, D. Deep Learning Based Automatic Grape Downy Mildew Detection. *Front. Plant Sci.* **2022**, *13*, 872107. [CrossRef]

20. Dai, G.; Hu, L.; Fan, J.; Yan, S.; Li, R. A Deep Learning-Based Object Detection Scheme by Improving YOLOv5 for Sprouted Potatoes Datasets. *IEEE Access* **2022**, *10*, 85416–85428. [CrossRef]
21. Gao, G.; Wang, S.; Shuai, C.; Zhang, Z.; Zhang, S.; Feng, Y. Recognition and Detection of Greenhouse Tomatoes in Complex Environment. *Traitement Signal.* **2022**, *39*, 291–298. [CrossRef]
22. Ye, Z.; Guo, Q.; Wei, J.; Zhang, J.; Zhang, H.; Bian, L.; Guo, S.; Zheng, X.; Cao, S. Recognition of terminal buds of densely-planted Chinese fir seedlings using improved YOLOv5 by integrating attention mechanism. *Front. Plant Sci.* **2022**, *13*, 991929. [CrossRef]
23. Li, D.H.; Zhao, H.; Yu, X. Overlapping green apple recognition based on improved spectral clustering. *Spectrosc. Spect. Anal.* **2019**, *39*, 2974–2981.
24. Uijlings, J.R.; Van De Sande, K.E.; Gevers, T.; Smeulders, A.W. Selective search for object recognition. *Int. J. Comput. Vis.* **2013**, *104*, 154–171. [CrossRef]
25. Miao, R.; Shan, Z.; Zhou, Q.; Wu, Y.; Ge, L.; Zhang, J.; Hu, H. Real-time defect identification of narrow overlap welds and application based on convolutional neural networks. *J. Manuf. Syst.* **2022**, *62*, 800–810. [CrossRef]

horticulturae

MDPI

Article

GA-YOLO: A Lightweight YOLO Model for Dense and Occluded Grape Target Detection

Jiqing Chen [1,2], Aoqiang Ma [1,*], Lixiang Huang [1], Yousheng Su [1], Wenqu Li [1], Hongdu Zhang [1] and Zhikui Wang [1]

1 College of Mechatronic Engineering, Guangxi University, Nanning 530007, China
2 Guangxi Manufacturing System and Advanced Manufacturing Technology Key Laboratory, Nanning 530007, China
* Correspondence: aoqiangma@163.com; Tel.: +86-178-6888-2739

Abstract: Picking robots have become an important development direction of smart agriculture, and the position detection of fruit is the key to realizing robot picking. However, the existing detection models have the shortcomings of missing detection and slow detection speed when detecting dense and occluded grape targets. Meanwhile, the parameters of the existing model are too large, which makes it difficult to deploy to the mobile terminal. In this paper, a lightweight GA-YOLO model is proposed. Firstly, a new backbone network SE-CSPGhostnet is designed, which greatly reduces the parameters of the model. Secondly, an adaptively spatial feature fusion mechanism is used to address the issues of difficult detection of dense and occluded grapes. Finally, a new loss function is constructed to improve detection efficiency. In 2022, a detection experiment was carried out on the image data collected in the Bagui rural area of Guangxi Zhuang Autonomous Region, the results demonstrate that the GA-YOLO model has an mAP of 96.87%, detection speed of 55.867 FPS and parameters of 11.003 M. In comparison to the model before improvement, the GA-YOLO model has improved mAP by 3.69% and detection speed by 20.245 FPS. Additionally, the GA-YOLO model has reduced parameters by 82.79%. GA-YOLO model not only improves the detection accuracy of dense and occluded targets but also lessens model parameters and accelerates detection speed.

Keywords: picking robot; computer vision; grape detection; GA-YOLO; dense and occluded target; lightweight model

Citation: Chen, J.; Ma, A.; Huang, L.; Su, Y.; Li, W.; Zhang, H.; Wang, Z. GA-YOLO: A Lightweight YOLO Model for Dense and Occluded Grape Target Detection. *Horticulturae* **2023**, *9*, 443. https://doi.org/10.3390/horticulturae9040443

Academic Editor: Stefano Poni

Received: 7 March 2023
Revised: 23 March 2023
Accepted: 24 March 2023
Published: 28 March 2023

1. Introduction

Grapes, known as the queen of fruits, have high economic value. The short fruit period of grapes means that timely picking is essential for quality. Currently, hand grape harvesting is the most common method, which takes a lot of time and labor. With the transfer of rural labor from agriculture to non-agricultural industries, the rural surplus labor is gradually decreasing [1]. Therefore, developing grape-picking robots has important research prospects. At present, picking robots mainly rely on the vision system to realize the location of fruits. Accurately detecting the location of the fruit is the key to achieving picking [2]. Especially in the complex environment of grape orchards, is disturbed by factors such as illumination change, leaf occlusion, and fruit overlapping, which bring huge challenges to picking robots.

Traditional fruit detection methods, such as support vector machine [3], template matching [4], edge detection [5], and threshold segmentation [6], mainly extract inherent features, such as geometric shape [7], color [8–10], spectral information [11], texture [12] and edge [13], to realize the detection of the grape region. Liu et al. [14] used the least square method to fit the elliptic boundary of pomelo, to realize the segmentation of pomelo. Lin et al. [4] proposed a local template matching algorithm and trained a new vector machine classifier by using color and texture, which can detect tomatoes, pumpkins, mangoes, and oranges. Nazari et al. [15] designed an RGB classifier based on the color difference between red grapes and the background, which can segment red grapes. Pérez-Zavala et al. [16] extracted the edge gradient information and surface texture information of grapes as classification features and used the support vector machine classifier to realize the segmentation of grapes and the background. Behroozi-Khazaei et al. [8] put forward a method combining an artificial neural network and genetic algorithms, which can overcome the problem that greens grapes are similar to the background. Traditional grape detection methods can achieve good segmentation results when only a few fruits with a specified color and shape. At the same time, traditional image processing techniques rely on high-quality images and require complex artificial features. However, when there are complex scenes, such as scenes with changing illumination, scenes with dense fruits, and scenes with hidden fruits, the performance of fruit detection becomes poor. Under the circumstances, multiple overlapping grapes may be detected as one.

In the recent ten years, with the wide application of deep learning, great breakthroughs have been made in the object detection field [17–22]. Gao et al. [23] divided the blocked apples into three categories, including apples occluded by leaves, apples occluded by branches, and apples occluded by other apples, and used the Faster R-CNN algorithm to detect the occluded apples. Tu et al. [24] proposed a multi-scale feature fusion MS-FRCNN algorithm, which combined the semantic information of the deep network and the location information of the shallow network to improve the detection accuracy in the case of dense passion fruit. Mai et al. [25] increased the single classifier in Faster-RCNN to three classifiers, which effectively enhanced the detection performance of dense fruit targets. Ding et al. [26] improved the SSD model by using the receptive field block and attention mechanism, which effectively reduced the missed detection rate of occluded apples. Behera et al. [27] changed IOU to MIOU in the loss function of Fast RCNN, which improved the recognition performance of occluded and dense fruits. Tu et al. [24] and Ding et al. [26] improved the feature fusion module of the model, and Behera et al. [27] improved the loss function to solve the issue of difficult recognition of occluded and dense targets. However, due to the slow detection speed and a large number of parameters, the above models are difficult to deploy on the mobile end of harvesting robots.

In order to solve the issues of large parameters and slow detection speed, some scholars have studied in the field of lightweight. Generally speaking, the detection speed increases with the decrease in the model parameters. The main methods to reduce the parameters are replacing the convolution module and reducing the convolution layer [28–31]. Mao et al. [32] proposed the Mini-YOLOv3 model, which used depthwise separable convolution and point group convolution to decrease the parameters. A lightweight YOLOv4 model was proposed by Zhang et al. [33], the backbone network Darknet-53 of YOLOv4 is replaced with the GhostNet network and the basic convolution is replaced with a depthwise separable convolution in the neck and head. Ji et al. [34] took YOLOVX-Tiny as the baseline, adopted a lightweight backbone network, and proposed a method for apple detection based on Shufflenetv2-YOLOX. Fu et al. [35] used 1×1 convolution to decrease the parameters of the original model and proposed the DY3TNet model to detect kiwifruit. Li et al. [36] reduced the calculations and parameters by introducing deep separable convolution and ghost modules. Liu et al. [37] proposed the YOLOX-RA model, which pruned part of the network structure in the backbone network and used depth separable convolution in the neck network. Cui et al. [38] changed the backbone network from CSPdarknet-Tiny to ShuffleNet in YOLOv4-tiny and reduced the three detection heads to one detection head. Zeng et al. [39] replaced CSPdarknet with Mobilenetv3 and compressed the neck network of YOLOv5s by pruning technology [40].

Although these models achieve lightweight, the detection accuracy suffers. In the vineyard, clusters of grapes grow densely and overlap each other, and the huge leaves easily cover the grapes. The complex growing environment leads to a low recall rate of the deep learning detection model for grape detection. In addition, the model parameters with high detection accuracy are redundant, which makes it difficult to deploy to the mobile end of the picking robots. The existing detection model can hardly meet the two advantages of detection accuracy and detection speed. To sum up, our research objective is to solve the problem that targets are difficult to identify while ensuring the accuracy of model detection and reducing the parameters of the model. In this paper, a GA-YOLO model with fast detection speed, small parameters, and a low missed detection rate is proposed for dense and occluded grapes.

In short, our innovations are as follows:

(1) A new backbone network SE-CSPGhostnet is designed, which greatly reduces the parameters.
(2) ASFF mechanism is used to address the issues of difficult detection of occluded and dense targets, and the model's detection accuracy is raised.
(3) A novel loss function is constructed to improve detection efficiency.

The architecture of this paper is as follows: Section 1 introduces the background, significance, and current status. Section 2 introduces dataset collection, annotation, and augmentation. Section 3 introduces the GA-YOLO algorithm. Section 4 contains the experimental process, the comparison of model performance, and the analysis of the results. Section 5 describes the use of human–computer interaction interface. Section 6 discusses the experimental results and points out the limitations of the algorithm. Section 7 concludes the paper and provides future research plans.

In the paper, the full names and acronyms are displayed in Table 1.

Table 1. The acronyms and full names.

Acronyms	Full Name
ASFF	Adaptively Spatial Feature Fusion
CBL	Convolution, Batch normalization and Leaky Relu activation function
CBM	Convolution, Batch normalization and Mish activation function
CSP	Cross Stage Partial
FLOPs	Floating point operations per second
FPNet	Feature Pyramid Network
FPS	Frames Per Second
GBM	Ghost convolution, Batch normalization and Mish activation functions
IOU	Intersection over Union
mAP	Mean Average Precision
PANet	Path Aggregation Network
RCNN	Regions with CNN features
Res element	Residual element
SENet	Squeeze-and-Excitation Networks
SPP	Spatial Pyramid Pooling
SSD	Single Shot MultiBox Detector
YOLO	You Only Look Once

2. Datasets

2.1. Collection of Datasets

The study's grape datasets were collected from 21 June 2022 to 26 June 2022 in Bagui Garden, Nanning City, Guangxi Zhuang Autonomous Region, including three varieties of grapes: "Kyoho", "Victoria" and "Red Fuji". We used Daheng Industrial Camera MER-132-43U3C-L for the acquisition of datasets. All images were acquired under natural lighting at 8:30 am, 11:30 noon, 2:30 pm, and 5:30 pm on sunny and overcast days. The distance from the camera lens to the grapes is 0.5 m~1.2 m. The camera's shooting angles include flat, up, and down. The camera is shown in Figure 1, and the basic parameters of the industrial camera are displayed in Table 2.

In order to avoid the overfitting of the network model caused by the single feature of the datasets, 200 "Kyoho Grapes" images, 200 "Red Fuji Grapes" images, and 200 "Victoria Grapes" images were collected in consideration of different light intensity, different occlusion degree, and different fruit sparseness. Figure 2 shows the images of three grape varieties, and Table 3 shows the number of grape images in different collection conditions.

Figure 1. Daheng Industrial Camera: (**a**) overall camera; (**b**) camera base; (**c**) camera lens.

Table 2. The basic parameters of the industrial camera.

Parameter	Value
Model	MER-132-30UC
Frame rate	30 fps
Sensor type	1/3″ CCD
Spectrum	black/color
Data Interface	USB2.0
Working temperature	0–60 °C
Working humidity	10–80%
Resolution ratio	1292 × 964

Figure 2. Three grape varieties: (**a**) Kyoho; (**b**) Red Fuji; (**c**) Victoria.

Table 3. Number of grapes images in different collection conditions. ("√" indicates "yes"; "×" indicates "no").

Varieties	Kyoho				Red Fuji				Victoria			
Is the number of grape clusters more than twelve?	√		×		√		×		√		×	
Is there occlusion?	√	×	√	×	√	×	√	×	√	×	√	×
Number of images	100	100	50	50	50	50	100	100	50	50	50	50

2.2. *Annotation of Datasets*

This paper uses labelImg software [41] for labeling, annotation format is Pascal VOC. LabelImg software is shown in Figure 3a, and the label format corresponding to the labeled picture is shown in Figure 3b. Image labeling is based on the following principles:

(1) unripe grapes are not labeled; (2) grapes falling on the ground will not be labeled; (3) grapes with an occlusion area exceeding $\frac{4}{5}$ are not labeled; (4) grape images that are blurred, but the grape area is larger, it is also labeled; (5) when labeling, ensure that the label box and the grape area overlap to the maximum.

(a) (b)

Figure 3. (**a**) LabelImg software; (**b**) XML label file.

2.3. *Augmentation of Datasets*

Data augmentation has the advantages of saving time for making labels, preventing model overfitting, and improving model generalization ability. The augmentation methods are shown in Figure 4, which contains 14 augmentation methods such as scaling, cropping, rotation, brightness change, saturation change, contrast change, blurring process, and mosaic data augmentation. Finally, 600 × 14 = 8400 valid images are obtained. According to the ratio of 7:2:1, the photos are separated into training set, validation set, and test set.

Figure 4. Image augmentation method: (**a**) original image; (**b**) crop and zoom; (**c**) rotate by 180 degrees; (**d**) flip horizontally; (**e**) vertical turnover; (**f**) fuzzy median value; (**g**) Gaussian blur; (**h**) 50% increase in saturation; (**i**) 50% reduction in saturation; (**j**) 50% increase in brightness; (**k**) 50% reduction in brightness; (**l**) 50% increase in contrast; (**m**) 50% reduction in contrast; (**n**) mosaic data enhancement method.

3. Methodologies

3.1. YOLOv4 and GA-YOLO

The YOLOv4 model's structure is shown in Figure 5a. The GA-YOLO model is improved on the basis of the YOLOv4 [22], as shown in Figure 5b. We propose a new backbone network SE-CSPGhostnet and incorporate the ASFF mechanism into the head network. Furthermore, a new loss function is used to improve the detection performance.

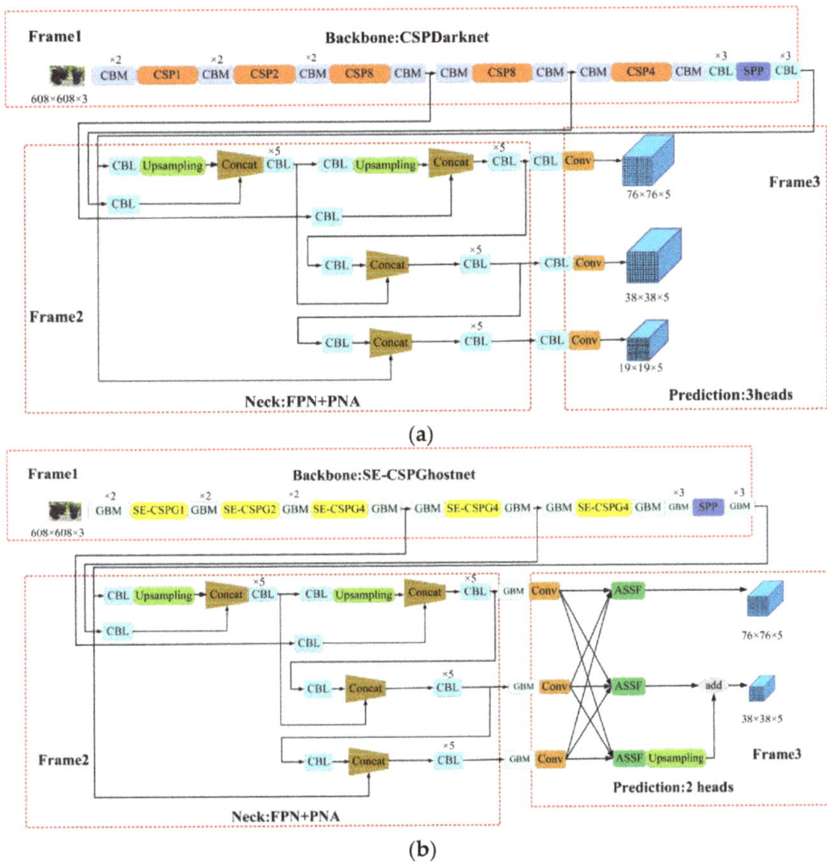

Figure 5. (**a**) YOLOV4 network model structure (**b**) GA-YOLO network model structure.

In Figure 5a, the YOLOv4 network model consists of modules such as SPP, CBM, CBL, and CSP. Among them, Spatial Pyramid Pooling [42] (SPP) fixes feature maps of any size as feature vectors of the same length through a pooling of three scales. CBL contains convolution, Batch normalization, and Leaky Relu activation functions, which are used in the latter position of the backbone network to extract features. The CBM module is composed of convolution, Batch normalization, and Mish activation functions, which are used in the front position of the backbone network to extract features. We changed the convolution in CBM to ghost convolution and proposed the GBM module. In order to reduce parameters of model, the CBM and CBL modules in the backbone network are changed to GBM modules.

Meanwhile, the CBL modules at the junction of the neck network and the head network were changed to GBM modules. CBM, CBL, GBM, and SPP are shown in Figure 6.

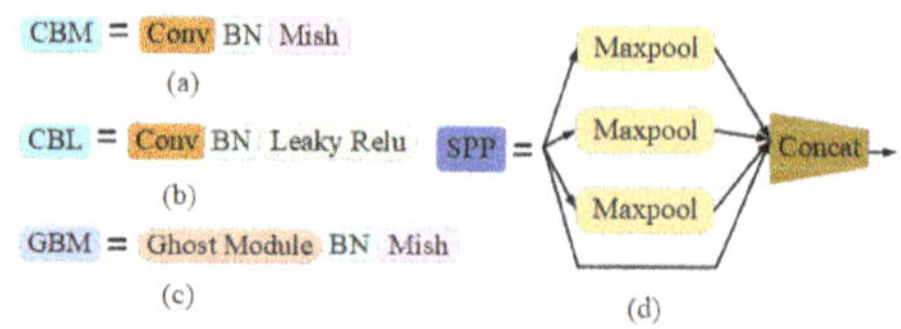

Figure 6. (**a**) CBM module (**b**) CBL module (**c**) GBM module (**d**) SPP module.

The CBM uses ordinary convolution, and convolution process is shown in Figure 7a. GBM uses ghost convolution [28], which greatly reduces the number of parameters, as shown in Figure 7b. The primary distinction between ghost convolution and ordinary convolution is that ghost convolution has two convolution processes. Firstly, $\frac{n}{s}$ intermediate feature maps are obtained by ordinary convolution. Then, the intermediate feature map is convoluted with a convolution kernel of $d \times d$ size to obtain $(s-1) \times \frac{n}{s}$ feature maps. Finally, the $\frac{n}{s}$ intermediate feature maps acquired in the first step and the $(s-1) \times \frac{n}{s}$ feature maps acquired in the second step are superimposed on the channel dimension to obtain a total of n feature maps. The parameter quantity of GBM is shown in Formula (1). In contrast, the ordinary convolution in Figure 7a is a direct convolution to obtain n output feature maps. The parameter quantity of CBM is shown in Formula (2). Obviously, the amount of final feature maps of GBM convolution and ordinary convolution is the same. The ratio of parameters of CBM and GBM is shown in Formula (3). Through calculation and analysis, theoretically, the parameter quantity of GBM is $\frac{1}{s}$ of that of CBM.

$$P_1 = h' \times w' \times \frac{n}{s} \times k \times k \times c + (s-1) \times h' \times w' \times \frac{n}{s} \times d \times d \tag{1}$$

$$P_2 = h' \times w' \times n \times k \times k \times c \tag{2}$$

$$r_s = \frac{P_2}{P_1} \approx s \tag{3}$$

where, h' represents the length of output feature map; w' represents the width of output feature map; n represents the number of channels for output feature map; k represents the size of the convolution kernel; c represents the channel number of convolution kernel; s represents the ratio of the number of channels of output feature map to the number of channels of input feature map; d represents the size of convolution kernel in the second convolution of ghost convolution.

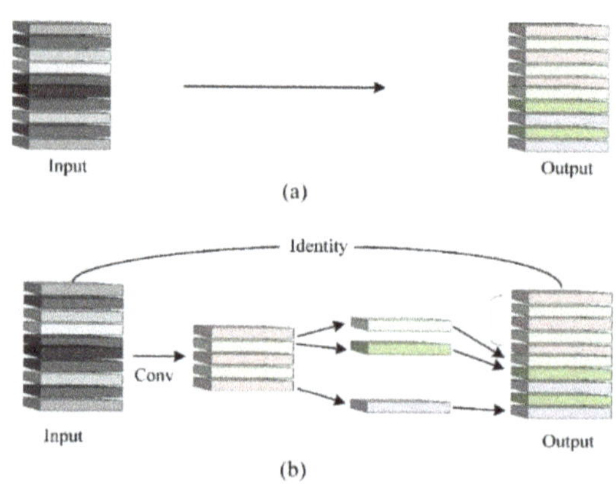

Figure 7. Two convolution methods: (**a**) ordinary convolution (**b**) ghost convolution.

3.2. Improvement of GA-YOLO Backbone Network

The role of backbone network is to extract features, and its structure is shown in Figure 5. We made three improvements to YOLOv4 backbone network CSPdarknet to obtain the GA-YOLO backbone network CSPGhostnet: (1) change the CBM and CBL to GBM; (2) change the CSP structure to SE-CSPG structure; (3) reduce the number of iterations of the SE-CSPG module. The above three improved methods all greatly reduce parameters and calculations.

The CSP structure [43] of CSPdarknet has a Res unit component that iterates X times, as shown in Figure 8a. In Res unit, the CBM module is replaced with a GBM module. At the same time, the insertion of Squeeze-and-Excitation Networks (SENet) improves the performance of the Res element to solve the problem of gradient degradation. After adding skip connection and Res element in SE-CSPG, the shallow feature information is integrated into the deep feature information, which improves the generalization performance of the model.

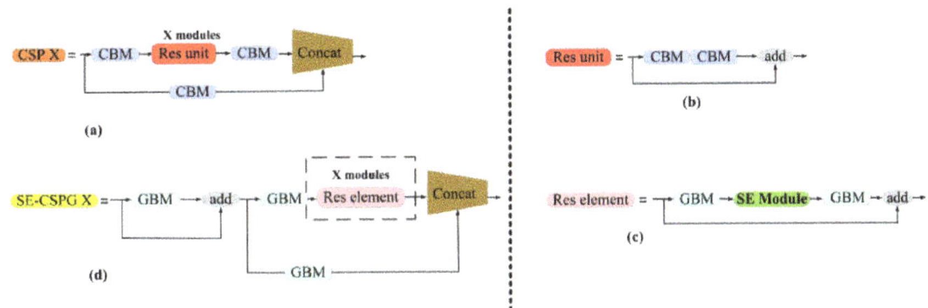

Figure 8. (**a**) CSP structure (**b**) Res unit structure (**c**) SE-CSPG structure (**d**) Res element structure.

The attention mechanism can correct the features, make the network focus on important local information. Useless feature information is filtered out, so as to simplify the model and accelerate the calculation. SENet [44] mainly studies the relationship between channels and realizes the effect of adaptively correcting channel characteristics. SENet is a typical representative of the channel attention mechanism, as shown in Figure 9a.

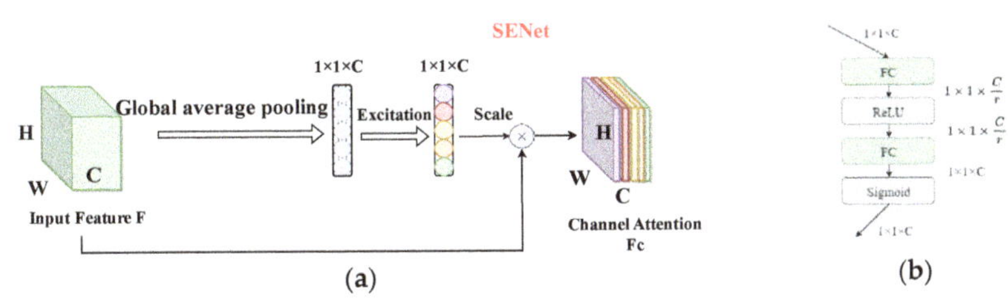

Figure 9. (**a**) SENet (**b**) the process of Excitation.

The input feature map F compresses the two-dimensional feature $H \times W$ of each channel into a real number through global average pooling. After this compression operation, the size of the feature map is converted from the original $H \times W \times C$ to $1 \times 1 \times C$. The Excitation operation is performed on the obtained feature map F_g to generate a weight value for each feature channel. The Excitation operation is to use two fully connected layers to build the correlation between channels, as shown in Figure 9b. The normalized weight is obtained after the activation function Sigmoid. The F_e represents the importance of the channel, and it is weighted to the features of each channel in the F to obtain the channel attention feature map F_c.

3.3. Introduction of GA-YOLO Neck Network

The neck network is shown in Frame 2 of Figure 5. The role of the neck network is to fuse the features of different feature layers. Feature Pyramid Networks (FPN) [45] and Path Aggregation Networks [46] (PAN) are used as the feature fusion module, making full use of the semantic information of high-dimensional feature maps and the location information of low-dimensional feature maps. The feature fusion of neck network improves the detection accuracy of dense and occluded targets.

3.4. Improvement of GA-YOLO Head Network

The structure of the head network is shown in Frame 3 of Figure 5. The role of the head network is to predict the class and location. In the head network, Adaptively Spatial Feature Fusion [47] (ASFF) is added to the front of the prediction head. ASFF can adaptively learn the spatial weight of each scale feature map fusion, which is used to solve the problem of inconsistent scales in spatial feature fusion. The structure of ASFF is shown in Figure 10.

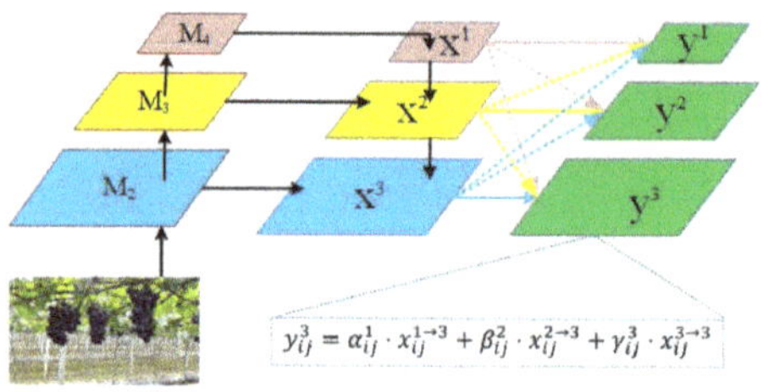

Figure 10. ASFF structure diagram.

Where, M_1, M_2 and M_3 represent the feature map obtained by convolution of the backbone network; x^1, x^2 and x^3 represent the feature map after PANet feature fusion; y^1, y^2 and y^3 represent the feature maps after ASFF mechanism processing.

In Figure 10, the learning process of spatial weight of ASFF is shown in Formula (4). Where, x^l represents the l-layer feature map and $x_{ij}^{s\to t}$ represents the feature map from s-layer resize to t-layer. α_{ij}^l,β_{ij}^l,γ_{ij}^l are the learned spatial weights, which represent the importance of the pixel at the coordinate (i, j) in the feature map of the lth layer. Meanwhile, α_{ij}^l,β_{ij}^l,γ_{ij}^l satisfy $\alpha_{ij}^l + \beta_{ij}^l + \gamma_{ij}^l = 1$ and $\alpha_{ij}^l, \beta_{ij}^l, \gamma_{ij}^l \in [0,1]$. α_{ij}^l,β_{ij}^l,γ_{ij}^l can be obtained according to Formula (5). In Formula (5), $\lambda_\alpha^l, \lambda_\beta^l, \lambda_\gamma^l$ are control parameters, which can be learned by back-propagation of the network.

$$y_{ij}^3 = \alpha_{ij}^1 \cdot x_{ij}^{1\to3} + \beta_{ij}^2 \cdot x_{ij}^{2\to3} + \gamma_{ij}^3 \cdot x_{ij}^{3\to3} \tag{4}$$

$$\alpha_{ij}^l = \frac{e^{\lambda_{\alpha_{ij}}^l}}{e^{\lambda_{\alpha_{ij}}^l} + e^{\lambda_{\beta_{ij}}^l} + e^{\lambda_{\gamma_{ij}}^l}} \tag{5}$$

3.5. Improvement of GA-YOLO Loss Function

The YOLOv4 loss function before improvement includes three sections: confidence loss, rectangular box loss and classification loss, as shown in Formula (6).

$$loss = a \cdot loss_{conf} + b \cdot loss_{box} + c \cdot loss_{clc} \tag{6}$$

Three improvements to the loss function are made: (1) Since there is only one class of target to be detected, set $c = 0$, which removes the ineffective classification loss. (2) Increase the weight of confidence loss, take $a = 0.6$, $b = 0.4$. (3) Confidence loss consists of target confidence loss and background confidence loss. The weight of target confidence loss is increased, taking λ_{obj} = 0.7, λ_{noobj} = 0.3. The loss of improved confidence is shown in Formula (7).

$$\begin{aligned} loss_{conf} = \lambda_{obj} \sum_{i=0}^{S^2} \sum_{j=0}^{B} 1_{i,j}^{obj} [\hat{C}_i^j \log(C_i^j) + (1-\hat{C}_i^j)\log(1-C_i^j)] \\ +\lambda_{noobj} \sum_{i=0}^{S^2} \sum_{j=0}^{B} 1_{i,j}^{noobj} [\hat{C}_i^j \log(C_i^j) + (1-\hat{C}_i^j)\log(1-C_i^j)] \end{aligned} \quad (7)$$

In Formulas (6) and (7), some proper nouns are defined as follows: confidence indicates the confidence degree of the predicted rectangular box containing the target, and binary cross entropy loss is employed for calculating confidence loss. S^2 indicates the number of divided grids in image, B indicates the number of prior frames in each grid, λ_{obj} indicates the weight factor of the target's confidence loss and λ_{noobj} indicates the weight factor of the background's confidence loss. $\hat{C}_i^j$ is the label value of the prediction box's confidence and C_i^j is the predicted value of the prediction box's confidence. $1_{i,j}^{obj}$ indicates that if the ith grid's jth prediction box has a target, its value is 1, otherwise, it is 0. $1_{i,j}^{noobj}$ indicates that there is no target in the ith grid's jth prediction box, and its value is 1, otherwise it is 0.

In Formula (6), rectangular box loss is employed to calculate the position error between the predicted box and the ground-truth box, including the error loss of the central point coordinate and the height and width of the rectangular box, which is calculated by using CIOU loss function [48], as shown in Formula (8).

$$loss_{box} = \sum_{i=0}^{S^2} \sum_{j=0}^{B} 1_{i,j}^{obj} [1 - IOU + \frac{\rho^2}{c^2} + \frac{\frac{16}{\pi^4}(arctan\frac{w^{gt}}{h^{gt}} - arctan\frac{w}{h})^4}{1 - IOU + \frac{4}{\pi^2}(arctan\frac{w^{gt}}{h^{gt}} - arctan\frac{w}{h})^2}] \quad (8)$$

In Formula (8), IOU represents the ratio of the area of the intersection region between ground-truth rectangular box and predicted rectangular box to the area of the merged region. ρ represents the distance between the central point of prediction rectangular box and the central point of ground-truth rectangular box. c represents the length of the diagonal of the external rectangular box of the prediction rectangular box and the ground-truth rectangular box. w^{gt} and h^{gt} represent the width and height of the ground-truth rectangular box. w and h represent the width and height of the prediction rectangular box.

4. Results of Experiment

4.1. Experimental Details

In order to confirm that the GA-YOLO substantially improve the detection performance, 8400 grape datasets are used to conduct experiments. The experimental hardware and software configuration parameters are displayed in Table 4.

Table 4. The basic parameters of the industrial camera.

Hardware/Software	Configuration/Version
CPU	Intel(R) Xeon(R) CPU E5-2680
GPU	Tesla M40 24 G × 4
Memory	DDR4 64G KF3200C16D4/8GX
Hard disk	SSD 980 500 G
Operating system	Ubuntu20.04.1
Python	3.9
Pytorch	1.8.1
CUDA	10.0.3

To ensure the fairness of the experimental comparisons, all models are trained under the same hardware condition and the same initial training parameters. The learning rate is adopted by means of the cosine annealing decay method. The weights are saved every

10 epochs during the training process. The specific training initial parameters are displayed in Table 5.

Table 5. Initial training parameters.

Parameter	Form/Value
Init-learning rate	0.01
Min-learning rate	Init-lr × 0.01
Optimizer-class	SGD
momentum	0.937
Lr-decay-class	Cos
Weight decay	0.0005
Num-works	4
Batch size	64
Epoch	50

4.2. Metrics for Evaluation

FPS refers to the number of images that can be detected per second, which is employed to assess the network model's detecting speed. Parameters represent the volume of parameters that require training in the network model. Weights represent the size of the weight file obtained by the final training of the network model. Parameters and weights are employed to evaluate the size of the network model, and the size of the weights is generally four times the size of the parameters. The smaller the parameters and weights, the easier the model to be deployed to the mobile terminal of the picking robot. Floating-point operations per second (FLOPs) are employed to evaluate the calculation effort of the model. The precision rate, recall rate, F_1 score, and AP are employed to evaluate the accuracy of the target detection method.

The precision rate indicates the ratio of being a positive sample among predicted positive samples, as shown in Formula (9):

$$Precision = \frac{TP}{TP + FP} \times 100\% \quad (9)$$

The *recall* rate indicates the ratio of correctly predicted positive samples to labeled positive samples, as shown in Formula (10):

$$Recall = \frac{TP}{TP + FN} \times 100\% \quad (10)$$

The harmonic mean of the *precision* rate and *recall* rate is the F_1 score, as shown in Formula (11):

$$F_1 = \frac{2 \times Precision \times Recall}{Precision + Recall} \quad (11)$$

The two indicators of precision rate and recall rate show a negative correlation. Therefore, to comprehensively assess the quality of the algorithm, the PR curve is usually drawn with the recall rate as the horizontal axis and with the precision rate as the vertical axis. The area below the PR curve is average precision (AP) value, as shown in Formula (12):

$$AP = \int_0^1 p(r)dr \quad (12)$$

4.3. Comparison of Network Models

4.3.1. Calculation Volume, Parameter Volume, and the Size of Weight File

The GA-YOLO network model is compared with mainstream detection network models such as Faster RCNN, YOLOv3, YOLOv4, SSD, YOLOv4-MobileNetv1, YOLOv4-MobileNetv2, YOLOv4-MobileNetv3, YOLOv4-tiny, YOLOv5s, YOLOv5m, YOLOv5l, and YOLOv5x.

On account of the large number of network models engaged in the comparison, the models are divided into the light network model (0 < FLOPs ≤ 50 G), medium network model (50 < FLOPs ≤ 100 G), and large network model (FLOPs > 100 G) according to the calculation volume of the network model. Therefore, in Table 6, the light network model contains YOLOv4-MobileNetv1, YOLOv4-MobileNetv2, YOLOv4-MobileNetv3, YOLOv4-tiny, GA-YOLO, and YOLOv5s, the medium network model contains YOLOv3, YOLOv4, and YOLOv5m, and the large network model contains Faster RCNN, SSD, YOLOv5l, and YOLOv5x. The comparison results on calculation volume, parameter volume, and the size of the weight file are displayed in Table 6.

Table 6. Comparison of calculation volume, parameter volume, and weight file of different network models.

Network Models	FLOPs (G)	Parameters (M)	Weights (M)
Faster RCNN	252.676	136.689	108
SSD	115.513	23.612	90.6
YOLOv3	65.520	61.524	235
YOLOv4	59.7758	63.938	244
YOLOv4-MobileNetv1	21.285	14.267	57.1
YOLOv4-MobileNetv2	16.185	12.376	49.4
YOLOv4-MobileNetv3	14.999	11.304	53.6
YOLOv4-tiny	16.438	7.057	28.4
GA-YOLO	13.860	11.003	32.5
YOLOv5s	16.377	8.064	32.1
YOLOv5m	50.404	21.056	80.6
YOLOv5l	114.240	46.631	178
YOLOv5x	217.323	87.244	333
Faster RCNN	252.676	136.689	108

The data are analyzed in Table 6. Firstly, the calculation volume, parameter volume and weight file size of the GA-YOLO model are 13.860 G, 11.003 M, and 32.5 M, respectively, which are 76.81%, 82.79%, and 86.68% lower than YOLOv4. Secondly, GA-YOLO is 34.88%, 14.37%, 7.59%, 15.68%, and 15.68% lower in calculation volume than light networks such as YOLOv4-MobileNetv1, YOLOv4-MobileNetv2, YOLOv4-MobileNetv3, YOLOv4-tiny, and YOLOv5s, respectively. At the same time, GA-YOLO is 22.88%, 11.09%, and 2.66%, lower in parameter volume than light networks such as YOLOv4-MobileNetv1, YOLOv4-MobileNetv2, and YOLOv4-MobileNetv3, respectively. Again, GA-YOLO is at least 78.85%, 47.74%, and 59.68% lower than medium-sized networks such as YOLOv3, YOLOv4, and YOLOv5m on calculation volume, parameter volume and the size of the weight file. Finally, GA-YOLO is at least 87.87%, 53.40%, and 64.13% lower than large networks such as Faster RCNN, SSD, YOLOv5l, and YOLOv5x on GFLOPs, params, and weights. This shows that the use of ghost convolution greatly decreases the volume of parameters and calculations.

4.3.2. Comparison of Convergence Speed

In order to confirm that the training convergence speed of the GA-YOLO has been improved after being lightweight, it is compared with other network models in Table 6. The loss value of each epoch of training is recorded, and the loss value change graph is drawn, as shown in Figure 11. Where the horizontal and vertical axes are epoch and loss values, respectively. For the convenience of comparison, we draw network models with similar convergence speed and loss value in one graph. The statistics of the convergent algebra are shown in Table 7.

According to Figure 10 and Table 7, the SSD has the slowest convergence speed, and it converges at the 50th epoch. The convergence speed of Faster RCNN and YOLOv5s is also relatively slow, reaching convergence after the 35th epoch and 40th epoch, respectively. YOLOv4-MobileNetv1, YOLOv4-MobileNetv2, and YOLOv4-MobileNetv3 have basically the same convergence speed, and they all reach convergence around the 30th epoch. YOLOv4-tiny has the fastest convergence speed and has basically converged in the 12th epoch. GA-YOLO completed the convergence at the 15th epoch, which is about 7 epochs faster than the YOLOv4 network before the improvement, which shows that GA-YOLO saves the training time of the model.

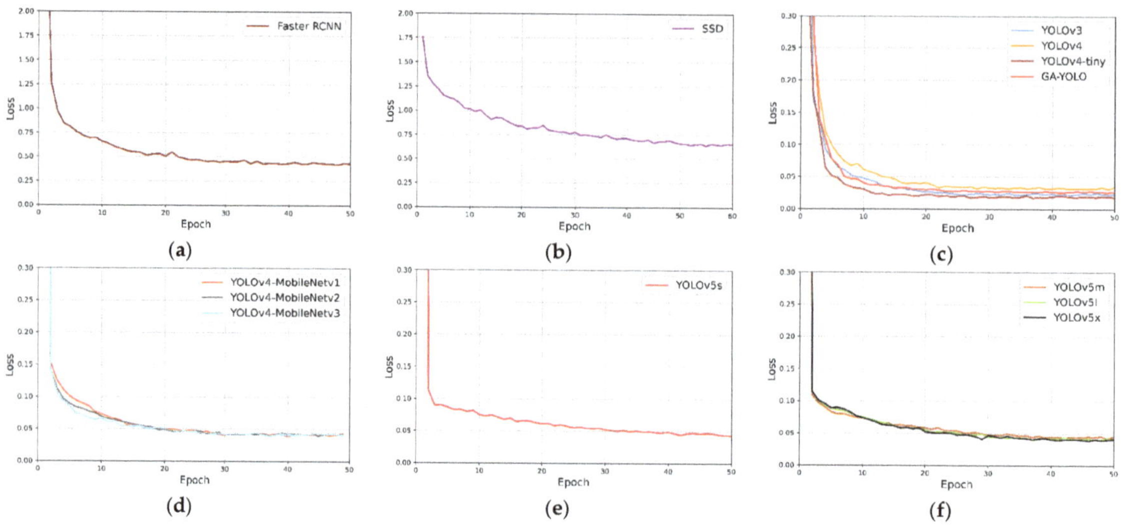

Figure 11. Loss convergence graphs of different network models. (**a**) Loss convergence graph of Faster RCNN; (**b**) Loss convergence graph of SSD; (**c**) Loss convergence graphs of YOLOv3, YOLOv4, YOLOv4-tiny and GA-YOLO; (**d**) Loss convergence graphs of YOLOv4-MobileNetv1, YOLOv4-MobileNetv2 and YOLOv4-MobileNetv3; (**e**) Loss convergence graphs of YOLOv5s; (**f**) Loss convergence graphs of YOLOv5m, YOLOv5l and YOLOv5x.

Table 7. Convergence epochs of different network models.

Network Models	Epoch	Network Models	Epoch
Faster RCNN	35	SSD	50
YOLOv3	25	YOLOv4	22
YOLOv4-tiny	13	GA-YOLO	15
YOLOv5s	40	YOLOv5m	32
YOLOv5l	33	YOLOv5x	35
YOLOv4-MobileNetv1	30	YOLOv4-MobileNetv2	30
YOLOv4-MobileNetv3	30		

4.3.3. Ablation Experiment

The ablation experiment aims to verify the role played by the SE-CSPGhostnet backbone network module, ASFF module, and improved loss function in the GA-YOLO network model. The definitions are as follows: YOLOv4-a indicates that the SE-CSPGhostnet backbone network is employed on the basis of YOLOv4. YOLOv4-b indicates that the ASFF module is employed on the basis of YOLOv4-a. GA-YOLO indicates that an improved loss function is employed on the basis of YOLOv4-b. The comparison of the mAP and F_1 values of the grape detection results of the ablation experiment is displayed in Table 8. Where, × indicates that the improved module of the corresponding column is not used. Conversely, √ indicates that the improved module of the corresponding column was adopted.

In Table 8, the mAP and F_1 score of YOLOv4-a are 92.24% and 90.20%, respectively, which are 0.94% and 0.96% lower than that of YOLOv4, respectively. This demonstrates that after YOLOv4 is lightweight, the detection accuracy is only slightly affected. However, as displayed in Table 6, the GA-YOLO network model reduces the volume of calculation, the volume of parameter, and the weight file by 76.81%, 82.79%, and 86.68%, respectively, compared with YOLOV4. The loss of accuracy is acceptable relative to the improvement in the volume of parameters and calculations. The mAP and F_1 scores of YOLOv4-b are 95.22% and 93.21%, respectively, which are 2.98% and 3.01% superior to that of YOLOv4-a, respectively. This is because the ASFF module improves the detection accuracy of the model for dense targets. ASFF enables the network to filter out contradictory and useless information, thereby retaining only useful information for combination, which solves the issue of poor detection accuracy of dense targets. The mAP and F_1 scores of GA-YOLO are 96.87% and 94.78%, respectively, which are 1.65% and 1.57% higher than that of YOLOv4-b, respectively. This demonstrates that application of an improved loss function improves the detection accuracy. In fact, the original loss function meets the highest accuracy conditions for the detection of 80 classes of targets in the MS COCO dataset but does not meet the highest accuracy conditions for single-target detection. Therefore, the improvement of the loss function is effective.

Table 8. The grape detection results of the four network models.

Network Model	SE-CSPGhostblock	ASFF	Improved Loss Function	mAP (%)	F_1
YOLOv4	×	×	×	93.18	91.16
YOLOv4-a	√	×	×	92.24	90.20
YOLOV4-b	√	√	×	95.22	93.21
GA-YOLO	√	√	√	96.87	94.78

4.3.4. Comparison of Detection Performance

In order to express the performance of the 13 models more intuitively, we draw the PR curves of the 13 network models, as depicted in Figure 12a. Where, the horizontal and vertical axes are the recall rate and the precision rate, respectively. It is evident from Formula (9) that the mAP value of the network model is the area enclosed by the PR curve and the axis of coordinates. The mAP values of the 13 network models are shown in Figure 12b. In the meantime, the parameters such as the precision rate, the recall rate, F_1 score and FPS of 13 network models are listed in Table 9.

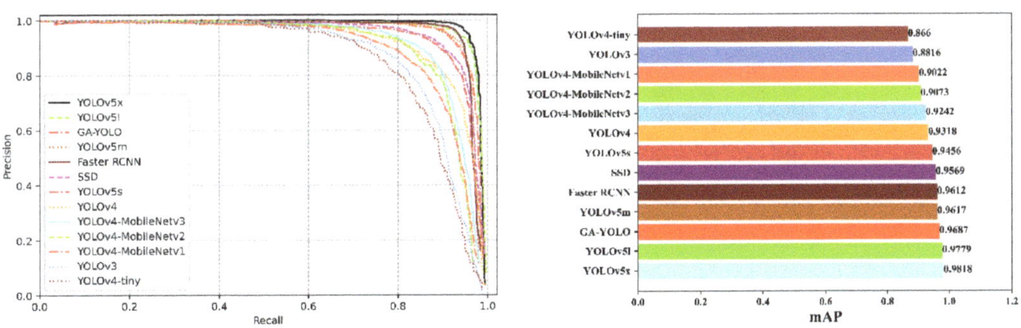

(a) (b)

Figure 12. Experimental results (**a**) PR curve; (**b**) average precision chart.

In Figure 12a, the relation of the area enclosed by the PR curves of each model can be clearly seen. mAP is the average of multiple category AP, so in single object detection, it is equal to the value of AP.

In Figure 12b, the mAP of the GA-YOLO model is 96.87%, which is 0.92% and 1.31% lower than that of the YOLOv5l model and YOLOv5x model, respectively. However, according to Table 9, the detection speed of the GA-YOLO model is 55.867 FPS, which is 35.802 FPS and 43.293 FPS higher than the YOLOv5l model and YOLOv5x model, respectively. At the same time, in Table 6, the calculation volume, parameter volume, and weight file of the GA-YOLO model are 13.860 G, 11.003 M, and 32.5 M, respectively, which are 93.62%, 87.39%, and 90.24% lower than the YOLOv5x model, respectively. Therefore, as a lightweight model, the GA-YOLO model is more appropriate for application to grape-picking robots when it comes to storage space and detection speed.

Table 9. Comparison of calculation volume, parameter volume, and weight file of different network models.

Network Models	F_1	Precision	Recall	FPS
Faster RCNN	0.9310	0.9491	0.9136	10.375
SSD	0.9221	0.9534	0.8929	54.858
YOLOv3	0.8570	0.9636	0.8150	27.254
YOLOv4	0.9116	0.9426	0.8826	35.622
YOLOv4-MobileNetv1	0.8870	0.9595	0.8247	33.512
YOLOv4-MobileNetv2	0.8943	0.9651	0.8331	27.917
YOLOv4-MobileNetv3	0.9023	0.9399	0.8675	25.859
YOLOv4-tiny	0.8495	0.9301	0.7818	121.374
GA-YOLO	0.9478	0.9533	0.9422	55.867
YOLOv5s	0.9292	0.9517	0.9078	44.430
YOLOv5m	0.9378	0.9482	0.9278	29.160
YOLOv5l	0.9550	0.9697	0.9407	12.574
YOLOv5x	0.9567	0.9675	0.9462	20.065

4.3.5. Object Detection Experiment in Actual Natural Environment

For the sake of further confirm the detection accuracy and robustness of the GA-YOLO, this paper conducts object detection experiments in the actual vineyard environment. The grape image with leaf occlusion, illumination change, and dense targets is selected for detection experiment. The grape image to be tested is shown in Figure 13.

Figure 13. Image of grapes to be tested in the actual natural environment.

In Section 4.3.1, network models are divided into three classes according to the volume of calculation: light network models, medium network models, and large network models. The detection results of light networks, medium networks, and large networks are shown in Figures 14–16, respectively. Meanwhile, the number of grape clusters detected by 13 network models is counted as displayed in Table 10.

Table 10. Number of grape clusters detected by 13 network models.

Network Models	Cluster	Network Models	Cluster
Faster RCNN	24	SSD	19
YOLOv3	16	YOLOv4	17
YOLOv4-tiny	15	GA-YOLO	22
YOLOv5s	17	YOLOv5m	20
YOLOv5l	24	YOLOv5x	24
YOLOv4-MobileNetv1	16	YOLOv4-MobileNetv2	16
YOLOv4-MobileNetv3	18		

Figure 14. Actual grape detection results of the light network model: (**a**) YOLOv4-MobileNetv1; (**b**) YOLOv4-MobileNetv2; (**c**) YOLOv4-MobileNetv3; (**d**) YOLOv4-tiny; (**e**) GA-YOLO; (**f**) YOLOv5s. The yellow ellipse represents the missed detection of grapes.

Figure 15. Actual detection grape results of the medium network models: (**a**) YOLOv3 (**b**) YOLOv4 (**c**) YOLOv5m. The yellow ellipse represents the missed detection of grapes.

In Figure 14, the GA-YOLO network model detects 22 clusters of grapes, which are 6 clusters, 6 clusters, 4 clusters, 7 clusters, and 5 clusters more than YOLOv4-MobileNetv1, YOLOv4-MobileNetv2, YOLOv4-MobileNetv3, YOLOv4-tiny, and YOLOv5s, respectively, indicating that it significantly outperforms other light networks in occluded and dense targets detection performance. In Figure 15, medium network models such as YOLOv3, YOLOv4, and YOLOv5m detected 16 clusters, 17 clusters, and 20 clusters of grapes, respectively. They have the phenomenon of missing detection in occluded target detection, which may be due to the defects of their feature fusion module. In Figure 16 and Table 10, large networks such as YOLOv5l and YOLOv5x detect the highest number of grape clusters, which detect 24 clusters. This is because large networks have deeper convolutional layers, which can extract richer features. At the same time, the Faster RCNN network model detects 24 clusters. Although it successfully detects most of the dense targets, it incorrectly detects the leaves as grapes in the first red oval on the left. This shows that Faster RCNN has the risk of false detection, which will reduce the picking efficiency of the robot. Among the 13 models, the GA-YOLO model can not only meet the demands of lightweight and real-time grape detection performance but also ensure the accuracy of grape detection.

Figure 16. Actual grape detection results of the large network models: (**a**) Faster RCNN (**b**) SSD (**c**) YOLOv5l (**d**) YOLOv5x. The yellow circle represents the missed detection of grapes. The yellow circle represents the missed detection of grapes.

5. Interactive Interface

In order to make it convenient for non-professionals to use the detection model, an interactive interface based on PyQt5, as shown in Figure 17. The interface includes a detection toolbar, image detection results, and text information. There are three detection models built into the detection interface: YOLOV4, GA-YOLO, and YOLOv5s. We can choose any model for testing. The detection modes include image detection and video detection. Function buttons include start, pause, and exit systems. The text information includes four parameters: AP, FPS, precision, and recall. The running process of the whole interactive interface is divided into three steps: Step 1: Select the detection model and detection mode. Step 2: Click the Start button and call the trained weights to detect the grape targets. Step 3: save the detected results to the hard disk of the local computer.

Figure 17. Human–machine interaction.

6. Discussion of Experiment

The problems of agricultural health monitoring [49–55] and harvesting [56,57] have always been hot spots of scientific research. In particular, the deep learning algorithm has become the mainstream research algorithm of the vision system of fruit-picking robots. Compared with the Faster RCNN algorithm [20], the YOLO algorithm [17–19,22] has the advantage of high speed because it unifies regression and classification into one stage. In recent years, some scholars [28–39] have applied the YOLO algorithm to the visual detection of fruit-picking robots, which provides technical help to solve the picking problem in agriculture. However, the YOLO algorithm still has some shortcomings, such as large parameters and low detection accuracy of occluded targets, which are exactly what we want to solve.

In fact, in recent years, some scholars have begun to study the lightweight model while ensuring the detection accuracy of complex objects. Zhao et al. [58] changed the backbone network in YOLOV4 from CSPdarknet53 to MobileNet53 to obtain a lightweight model, and at the same time, the deformable convolution was used to achieve dense target detection. Betti [59] and others pruned Darknet53 and compressed the backbone network from 53 layers to 20 layers. In addition, YOLO-S replaces the maximum pooling with cross-border convolution, which reduces the information loss in the transmission process and improves the detection accuracy of small targets. Huang et al. [60] proposed a GCS-YOLOv4-tiny model based on YOLOV4-Tiny. In this model, grouping convolution is used to reduce the parameter of the model by 1.7 M, and the attention mechanism is used to improve the mAP of F. margarita to 93.42%. Sun et al. [61] designed a shuffle module, lightened YOLOv5s and obtained the YOLO-P model. What is more, the YOLO-P model adopts a Hard-Swish activation function and CBAM attention mechanism. The research methods of the above scholars mainly use lightweight modules to partially replace the original network to achieve the purpose of reducing parameters. Meanwhile, methods such as the replacement of activation functions and the addition of attention mechanisms ensure the detection accuracy of the model for occlusions and dense objects. To conclude, using lightweight convolution modules (such as depth separable convolution, group convolution, etc.), and replacing backbone networks are the most frequently used to reduce the parameters of the model. Mao et al. [32], Fu et al. [35], Li et al. [36], and Liu et al. [37] used depth separable convolution to reduce the parameters of the model by 77%, 18.18%, 49.15%, and 10.03%, respectively. Moreover, Zhang et al. [33], Cui et al. [38], Zeng et al. [39], and Zhao et al. [58] replaced the backbone network to reduce the parameters of the model by 82.3%, 52.3%, 78%, and 82.81%, respectively. Replacing backbone networks can reduce more parameters than using lightweight convolution modules, but the accuracy drops even more. Similar to using deep separable convolution to replace ordinary convolution, we use a ghost module to replace ordinary convolution, which reduces the parameters of the model by 82.79%, and the accuracy loss is less affected than replacing the backbone network. In order to solve the problem of decreasing accuracy, attention mechanisms and improving loss function are common methods, which have been adopted by most researchers [26,27,33,34,36–38,60,61]. In addition to these two improved methods, we adopt the ASFF method [42] in the head network to effectively improve the detection accuracy of the model. ASFF performs spatial filtering on the feature maps at all levels, thus retaining only useful information for combination. GA-YOLO is proposed under the guidance of similar design ideas. The GA-YOLO model is of great significance for improving the picking speed and picking quality (low missing picking rate) of the picking robot.

The model proposed in this paper mainly aims at the target detection of dense and occluded grapes. The model can also be used for other fruits in the same growth state (clusters) such as tomatoes, bananas, and strawberries. According to the ablation experiments in Section 4, we found that the detection accuracy of the model decreased by 0.94% after the model was lightened by 82.79%. Yet, we can add an ASFF module and improve the loss function to heighten accuracy. The model is lightweight, which is of great significance to solve the deployment problem of the mobile end of the model. In addition, the recall rate of the GA-YOLO model and other target detection models is lower than the precision rate, which shows that the problem of missed detection is puzzling grape detection. By lowering the confidence threshold for prediction, it is easier for the model to detect grapes and reduce the missed detection rate. However, this will increase the risk of false detection, so subsequent debugging of the model is required. Finally, it may be possible to increase the size of the input image to obtain more abundant location features and semantic features to reduce the missed detection rate, but this method will increase the number of parameters of the model, so it is necessary to find the optimal input image size.

There are still some problems to be considered when picking grapes by picking robots.

(1) We need to distinguish the maturity of grapes to avoid picking immature grapes.
(2) The detection of grape clusters is only a part of picking steps, and we also need to realize the detection of picking points. Some scholars [62–64] have developed the detection algorithm of grape picking points, the position errors between most predicted picking points and real points are within 40 pixels. However, the detected grapes are not in dense and shaded conditions, and the detection accuracy is low, so there is much room for improvement. The occlusion problem is not only solved by visual models but also requires appropriate planting strategies, such as farmers paying attention to thinning leaves and fruits when planting.
(3) The picking robot can work 24 h a day, so it is necessary to obtain the grape dataset at night. In fact, when the fruit is picked at night, it will not be exposed to the sun to cause water loss, so the quality of the fruit will be better. In addition, a richer dataset can increase the robustness of GA-YOLO.

Overall, the development of deep learning-based methods for fruit detection in agricultural settings has shown great promise in recent years. Other deep learning models (such as Faster RCNN and SSD, etc.), have their own unique advantages in specific fruit detection. In the future, we can also combine the design ideas of these models to better solve the identification problem of tropical fruits.

7. Conclusions

The goal of this paper is to decrease the parameters and calculations and raise the detection accuracy of the model. In this research, a lightweight network model named GA-YOLO was proposed. This model uses a backbone network of SE-CSPGhost, which reduces the parameter amount of the original model by 82.79% and improves the detection speed of the model by 20.245 FPS. This lightweight approach is of great significance for model deployment to mobile terminals. At the same time, although the lightweight model reduces the detection accuracy of dense and occluded grapes by 0.94%. By adding the attention mechanism and ASFF mechanism, and improving the loss function, the accuracy rate is increased by 3.69%. In short, the parameter quantity of the GA-YOLO model is 11.003 M, the mAP is 96.87%, the detection speed is 20.245 FPS and the F_1 value is 94.78%. Compared with YOLOv4 and the other 11 commonly used models, the GA-YOLO has the advantages of high detection accuracy and low model parameters. It has excellent comprehensive performance and can meet the precision and speed requirements of picking robots. Finally, we use PyQt5 to design a human–computer interaction interface to facilitate the use of the GA-YOLO model by non-professionals. In future research, we will consider the mobile deployment of the model, and deploy the GA-YOLO model on small computing devices (Raspberry Pie, developed by the Raspberry Pie Foundation in Cambridge, England; Jetson Nano, developed by the NVIDIA Corporation in Santa Clara, CA, USA; Intel NCS 2, developed by the Intel Corp in Santa Clara, CA, USA), using the deep learning inference framework NCNN and TensorRT. In addition, we will consider collecting grape datasets under night illumination and training a widely used GA-YOLO model.

Author Contributions: Conceptualization, J.C. and A.M.; methodology, A.M. and L.H.; software, L.H. and A.M.; validation, Y.S.; formal analysis, W.L.; investigation, H.Z.; resources, J.C.; data curation, Z.W.; writing—original draft preparation, A.M.; writing—review and editing, A.M. and L.H; visualization, A.M. and L.H.; supervision, J.C.; project administration, A.M.; funding acquisition, J.C. All authors have read and agreed to the published version of the manuscript.

Funding: This research was funded by the National Nature Science Foundation of China (grant number 62163005), the Natural Science Foundation of Guangxi Province (grant number 2022GXNSFAA035633).

Institutional Review Board Statement: Not applicable.

Informed Consent Statement: Not applicable.

Data Availability Statement: Not applicable.

Conflicts of Interest: The authors affirm that they have no known financial or interpersonal conflict that would have appeared to have an impact on the research presented in this study.

References

1. Chang, M.; Liu, J.; Shi, H.; Guo, T. The Effect of Off-Farm Employment on Agricultural Production Efficiency: Micro Evidence in China. *Sustainability* **2022**, *14*, 3385. [CrossRef]
2. Li, Y.; Feng, Q.; Li, T.; Xie, F.; Liu, C.; Xiong, Z. Advance of Target Visual Information Acquisition Technology for Fresh Fruit Robotic Harvesting: A Review. *Agronomy* **2022**, *12*, 1336. [CrossRef]
3. Liu, S.; Whitty, M. Automatic grape bunch detection in vineyards with an SVM classifier. *J. Appl. Log.* **2015**, *13*, 643–653. [CrossRef]
4. Lin, G.; Tang, Y.; Zou, X.; Cheng, J.; Xiong, J. Fruit detection in natural environment using partial shape matching and probabilistic Hough transform. *Precis. Agric.* **2020**, *21*, 160–177. [CrossRef]
5. Lu, J.; Sang, N. Detecting citrus fruits and occlusion recovery under natural illumination conditions. *Comput. Electron. Agric.* **2015**, *110*, 121–130. [CrossRef]
6. Wei, X.; Jia, K.; Lan, J.; Li, Y.; Zeng, Y.; Wang, C. Automatic method of fruit object extraction under complex agricultural background for vision system of fruit picking robot. *Optik* **2014**, *125*, 5684–5689. [CrossRef]
7. Luo, L.; Liu, W.; Lu, Q.; Wang, J.; Wen, W.; Yan, D.; Tang, Y. Grape berry detection and size measurement based on edge image processing and geometric morphology. *Machines* **2021**, *9*, 233. [CrossRef]
8. Behroozi-Khazaei, N.; Maleki, M.R. A robust algorithm based on color features for grape cluster segmentation. *Comput. Electron. Agric.* **2017**, *142*, 41–49. [CrossRef]
9. Reis, M.J.; Morais, R.; Peres, E.; Pereira, C.; Contente, O.; Soares, S.; Valente, A.; Baptista, J.; Ferreira, P.; Cruz, J.B. Automatic detection of bunches of grapes in natural environment from color images. *J. Appl. Log.* **2012**, *10*, 285–290. [CrossRef]
10. Luo, L.; Tang, Y.; Zou, X.; Wang, C.; Zhang, P.; Feng, W. Robust grape cluster detection in a vineyard by combining the AdaBoost framework and multiple color components. *Sensors* **2016**, *16*, 2098. [CrossRef]
11. Feng, J.; Zeng, L.; He, L. Apple fruit recognition algorithm based on multi-spectral dynamic image analysis. *Sensors* **2019**, *19*, 949. [CrossRef] [PubMed]
12. Syazwani, R.W.N.; Asraf, H.M.; Amin, M.M.S.; Dalila, K.N. Automated image identification, detection and fruit counting of top-view pineapple crown using machine learning. *Alex. Eng. J.* **2022**, *61*, 1265–1276. [CrossRef]
13. Xiang, R.; Ying, Y.; Jiang, H.; Rao, X.; Peng, Y. Recognition of overlapping tomatoes based on edge curvature analysis. *Nongye Jixie Xuebao Trans. Chin. Soc. Agric. Mach.* **2012**, *43*, 157–162.
14. Liu, T.-H.; Ehsani, R.; Toudeshki, A.; Zou, X.-J.; Wang, H.-J. Identifying immature and mature pomelo fruits in trees by elliptical model fitting in the Cr–Cb color space. *Precis. Agric.* **2019**, *20*, 138–156. [CrossRef]
15. Nazari, A.; Mohammadzamani, D.; Javidan, S.M. Using Image Processing Technology for Detection of Red Grape Fruits in natural conditions. *Res. Sq.* 2023; *in preprint*. [CrossRef]
16. Pérez-Zavala, R.; Torres-Torriti, M.; Cheein, F.A.; Troni, G. A pattern recognition strategy for visual grape bunch detection in vineyards. *Comput. Electron. Agric.* **2018**, *151*, 136–149. [CrossRef]
17. Redmon, J.; Divvala, S.; Girshick, R.; Farhadi, A. You only look once: Unified, real-time object detection. In Proceedings of the IEEE Conference on Computer Vision and Pattern Recognition, Las Vegas, NV, USA, 27–30 June 2016; pp. 779–788.
18. Redmon, J.; Farhadi, A. YOLO9000: Better, faster, stronger. In Proceedings of the IEEE Conference on Computer Vision and Pattern Recognition, Honolulu, HI, USA, 21–26 July 2017; pp. 7263–7271.
19. Redmon, J.; Farhadi, A. Yolov3: An incremental improvement. *arXiv* **2018**, arXiv:1804.02767.
20. Ren, S.; He, K.; Girshick, R.; Sun, J. Faster r-cnn: Towards real-time object detection with region proposal networks. *Adv. Neural Inf. Process. Syst.* **2015**, *28*, 1137–1149. [CrossRef]
21. Liu, W.; Anguelov, D.; Erhan, D.; Szegedy, C.; Reed, S.; Fu, C.Y.; Berg, A.C. *Ssd: Single Shot Multibox Detector, Proceedings of the European Conference on Computer Vision, Amsterdam, The Netherlands, 8–16 October 2016*; Springer: Berlin/Heidelberg, Germany, 2016; pp. 21–37.
22. Bochkovskiy, A.; Wang, C.-Y.; Liao, H.-Y.M. Yolov4: Optimal speed and accuracy of object detection. *arXiv* **2020**, arXiv:2004.10934.

23. Gao, F.; Fu, L.; Zhang, X.; Majeed, Y.; Li, R.; Karkee, M.; Zhang, Q. Multi-class fruit-on-plant detection for apple in SNAP system using Faster R-CNN. *Comput. Electron. Agric.* **2020**, *176*, 105634. [CrossRef]
24. Tu, S.; Pang, J.; Liu, H.; Zhuang, N.; Chen, Y.; Zheng, C.; Wan, H.; Xue, Y. Passion fruit detection and counting based on multiple scale faster R-CNN using RGB-D images. *Precis. Agric.* **2020**, *21*, 1072–1091. [CrossRef]
25. Mai, X.; Zhang, H.; Jia, X.; Meng, M.Q.-H. Faster R-CNN with classifier fusion for automatic detection of small fruits. *IEEE Trans. Autom. Sci. Eng.* **2020**, *17*, 1555–1569. [CrossRef]
26. Ding, X.; Li, Q.; Wang, X.; Chen, L.; Son, J.; Song, J.Y. Apple Detection Algorithm based on an Improved SSD. *J. Inst. Internet Broadcast. Commun.* **2021**, *21*, 81–89.
27. Behera, S.K.; Rath, A.K.; Sethy, P.K. Fruits yield estimation using Faster R-CNN with MIoU. *Multimed. Tools Appl.* **2021**, *80*, 19043–19056. [CrossRef]
28. Han, K.; Wang, Y.; Tian, Q.; Guo, J.; Xu, C.; Xu, C. Ghostnet: More features from cheap operations. In Proceedings of the IEEE/CVF Conference on Computer Vision and Pattern Recognition, Seattle, WA, USA, 13–19 June 2020; pp. 1580–1589.
29. Howard, A.; Sandler, M.; Chu, G.; Chen, L.C.; Chen, B.; Tan, M.; Wang, W.; Zhu, Y.; Pang, R.; Le, Q.V.; et al. Searching for mobilenetv3. In Proceedings of the IEEE/CVF International Conference on Computer Vision, Seoul, Republic of Korea, 27 October–2 November 2019; pp. 1314–1324.
30. Sandler, M.; Howard, A.; Zhu, M.; Zhmoginov, A.; Chen, L.-C. Mobilenetv2: Inverted residuals and linear bottlenecks. In Proceedings of the IEEE Conference on Computer Vision and Pattern Recognition, Salt Lake City, UT, USA, 18–23 June 2018; pp. 4510–4520.
31. Zhang, X.; Zhou, X.; Lin, M.; Sun, J. Shufflenet: An extremely efficient convolutional neural network for mobile devices. In Proceedings of the IEEE Conference on Computer Vision and Pattern Recognition, Salt Lake City, UT, USA, 18–23 June 2018; pp. 6848–6856.
32. Mao, Q.-C.; Sun, H.-M.; Liu, Y.-B.; Jia, R.-S. Mini-YOLOv3: Real-time object detector for embedded applications. *IEEE Access* **2019**, *7*, 133529–133538. [CrossRef]
33. Zhang, C.; Kang, F.; Wang, Y. An Improved Apple Object Detection Method Based on Lightweight YOLOv4 in Complex Backgrounds. *Remote Sens.* **2022**, *14*, 4150. [CrossRef]
34. Ji, W.; Pan, Y.; Xu, B.; Wang, J. A real-time Apple targets detection method for picking robot based on ShufflenetV2-YOLOX. *Agriculture* **2022**, *12*, 856. [CrossRef]
35. Fu, L.; Feng, Y.; Wu, J.; Liu, Z.; Gao, F.; Majeed, Y.; Al-Mallahi, A.; Zhang, Q.; Li, R.; Cui, Y. Fast and accurate detection of kiwifruit in orchard using improved YOLOv3-tiny model. *Precis. Agric.* **2021**, *22*, 754–776. [CrossRef]
36. Li, S.; Zhang, S.; Xue, J.; Sun, H. Lightweight target detection for the field flat jujube based on improved YOLOv5. *Comput. Electron. Agric.* **2022**, *202*, 107391. [CrossRef]
37. Liu, B.; Luo, L.; Wang, J.; Lu, Q.; Wei, H.; Zhang, Y.; Zhu, W. An improved lightweight network based on deep learning for grape recognition in unstructured environments. *Inf. Process. Agric.* 2023; in press.
38. Cui, M.; Lou, Y.; Ge, Y.; Wang, K. LES-YOLO: A lightweight pinecone detection algorithm based on improved YOLOv4-Tiny network. *Comput. Electron. Agric.* **2023**, *205*, 107613. [CrossRef]
39. Zeng, T.; Li, S.; Song, Q.; Zhong, F.; Wei, X. Lightweight tomato real-time detection method based on improved YOLO and mobile deployment. *Comput. Electron. Agric.* **2023**, *205*, 107625. [CrossRef]
40. He, Y.; Liu, P.; Wang, Z.; Hu, Z.; Yang, Y. Filter pruning via geometric median for deep convolutional neural networks acceleration. In Proceedings of the IEEE/CVF Conference on Computer Vision and Pattern Recognition, Long Beach, CA, USA, 15–20 June 2019; pp. 4340–4349.
41. Tzutalin, D. LabelImg: LabelImg is a Graphical Image Annotation Tool and Label Object Bounding Boxes in Images. Available online: https://github.com/tzutalin/labelImg (accessed on 23 March 2023).
42. He, K.; Zhang, X.; Ren, S.; Sun, J. Spatial pyramid pooling in deep convolutional networks for visual recognition. *IEEE Trans. Pattern Anal. Mach. Intell.* **2015**, *37*, 1904–1916. [CrossRef] [PubMed]
43. Wang, C.-Y.; Liao, H.Y.M.; Wu, Y.H.; Chen, P.Y.; Hsieh, J.W.; Yeh, I.H. CSPNet: A new backbone that can enhance learning capability of CNN. In Proceedings of the IEEE/CVF Conference on Computer Vision and Pattern Recognition, Seattle, WA, USA, 13–19 June 2020; pp. 390–391.
44. Hu, J.; Shen, L.; Sun, G. Squeeze-and-excitation networks. In Proceedings of the IEEE Conference on Computer Vision and Pattern Recognition, Salt Lake City, UT, USA, 18–23 June 2018; pp. 7132–7141.
45. Lin, T.-Y.; Dollár, P.; Girshick, R.; He, K.; Hariharan, B.; Belongie, S. Feature pyramid networks for object detection. In Proceedings of the IEEE Conference on Computer Vision and Pattern Recognition, Honolulu, HI, USA, 21–26 July 2017; pp. 2117–2125.
46. Liu, S.; Qi, L.; Qin, H.; Shi, J.; Jia, J. Path aggregation network for instance segmentation. In Proceedings of the IEEE Conference on Computer Vision and Pattern Recognition, Salt Lake City, UT, USA, 18–23 June 2018; pp. 8759–8768.
47. Liu, S.; Huang, D.; Wang, Y. Learning spatial fusion for single-shot object detection. *arXiv* **2018**, arXiv:1911.09516.
48. Zheng, Z.; Wang, P.; Ren, D.; Liu, W.; Ye, R.; Hu, Q.; Zuo, W. Enhancing geometric factors in model learning and inference for object detection and instance segmentation. *IEEE Trans. Cybern.* **2021**, *52*, 8574–8586. [CrossRef] [PubMed]
49. Campos, B.O.O.; Paredes, F.; Rey, J.C.; Lobo, D.; Galvis-Causil, S. The relationship between the normalized difference vegetation index, rainfall, and potential evapotranspiration in a banana plantation of Venezuela. *SAINS TANAH-J. Soil Sci. Agroclimatol.* **2021**, *18*, 58–64. [CrossRef]

50. Islam, F.; Hoq, M.N.; Rahman, C.M. Application of transfer learning to detect potato disease from leaf image. In Proceedings of the 2019 IEEE International Conference on Robotics, Automation, Artificial-intelligence and Internet-of-Things (RAAICON), Dhaka, Bangladesh, 29 November–1 December 2019; pp. 127–130.
51. Lee, T.-Y.; Yu, J.-Y.; Chang, Y.-C.; Yang, J.-M. Health detection for potato leaf with convolutional neural network. In Proceedings of the 2020 Indo–Taiwan 2nd International Conference on Computing, Analytics and Networks (Indo-Taiwan ICAN), Rajpura, India, 7–15 February 2020; pp. 289–293.
52. Olivares, B.O.; Calero, J.; Rey, J.C.; Lobo, D.; Landa, B.B.; Gómez, J.A. Correlation of banana productivity levels and soil morphological properties using regularized optimal scaling regression. *Catena* **2022**, *208*, 105718. [CrossRef]
53. Olivares, B.O.; Rey, J.; Lobo, D.; Navas-Cortés, J.; Gómez, J.; Landa, B. Fusarium wilt of bananas: A review of agro-environmental factors in the Venezuelan production system affecting its development. *Agronomy* **2021**, *11*, 986. [CrossRef]
54. Olivares, B.O.; Rey, J.C.; Perichi, G.; Lobo, D. Relationship of microbial activity with soil properties in banana plantations in Venezuela. *Sustainability* **2022**, *14*, 13531. [CrossRef]
55. Olivares, B.O.; Vega, A.; Calderón, M.A.R.; Montenegro-Gracia, E.; Araya-Almán, M.; Marys, E. Prediction of banana production using epidemiological parameters of black sigatoka: An application with random forest. *Sustainability* **2022**, *14*, 14123. [CrossRef]
56. Kuznetsova, A.; Maleva, T.; Soloviev, V. Detecting apples in orchards using YOLOv3 and YOLOv5 in general and close-up images. In *Advances in Neural Networks, Proceedings of the ISNN 2020 17th International Symposium on Neural Networks, ISNN 2020, Cairo, Egypt, 4–6 December 2020*; Springer: Berlin/Heidelberg, Germany, 2020; Volume 17, pp. 233–243.
57. Le, T.-T.; Lin, C.-Y. Deep learning for noninvasive classification of clustered horticultural crops—A case for banana fruit tiers. *Postharvest Biol. Technol.* **2019**, *156*, 110922. [CrossRef]
58. Zhao, S.; Zhang, S.; Lu, J.; Wang, H.; Feng, Y.; Shi, C.; Li, D.; Zhao, R. A lightweight dead fish detection method based on deformable convolution and YOLOV4. *Comput. Electron. Agric.* **2022**, *198*, 107098. [CrossRef]
59. Betti, A.; Tucci, M. YOLO-S: A Lightweight and Accurate YOLO-like Network for Small Target Selection in Aerial Imagery. *Sensors* **2023**, *23*, 1865. [CrossRef] [PubMed]
60. Huang, M.-L.; Wu, Y.-S. GCS-YOLOV4-Tiny: A lightweight group convolution network for multi-stage fruit detection. *Math. Biosci. Eng.* **2023**, *20*, 241–268. [CrossRef] [PubMed]
61. Sun, H.; Wang, B.; Xue, J. YOLO-P: An efficient method for pear fast detection in complex orchard picking environment. *Front. Plant Sci.* **2023**, *13*, 1089454. [CrossRef]
62. Xu, D.; Xia, F.; Suyin, Z. A Keypoint-Based Method for Grape Stems Identification. *SSRN* **2022**. [CrossRef]
63. Jin, Y.; Liu, J.; Wang, J.; Xu, Z.; Yuan, Y. Far-near combined positioning of picking-point based on depth data features for horizontal-trellis cultivated grape. *Comput. Electron. Agric.* **2022**, *194*, 106791. [CrossRef]
64. Zhao, R.; Zhu, Y.; Li, Y. An end-to-end lightweight model for grape and picking point simultaneous detection. *Biosyst. Eng.* **2022**, *223*, 174–188. [CrossRef]

horticulturae

MDPI

Article

A Comparative Analysis of the Grafting Efficiency of Watermelon with a Grafting Machine

Huan Liang [1], Juhong Zhu [1], Mihong Ge [1], Dehuan Wang [1], Ke Liu [1], Mobing Zhou [1], Yuhong Sun [1], Qian Zhang [2], Kai Jiang [3,*] and Xianfeng Shi [1,*]

[1] Wuhan Academy of Agricultural Sciences, Wuhan 430070, China; lianghuanconf@126.com (H.L.); hongye408@163.com (J.Z.); gmh917@126.com (M.G.); wdhuan1987@163.com (D.W.); cokelk@163.com (K.L.); zmb19682022@163.com (M.Z.); sunyh68@163.com (Y.S.)

[2] Research Center of Information Technology, Beijing Academy of Agriculture and Forestry Sciences, Beijing 100097, China; zhangq@nercita.org.cn

[3] Research Center of Intelligent Equipment, Beijing Academy of Agriculture and Forestry Sciences, Beijing 100097, China

* Correspondence: jiangk@nercita.org.cn (K.J.); shixf124@163.com (X.S.); Tel.: +86-10-5150-3504 (K.J.); +86-10-6538-6852 (X.S.)

Abstract: The rising age of the population in rural China and the labor intensity of grafting have resulted in a decrease in the number of grafters and a subsequent increase in their wages. Manual grafting can no longer satisfy the increasing demand for watermelon-grafted transplanting; thus, machine grafting will be an effective alternative. In order to accelerate the implementation of machine grafting in China, a comparative analysis between the automatic grafting machine (model 2TJGQ-800) and traditional hand grafting was conducted. The reliability and feasibility of machine grafting were evaluated through a comprehensive evaluation of the production capacity and grafting seedling quality. This study focuses on the grafting application of watermelon plug-tray seedlings. The scion and rootstock seeds were sown on 9 November 2022. Grafting experiments using an automatic grafting machine, skilled workers, and ordinary workers were conducted with the root-pruned one-cotyledon grafting method on 24 November 2022. The results showed that the machine grafting had a high uniformity and grafting speed. The grafting speed of the grafting machine was 774 plant·h^{-1} and 1.65–2.55-fold higher than the hand grafting. With training, workers can improve their grafting speed, but it will still be slower than machine grafting. In addition, there was no significant difference in the grafting survival rate between the machine grafting and hand grafting. However, using machine grafting, the success rate decreased from 100% to 90.07% and the rootstock regrowth rate increased from 18.44% to 72.69%. Incomplete rootstock cutting, clip supply failure, and grafting drop failure are the three main factors that result in machine grafting failure. In conclusion, the grafting machine has advantages in terms of grafting speed and uniformity. Upon improving the accuracy of the cutting mechanism and grafting success rate, it will be adopted by commercial nurseries.

Keywords: watermelon; grafting efficiency; grafting machine; one-cotyledon grafting; comparison

Citation: Liang, H.; Zhu, J.; Ge, M.; Wang, D.; Liu, K.; Zhou, M.; Sun, Y.; Zhang, Q.; Jiang, K.; Shi, X. A Comparative Analysis of the Grafting Efficiency of Watermelon with a Grafting Machine. *Horticulturae* **2023**, *9*, 600. https://doi.org/10.3390/horticulturae9050600

Academic Editor: Francesco Giuffrida

Received: 4 April 2023
Revised: 10 May 2023
Accepted: 11 May 2023
Published: 19 May 2023

1. Introduction

Grafting is an asexual propagation method that allows for two plant segments to be joined together, only for them to later grow and develop as a single composite plant. Grafting is widely implemented in cucurbit vegetable production to control soil-borne diseases caused by bacteria, fungi, oomycetes, viruses, or root-knot nematodes in Japan, China, Korea, and some parts of European and Asian countries [1,2]. Moreover, it can improve nutrient transport and resistance to abiotic stresses [3–5]. Grafting can also promote the growth and development of the grafted plants, increase the plant yield, and positively influence the fruit quality [6,7].

Despite the above advantages, their high cost is one main factor limiting the utilization of grafted seedlings on a large scale [8]. In nurseries, the cost of hand-grafted plants is estimated at approximately CNY 0.58 ¥ in China, compared to CNY 0.32 ¥ for non-grafted plants [9]. However, the costs of hand-grafted plants in Japan, Korea, and the USA were higher than those in China, where they can accumulate three to four times extra costs without grafting [10–12]. The cost of hand-grafted plants includes the cost of grafting operations, rootstock seeds costs, clip costs, and energy costs at grafted seedling healing, etc. It is estimated that the grafting process itself can amount to approximately one quarter of the total costs per grafted plant [13].

At present, grafting is mainly reliant on Chinese laborers. The hole-insertion method is the most widely used method for cucurbit grafting [14,15]. Its grafting speed is approximately 350 plants·h^{-1} per worker. This method requires a highly skilled grafter and similarly high-quality scion and rootstock seedlings. At the time of grafting, the diameter of the scion must be smaller than the diameter of the rootstock stem, so that the scion can be inserted into a hole made between the two cotyledons of the rootstock. A smaller scion stem is easier to soften after cutting. It is difficult to insert the slender and soft stem into the hole of the rootstock. Moreover, rootstock regrowth commonly occurs after grafting using the above grafting method, which can result in graft failure or a decrease in yield, as the rootstock competes with the scion for water and nutrients [16]. To solve this problem, labor is required to remove the rootstock regrowth and damage to seedlings can occur [17,18]. Furthermore, the aging agricultural population and intensive workloads of grafting have led to a decrease in the number of available grafters. However, grafting is a task that requires considerable time. With the increasing demand for grafted seedlings, hand-grafting—which requires a large number of laborers—can no longer meet this demand.

Grafting machines, which can reduce the demands on hand-grafting laborers, expand production capacities, and improve product uniformity, have been proposed worldwide for many years. Since the 1980s, when researchers in Japan and Korea first began to conduct melon-grafting machine technology research, many types of grafting machines have been launched [19–21]. In 2010, ISEKI & CO., LTD. launched the GR803-U grafting robot in Japan (700 Umaki-cho, Matsuyama-shi, Ehime-ken, 799-2692 Japan). The robot was designed with a seedling linear cutting mechanism. It needed two workers to complete the seedling supply. Its productivity reached 800 plants·h^{-1} and its survival rate reached 95% [22]. Helper Robotech Co., Ltd. launched the AFGR-800CS grafting robot in Korea (93, Hitech-ro, Jinrye-myeon, Gimhae-si, Gyeongsangnam-do, Korea). This grafting robot used the rotary cutting mechanism. It also needed two workers to complete the seedling supply. Its productivity reached 800 plants·h^{-1} and its survival rate reached 95% [23,24]. Atlantic Man. SRL developed the Gr300/3 grafting robot in Italy (42024 Castelnovo di Sotto (RE), Italy). A single worker could complete the seedling supply. Its production efficiency was 300 plants·h^{-1} and the survival rate of the grafted seedlings was 98% [25]. However, the high cost of these robots, the strict requirements for the agronomic traits of the scion and rootstock seedlings, and the complexity of their repairs are obstacles to their adoption in China [26]. China began researching grafting machines in the 1990s [27–29]. In 2005, China Agricultural University launched the 2JSZ-600 grafting robot, which used a double-arm bidirectional cutting mechanism. It adopted the operation mode of double-station feeding for the seedlings. Its productivity reached 600 plants·h^{-1} and its survival rate reached 95% [27]. In 2011, the first commercial grafting machine (TJ-800, the National Agricultural Intelligent Equipment Engineering Technology Research Center, Beijing, China) was launched [30]. This robot was designed with a rotary cutting machine, which could precisely adjust the cutting angle. Two workers could complete the seedling supply, the productivity reached 800 plants·h^{-1}, and the success rate reached 95% in the laboratory. In order to improve the machine grafting accuracy, the improved and upgraded model 2TJGQ-800 was developed [31,32]. The grafting matching model was designed, which could adjust the grafting angle of the rootstock or scion. The matching rate of the rootstock incision and the scion incision reached 98.03% [33,34]. The properties of this grafting machine are being continuously improved, but their current use for cucurbit is relatively low [35]. In China, the demand for melon grafting seedlings is approximately 50 billion, and millions of grafters working to produce grafting seedlings could meet this demand. The demand for machine grafting becomes more and more urgent with the reduction in grafters. However, the related research and development institutions have not given a reasonable and accurate answer to whether a grafting machine could replace manual grafting at the present stage. The lack of a feasibility analysis of machine grafting technology means that farmers and seedling nurseries cannot adopt machine grafting.

To ensure that machine grafting can be implemented in China as soon as possible, this study was conducted to examine the grafting efficiency and grafting seedling quality of a grafting machine compared with those of skilled grafters and unskilled grafters, based on the grafting machine 2TJGQ-800 (the National Agricultural Intelligent Equipment Engineering Technology Research Center, Beijing, China). This study focuses on the subject of watermelon plug-tray seedlings. The grafting was conducted using the root-pruned one-cotyledon grafting method. These results will highlight the existing problems of grafting machines and provide references for grafting machine improvement and optimization, contributing to the complete mechanization of grafting in China.

2. Materials and Methods

2.1. Seedling Production

The experiment was conducted at the Wuhan Agricultural Academy, Central China (30°27′ N, 114°20′ E, and altitude 22 m above sea level), between November and December 2022. The watermelon 'Zaojia84-24' (*Citrullus lanatus*, Xinjiang Seed Co., Ltd., Urumqi City, China) was used as a scion and the interspecific pumpkin 'Zhenzhuang' (Jingyan Yinong, Seed Sci-Tech Co., Ltd., Beijing, China) was used as the rootstock. The scion and rootstock seeds were sown into 98- and 72-cell plug trays filled with a mixed seedling substrate (peat moss and pearlite at a volume ratio of 3:1) on the same day, respectively. After the sowing, the plug trays were covered with plastic film and placed in a germination room at a temperature of 30 °C for two days to promote germination. Then, they were moved to a greenhouse with a temperature of 18–28 °C, a relative humidity of 60–85%, and ordinary light. The plants were fertilized with a water-soluble fertilizer (Product number: 20-10-20 + TE, 1000 times liquid, Shanghai Yongtong chemical Co., Ltd., Shanghai, China).

2.2. Device Description and Productive Process

The 2TJGQ-800 grafting machine is composed of a seedling loading platform, clamping and handling mechanism, cutting mechanism, clip-feeding mechanism, clip-sequencing and supply device, conveyor belt, and control system, as shown in Figure 1A. This machine needs two people to operate it and supply the seedlings. It then automatically completes the seedling supplying, holding, cutting, joining, and clipping procedures, as shown in Figure 1B. Its operators are required to pass the grafting machine operation training for more than 48 h, and be proficient in seedling supplying, cutting angle adjusting, clip handling, and blade replacing. The technical requirements for supplying artificial seedlings include adjusting the cotyledon orientation and stem height of the seedlings and keeping the seedlings upright.

The working process was as follows: (1) The seedlings (rootstock and scion) were manually taken out of the plug tray and placed on the seedling loading platform of the grafting machine. (2) The clamping and handling mechanism held the seedlings (rootstock and scion) and transported them to the cutting station. The cutting mechanism cut the seedlings. The rootstock cut off a cotyledon and growth point, and the scion was cut off at 15 mm from the hypocotyl. (3) The clamping and handling mechanism continued to transport the seedlings (rootstock and scion) to the docking station, and the cutting of the rootstock and scion was jointed. (4) The clip-feeding mechanism pushed out a grafting clip to fix the cuts of the rootstock and scion, and then the grafted seedlings were released and dropped onto the conveyor belt to complete a grafting cycle. The height and cotyledon direction of the seedlings needed to be controlled by a worker during the seedling supplying, and the uniformity of seedlings was selected manually for the grafting machine. In addition, the cutting angle of the rootstock and scion was adjusted to induce the cutting fit effect of the two to reach a perfect state.

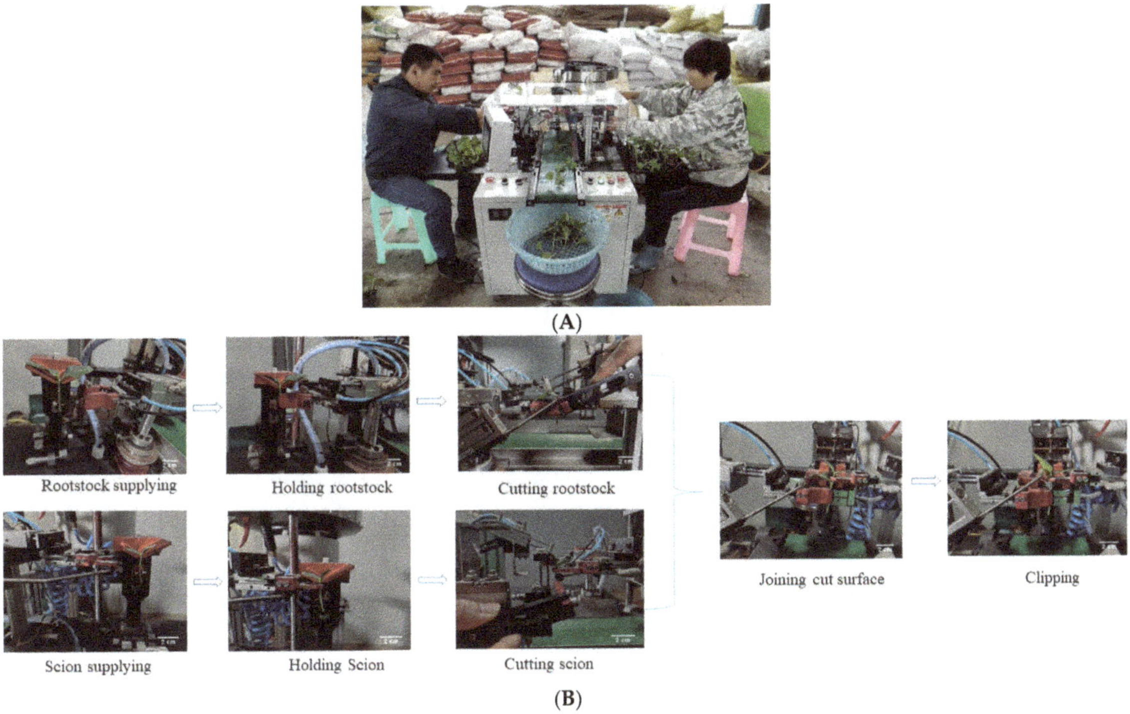

Figure 1. (**A**) 2TJGQ-800, launched by the National Agricultural Intelligent Equipment Engineering Technology Research Center in China, a semi-automatic grafting machine used in this study. (**B**) Machine grafting operation process.

2.3. Experimental Design

The plants were considered mature and ready for grafting when they had one true leaf, as shown in Figure 2A. The diameters in the area close to the cut varied from 1.58 to 1.75 mm for the scion and 2.71 to 2.80 mm for the rootstock.

Figure 2. (**A**) The morphology of rootstock and scion at grafting, and (**B**) grafting clip for experiments.

The experiments were set up as one-way experiments with three replicates, and each replicate had prepared 500 plants. The root-pruned one-cotyledon grafting was conducted by day 15 after the rootstock and scion seeds had been sown, which was on 24 November 2022 [36]. During the grafting operation, three grafting treatments were established: machine grafting using 2TJGQ-800 (machine), hand grafting by two unskilled grafters (unskilled,) and hand grafting by two skilled grafters (skilled). In addition, the clip was dedicated to the grafting machine. As shown in Figure 2B, it was a one-piece PVC plastic clip. The diameter of the steel wire was 0.7 mm. The preload pressure on the clip mouth was generated by the steel ring. The tail of the clip was squeezed to make the clip open. The procedure of clip was: (1) the tail of the clip was squeezed, (2) the incision of the grafted seedling was put into the clip mouth, and (3) the tail of the clip was released.

The experimental procedure was carried out as follows. First, six workers were chosen and divided into three groups. The first group chosen for the "machine" treatment needed to possess good command of a grafting machine. The operators was required to pass the grafting machine operation training for more than 48 h, and was proficient in seedling supplying, cutting angle adjusting, clip handling and blade replacing. Technical requirements for artificial seedling supplying include adjusting the cotyledon orientation and stem height of seedlings and keeping the seedlings upright. The second group chosen for the "skilled" treatment needed to have more than 2 years grafting experience. The third group chosen for the "unskilled" treatment needed to understand the grafting process and have a small amount of grafting experience (1–2 previous experiences). Secondly, the 500 rootstock plants and scion seedlings were cut off from the stem base. Then, the grafting was completed by the three groups of workers, respectively. Supplying the seedlings to the grafting machine needed two workers, one worker to supply the rootstock seedlings and the other worker to supply the scion seedlings. The hand-grafting work involved cutting, joining, and picking. The two workers cut 300 plants of the scion and rootstock seedlings, respectively. Additionally, they completed the joining and picking alone. Next, the grafting plants were transported into a 72-hole plug tray filled with a mixed seedling substrate.

2.4. *Grafting Seedling Healing and Cultivation*

Immediately after being transplanted, the plants were placed under a plastic film with a day/night cycle of 28 °C/18 °C and more than a 90% humidity in a low-light intensity (75 $\mu mol \cdot m^{-2} \cdot s^{-1}$, 12/12 h photoperiod) environment. The grafted plants were exposed to the air for 1–3 h per day until the scions were alive and had grown normally. After 10 days, the grafted seedlings were transferred to a greenhouse, following common practice.

2.5. *Grafting Efficiency*

The grafts obtained via the "machine" treatment were counted after 30 min of grafting. The time taken to graft 300 plants by hand was also counted. The measurement was conducted 3 times independently.

The grafting speed was investigated using the formula.

$$\text{Grafting speed (plants} \cdot \text{h}^{-1}) = (\text{the number of grafting plants/grafting time}) \quad (1)$$

After the grafting, the number of successfully grafted seedlings was calculated. The rootstock and scion incisions of the successfully grafted seedlings had to be joined together. After turning the grafted rootstocks upside down and shaking them, they were considered to be successfully grafted seedlings if the scion did not fall off. The successfully grafted seedlings are shown in Figure 3.

The grafting success rate was investigated using the formula.

$$\text{Grafting success rate (\%)} = (\text{the number of grafting success seedling/the number of grafting seedling}) \quad (2)$$

Figure 3. The successfully grafted seedlings.

2.6. Calculation of Hypocotyl Height, Incision Length of Scion and Rootstock

The hypocotyl height, incision length of the scion, and incision length of the rootstock were measured to analyze the uniformity of the grafting seedlings. After randomly selecting 100 plants of the rootstock and scion seedlings, the height of the hypocotyl and incision lengths of the scion and rootstock were measured using a digital display caliper. The hand-grafting treatment, height of the hypocotyl, and incision lengths of the scion and rootstock were measured after the workers had cut off the rootstock and scion.

When measuring the height of the hypocotyl and incision lengths of the scion and rootstock for the "machine" treatment, the feeding clip system was shut. During the grafting, the workers regularly supplied rootstock and scion seedlings. The cutting mechanism's normal operation and only cutting off the rootstock and scion seedlings without supply the clip, and that were used to measure the incision lengths of the scion and rootstock. This ensured the accuracy of the measurements and avoided a change in the incision lengths of the scion and rootstock as a result of clipping.

The coefficient of variation reflects the uniformity of the parameter. It was investigated using the formula.

$$\text{Coefficient of variation} = \text{standard deviation}/\text{mean}. \tag{3}$$

2.7. Grafted Survival and Rootstock Regrowth Rate Measurement

The survival rates of the grafted plants and the rootstock regrowth rates were assessed 10 days after the grafting [17,37]. The grafted plants were considered to have survived if the scion leaves and rootstock stem were turgid, whereas severely wilted scion leaves and stems of both the scion and rootstock were considered as graft failures. Rootstock regrowth (squash leaves) reflected whether the rootstock axillary bud, at the base of the rootstock cotyledon, was removed completely (Figure 4).

$$\text{Survival rate (\%)} = (\text{survival number}/\text{total number of grafted plants}) \times 100\%. \tag{4}$$

$$\text{Rootstock regrowth rate (\%)} = (\text{regrown rootstock number}/\text{total number of grafted plants}) \times 100\%. \tag{5}$$

Figure 4. The rootstock regrowth from grafting watermelon seedlings. The squash leaves indicated by the red arrow show the rootstock regrowth.

2.8. Statistical Analysis

All the data were statistically analyzed by analyses of variance, and the significant differences were determined based on Duncan's tests and were indicated by different letters ($p = 0.05$) using the SAS 9.0.2 software (SAS Institute Inc., Cary, NC, USA).

3. Results

3.1. Grafting Efficiency

The "machine", "unskilled", and "skilled" grafting treatments showed significant effects on the grafting speed (Figure 5A). The machine grafting speed was 774 plant·h^{-1}. Compared to the "unskilled" and "skilled" grafting treatments, the grafting speeds of the "machine" treatment were 1.65- and 2.55-fold higher. There was a significant difference in the grafting speeds between the hand-grafting treatments ("unskilled" and "skilled" grafting treatments). The grafting speed of the "skilled" grafting treatment increased by 40.53% compared to the "unskilled" grafting treatment. Thus, the grafting speed of skilled workers is more advantageous. From the perspective of grafting speed, machine grafting can compensate for the problem of insufficient labor in production. With the continuous upgrading of automatic technology, it will replace manual seedling supply and the grafting speed will be doubled, so the machine grafting speed can be accepted by users.

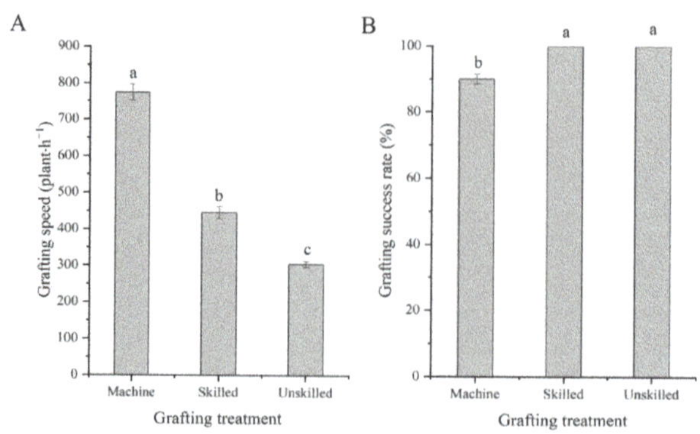

Figure 5. Grafting speed (**A**), and grafting success rate (**B**) were measured at grafting. Machine, unskilled, and skilled represent machine grafting with 2TJGQ-800, hand grafting by unskilled grafter, and hand grafting by skilled grafter, respectively. The different small letters indicate significant difference at $p = 0.05$ levels.

There was no significant difference in the grafting success rates between the hand-grafting treatments (Figure 5B), because one-cotyledon splice grafting is simple and manual grafting can accurately complete the grafting operation. The grafting success rate with hand grafting was 100%, whereas the grafting success rate significantly decreased when using machine grafting. The grafting success rate decreased from 100% to 90.07%. Through improving the standardization of the seedling cultivation and the operation accuracy of the grafting mechanism, the indicators of the machine grafting will be significantly improved.

The reason for the machine grafting failure was statistics. Incomplete rootstock cutting (Figure 6A), grafting seedlings drop failure (Figure 6B), and clip supply failure (Figure 6C) were the three main factors resulting in machine grafting failure. The failure rate caused by the three factors reached 88.89% for all the failures (Figure 6D). Among these, low rootstock positioning resulting in incomplete rootstock cutting accounted for 54.77%. Large rootstock cotyledons resulting in grafting seedlings getting stuck in the clip-feeding mechanism accounted for 16.95%. The clip supply failure accounted for 17.17%, because the attitude of the clip was not in a horizontal state before the clipping operation, resulting in the clip getting stuck in the slide. Other factors resulting in grafting failures accounted for 11.11%, including bent stems of the scion seedlings resulting in the scion stems being too short and the excessive cutting of the rootstock's two cotyledons due to the high position of the rootstock, both of which could not achieve the successful clipping of the grafting seedlings. It can be seen that accurately controlling the position of rootstock seedlings can solve most of the problems causing grafting failure. Achieving a stable and reliable clipping supply by improving and optimizing the clip-feeding mechanism is the second key factor to improving the grafting success rate. In addition, controlling the age of rootstock seedlings and artificially straightening the stem bending of scion seedlings are the keys to improving the accuracy of machine grafting and cutting operations.

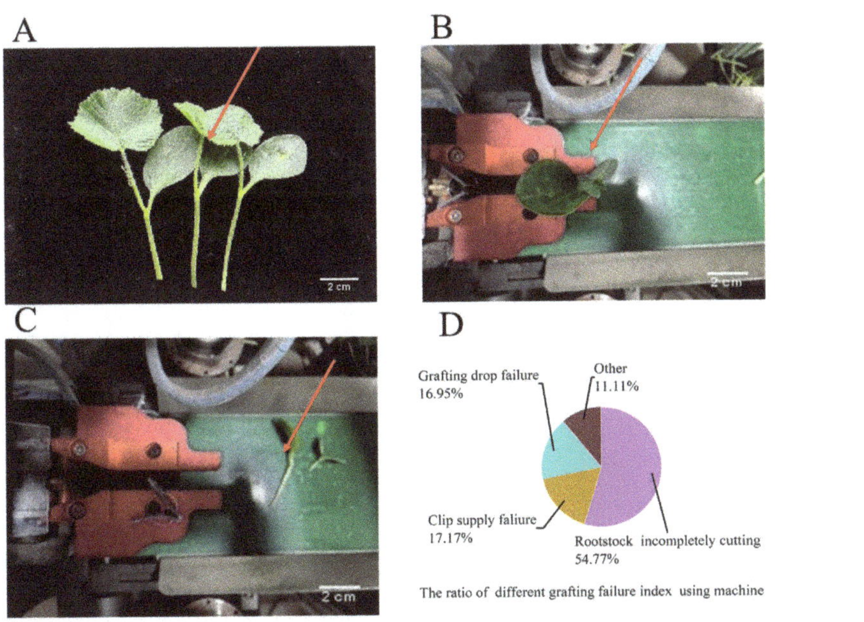

Figure 6. The picture of three main factors leading to machine grafting failure (**A–C**), and the ratio of different grafting failure indexes using a machine were measured (**D**). Pictures (**A–C**), respectively, represent incomplete rootstock cutting, grafting drop failure, and clip supply failure.

3.2. Measurement of the Scion Hypocotyl Height, Scion Incision Length, and Rootstock Incision Length at Grafting

The scion hypocotyl heights (Figure 7A), scion incision lengths (Figure 7B), and rootstock incision lengths (Figure 7C) of 100 plants were measured before the clipping. The coefficient of variation reflects the uniformity of the parameter. The smaller the coefficient of variation, the higher the uniformity. This result showed that the coefficients of variation for the scion hypocotyl height, incision length of the scion, and rootstock cutting carried out by the machine were 0.049, 0.089, and 0.068, respectively. The coefficients of variation for the parameters of the cutting carried out by the machine were the smallest. Compared to the "machine" grafting treatment, the coefficients of variation for the scion hypocotyl height treated by the "skilled" and "unskilled" grafting were 1.85- and 4.36-fold higher; those for the incision length of the scion were 1.22- and 1.76-fold higher; and those for the incision length of the rootstock were 1.22- and 1.87-fold higher, respectively.

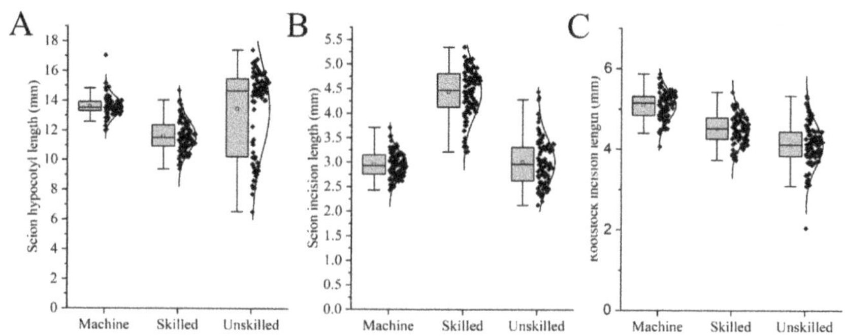

Figure 7. The scion hypocotyl height (**A**), scion incision length (**B**), and rootstock incision length (**C**) of 100 plants in each treatment were measured using a digital caliper at grafting. Machine, unskilled, and skilled represent machine grafting with 2TJGQ-800, hand grafting by unskilled grafter, and hand grafting by skilled grafter, respectively.

3.3. The Grafted Survival and Rootstock Regrowth Rate

The growth of the grafted watermelon seedlings after healing is shown in Figure 8A. There were no significant differences in the grafting survival rates among the "machine", "unskilled", and "skilled" grafting treatments (Figure 8B). The grafting survival rate of the "machine" treatment was 94.91%. Inappropriate rootstock incision, a mismatch of rootstocks, and scion incision can cause the failure of grafted seedling survival. The grafting survival rates of the unskilled and skilled workers were 97.15% and 97.67%, with a difference of 0.52 percentage points. Manual grafting can accurately complete the grafting work of rootstocks and scions and there is rarely incomplete cutting and joining, so the survival rate of manual grafting is slightly higher than that of machines. However, with the improvement in the cutting accuracy and intelligence of the grafting machine, this grafting survival rate will be improved.

There were no significant differences in the rootstock regrowth rates between the hand-grafting treatments (Figure 8C). The rootstock regrowth rates were 17.95–18.92%. However, the rootstock regrowth rate of the machine grafting reached 72.69%, significantly higher than that of the hand grafting ("unskilled" and "skilled" grafting treatments). An incomplete removal of rootstock growth points is the main reason for the high rate of the machine grafting rootstock regrowth rate. During production, there were morphological differences in the rootstock seedlings. The robot cannot adjust the cutting parameters based on the rootstock seedlings' morphological features and the position of the rootstock seedling supply. Compared to the machine grafting, the regeneration rate of the artificially grafted rootstock was relatively well controlled, because artificial cutting is based on the base of the growth point and its cutting success rate is high.

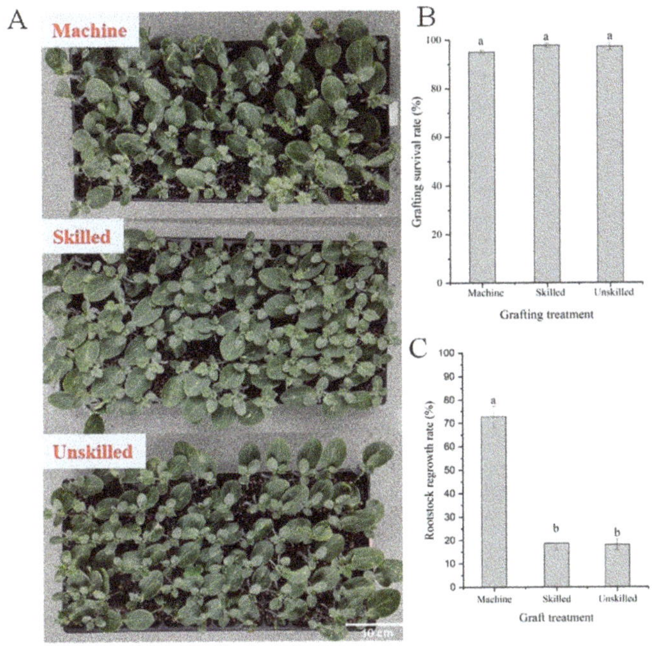

Figure 8. Growth of grafted watermelon seedlings after healing (**A**), grafting survival rate (**B**), and rootstock regrowth rate (**C**) were measured at 10 days after grafting. Machine, unskilled, and skilled represent machine grafting with 2TJGQ-800, hand grafting by unskilled grafter, and hand grafting by skilled grafter, respectively. The different small letters in picture (**B**,**C**) indicate significant difference at $p = 0.05$ levels.

It can be seen that the cutting angle consistency of the machine grafting was good and that the accuracy of the artificial seedling supply and cutting parameters were the keys to improving the grafting success rate and grafted seedlings' quality. Hand grafting had advantages in its grafting survival rates and rootstock regrowth rates, but it had no advantage for large scales of production.

4. Discussion

Manual grafting is a labor-intensive operation that is highly susceptible to human error. The development of a grafting machine has been considered as an effective alternative to manual grafting. Since the 1980s, researchers have conducted a significant amount of research on vegetable grafting machines [1]. Currently, there are some semi-automatic grafting machines that are relatively mature. However, compared to hand grafting, there were few advantages for grafting efficiency using machine grafting; as such, seedling companies in China cannot accept these machines at present. With the increasing demand for grafted seedlings, the demand for a grafting machine is becoming increasingly urgent. In order to ensure that a grafting machine can be implemented in China as soon as possible, this study was conducted to examine the grafting efficiency and grafting seedling quality of a grafting machine compared to those of skilled grafters and unskilled grafters, based on the grafting machine 2TJGQ-800.

4.1. Comparison of the Grafting Speed

One-cotyledon grafting is the simplest and most frequently used method when using a grafting machine [14,15]. The grafting machine in this experiment, 2TJGQ-800, also uses this method. During machine grafting, the grafting process includes feeding seedlings, cutting, joining the cut area of the rootstock and scion together, clipping, and transplantation. A simple grafting machine can produce 600 grafts per hour with two operators [38]. Due to improvements in grafting machines, the grafting machine in this study can produce 774 grafts per hour. The grafting speed was 1.65–2.55 fold higher than that of the hand grafting. Through training, workers can improve their grafting speed, but the grafting speed remains slower than machine grafting. Consequently, a grafting machine has an advantage over hand grafting in terms of speed.

In addition, manual grafting is labor intensive. During nursery production, grafters must work continuously for 8–10 h every day. With an increase in working time, the grafter will begin to feel tired and inattentive, and the grafting success rate and grafting speed will decrease. However, machine grafting is labor saving. During the machine grafting process, it only needs workers to supply the rootstock or scion seedlings to the machine. Manual grafting requires more work for the cutting, joining, and clipping of seedlings. Thus, the grafting success rate and grafting speed of a machine are stable and scarcely affected by working hours.

4.2. Comparison of the Grafted Quality

The grafting machine contains a clamping and handling mechanism, cutting mechanism, clip-feeding mechanism, clip-sequencing and supply device, conveyor belt, and control system. When grafting, the workers place the rootstock and scion seedling in the right position, and the cutting and clip mechanisms are operated based on the operating parameters. These devices ensure that the cutting position and angle are the same. However, as grafters generally cut seedlings based on experience, the cutting positions and angles of the seedlings were random. Thus, the accuracy and uniformity of the machine grafting were higher than those achieved via the hand grafting.

The results showed that the scion hypocotyl height, scion incision length, and rootstock incision length grafted by the machine were 13.5 mm, 2.8 mm, and 5.2 mm, respectively. The coefficients of variation for these parameters were 0.049, 0.089, and 0.068, respectively. Compared to the grafting by the skilled and unskilled grafters, the coefficients of variation for the parameters cut by the machine were the smallest. This highlights the standardized characteristics of a grafting machine. However, the scion incision length cutting completed by the machine was 2.8 mm. This is smaller and not optimized [39]. Thus, it is necessary to further adjust the cutting parameters. The setting of cutting operation parameters is crucial to improving the quality of machine grafting, and it can improve the adaptability of a grafting machine to seedlings.

4.3. Comparison of the Grafting Success Rate and the Grafting Survival Rate

In this study, compared to hand grafting, the grafting success rate of the machine grafting decreased from 100% to 90.07%. The grafting success rate was high because the working time of the grafter was under 2 h. The grafters were concentrated and able to correct any mistakes immediately. However, with an increase in working time, the grafter will begin to feel tired and inattentive, and the grafting success rate will decrease.

The machine grafting success rate was lower because the machine operation parameters were set before the grafting and could not be adjusted by changes in the seedling morphological characteristics. The reason for the machine grafting failure was statistics. The failure rate caused by incomplete rootstock cutting was 54.77%. This incomplete rootstock cutting may have been caused by the cutting parameters of the cutting mechanism being inaccurate, an unstable positioning of the seedlings supplied by the worker, or the cutter operation trajectory failing to completely meet the requirements for cutting the rootstock. Though optimizing the machine operating parameters, the machine grafting success rate would increase. Furthermore, the rootstock regrowth rate would decrease.

The grafting survival rate showed no significant differences between the machine grafting and hand grafting. The grafting survival rates for the "machine", "unskilled", and "skilled" grafting treatments were 94.91%, 97.15%, and 97.67%, respectively. The rootstock regrowth rate of the machine grafting reached 72.69%, which was significantly higher than that of the hand grafting (17.95–18.92%). This result is associated with the accuracy of the cutting device and also conforms to the actual situation. A higher rootstock regrowth rate needs extra labor input to remove it. Consequently, it is necessary to solve the problem of rootstock cutting precision.

4.4. Performance Comparison of Similar Grafting Machines

Machines similar to the one described in this study include GR803-U, launched by ISEKI Co., Ltd. in Japan [22] (700 Umaki-cho, Matsuyama-shi, Ehime-ken, 799-2692 Japan), and AFGR-800CS, launched by Helper Robotech Co., Ltd. in Korea (93, Hitech-ro, Jinrye-myeon, Gimhae-si, Gyeongsangnam-do, Korea) [23,24]. Both need two people to supply the seedlings; each job cycle can produce one individual grafting plant. In comparing the performance parameters, the grafting speeds of 2TJGQ-800, GR803-U (Figure 9A), and AFGR-800CS (Figure 9B) were 774 plants·h^{-1}, 800 plants·h^{-1}, and 800 plants·h^{-1}, respectively. The grafting survival rates were 94.07%, 95%, and 95%, respectively. The machine grafting speed depends on the proficiency of the operator. With an improvement in the operator's skill and experience, the grafting speed and efficiency can also be improved. Moreover, the cost of the 2TJGQ-800 grafting machine was CNY 150,000 ¥, amounting to approximately one quarter of that which the same kind of grafting machine costs. Therefore, the 2TJGQ-800 grafting machine has the advantage of market application in terms of both its performance and price.

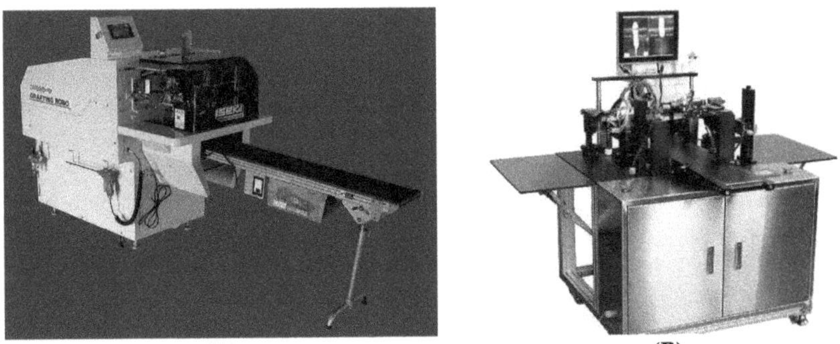

(A) (B)

Figure 9. Similar grafting equipment. (**A**) GR803-U, and (**B**) AFGR-800CS.

4.5. Estimation of Operation Cost

The intensive labor inputs of grafting by hand are the main reason for the high cost of grafted seedlings [15]. These labor costs are dependent on grafting speed, success, and the hourly wage of trained employees. The wage used for grafters was CNY 0.08 ¥·plant^{-1} and the wage used for machine operators was CNY 15 ¥·h^{-1}. For simplification purposes, a consistent production speed was assumed in the cost simulation. It is estimated that an experienced worker can graft 220 plants when working for 12 h every day. The grafting machine can produce 774 plants·h^{-1} when working for 24 h every day. Thirteen grafters working for 30 days can produce 1 million grafted seedlings. The cost of these grafters is CNY 82,368 ¥. Two machines working for 27 days can produce 1 million grafted seedlings. Six workers in three groups were required to take turns using the machine. The cost of machine operators is CNY 38,880 ¥. Comparatively, the cost of manual grafting is 1.12 times higher than that of machine grafting. In other words, producing 1 million grafted seedlings with a machine can save CNY 43,488 ¥ in labor costs. The cost of a grafting machine is approximately CNY 150,000 ¥. Compared to hand grafting, the labor costs saved in producing 3.45 million grafted seedlings was approximately equal to the cost of purchasing the grafting machine.

5. Conclusions

In conclusion, grafting machines have advantages in their grafting speed and uniformity. There were no significant differences in the grafting survival rates between the machine grafting and hand grafting. However, the lower rate of successful grafting and higher rate of rootstock regrowth with the machine grafting, compared to that of the hand grafting, are the main causes for the reduced adoption of grafting machines. In future, more research should be conducted to achieve higher accuracy and efficiency in machine grafting.

Firstly, the accuracy and adaptability of grafting machines will be upgraded. The grafting object is seedlings, which are variable. By improving the adaptability of the machine to the seedlings and the accuracy of the cutting device, the advantages of machine grafting will become apparent. It is necessary to add machine vision using machine vision technology and machine learning algorithms.

Secondly, it is necessary to develop seedling agricultural standards that match the grafting machine. This is beneficial for the standardized operation of machine grafting and realizing the efficient operation of a grafting machine.

Thirdly, a high-efficiency grafting machine is required. Compared to manual grafting using the hole-insertion method, which is mostly used in China, the production efficiency of grafting machines has little advantage. The innovative research and development of plant synchronous intelligent grafting mechanisms and intelligent control technology for the grafting parameters needed to break through the seedling morphological differentiation are the two difficult problems in future research.

Fourth, unmanned or flexible manpower grafted seedling production systems are under consideration. Grafting is a systematic project including a lot of links, such as sowing, grafting, healing, packing, handling, and transporting, etc. When grafting using workers, these seedling links are performed in the seedbed. In other words, grafting with a human needs less extra work. When using grafting robots, extra workers are needed to repeatedly handle the seedlings from the seedbed to the grafting robot. This process is very arduous and requires many workers to participate. Intelligent machinery and equipment will be needed to build an unmanned or flexible manpower grafted seedling production system.

Author Contributions: Conceptualization, H.L., K.J. and X.S.; methodology, H.L.; validation, J.Z., M.G. and D.W.; formal analysis, K.L.; investigation, H.L., J.Z., D.W., M.G. and K.L.; data curation, H.L., M.Z. and Y.S.; writing—original draft preparation, H.L. and K.J.; writing—review and editing, Q.Z., K.J. and X.S.; supervision, X.S.; funding acquisition, K.J. All authors have read and agreed to the published version of the manuscript.

Funding: This research was supported by the BAAFS Innovation Ability Project (KJCX20220403), the National Nature Science Foundation of China (Grant No. 32171898), the Knowledge Innovation Program of Wuhan-Shuguang Project (2022020801020418), the Key R&D projects in Hubei Province (2021BBA239), the Xiangyang Research and Development Project (2022ABA006997) and the China National Agricultural Research System (CARS-25-07).

Data Availability Statement: The data presented in this study are available upon demand from the correspondence author at (jiangk@nercita.org.cn).

Conflicts of Interest: The authors declare no conflict of interest.

References

1. Guan, W.; Zhao, X.; Hassell, R.; Thies, J. Defense mechanisms involved in disease resistance of grafted vegetables. *Horticulture* **2012**, *47*, 164–170. [CrossRef]
2. Maurya, D.; Pandey, A.K.; Kumar, V.; Dubey, S.; Prakash, V. Grafting techniques in vegetable crops: A review. *Int. J. Chem. Stud.* **2019**, *7*, 1664–1672.
3. Ulas, F.; Aydın, A.; Ulas, A.; Yetisir, H. Grafting for sustainable growth performance of melon (*Cucumis melo*) under salt stressed hydroponic condition. *Eur. J. Sustain. Dev.* **2019**, *8*, 201–210. [CrossRef]
4. Schwarz, D.; Rouphael, Y.; Colla, G.; Venema, J.H. Grafting as a tool to improve tolerance of vegetables to abiotic stresses. *Sci. Hortic.* **2010**, *127*, 162–171. [CrossRef]

5. Nawaz, M.A.; Imtiaz, M.; Kong, Q.; Fei, C.; Ahmed, W.; Huang, Y.; Bie, Z. Grafting: A technique to modify ion accumulation in horticultural crops. *Front. Plant Sci.* **2016**, *7*, 1457. [CrossRef]
6. Ertle, J.M.; Kubota, C. Watermelon Seedling Quality, Growth, and Development as Affected by Grafting and Chilling Exposure During Simulated Transportation. *HortScience* **2022**, *8*, 57. [CrossRef]
7. Ulas, A. Crossbreeding Rootstocks Improve Nitrogen Efficiency of Grafted Watermelon by Inducing Leaf Physiological and Root Morphological Responses. *Horticulturae* **2022**, *8*, 879. [CrossRef]
8. Sewoong, A.; Jong, H.B.; Ho, C.K.; Yurina, K. Production of Grafted Vegetable Seedlings in the Republic of Korea: Achievements, Challenges and Perspectives. *Hortic. Sci. Technol.* **2021**, *39*, 547–559.
9. Zhang, Y.; Yang, X.; Zhang, L.; Wei, J.; Liu, S.; Xu, L.; Wang, L. The cost analysis of vegetable seedling production in Shandong province. *China Veg.* **2014**, *9*, 49–51.
10. Singh, H.; Kumar, P.; Chaudhari, S.; Edelstein, M. Tomato Grafting: A Global Perspective. *HortScience* **2017**, *52*, 1328–1336. [CrossRef]
11. Djidonou, D.; Gao, Z.; Zhao, X. Economic analysis of grafted tomato production in sandy soils in northern Florida. *Horttechnology* **2013**, *23*, 613–621. [CrossRef]
12. Rysin, O.; Rivard, C.; Louws, F.J. Is vegetable grafting economically viable in the United States: Evidence from four different tomato production systems. *Acta Hortic.* **2015**, *1086*, 79–86. [CrossRef]
13. Pardo-Alonso, J.-L.; Carreño-Ortega, Á.; Martínez-Gaitán, C.-C.; Golasi, I.; Gómez Galán, M. Conventional Industrial Robotics Applied to the Process of Tomato Grafting Using the Splicing Technique. *Agronomy* **2019**, *9*, 880. [CrossRef]
14. Guan, W.; Zhao, X. Effects of Grafting Methods and Root Excision on Growth Characteristics of Grafted Muskmelon Plants. *HortTechnology* **2015**, *25*, 706–713. [CrossRef]
15. Devi, P.; Lukas, S.; Miles, C. Advances in Watermelon Grafting to Increase Efficiency and Automation. *Horticulturae* **2020**, *6*, 88. [CrossRef]
16. Guan, W.; Zhao, X. Techniques for Melon Grafting. *Acta Hortic.* **2016**, *1140*, 335–336. [CrossRef]
17. Liu, C.; Lin, W.; Feng, C.; Wu, X.; Fu, X.; Xiong, M.; Bie, Z.; Huang, Y. A New Grafting Method for Watermelon to Inhibit Rootstock Regrowth and Enhance Scion Growth. *Agriculture* **2021**, *11*, 812. [CrossRef]
18. Daley, S.L.; Hassell, R.L. Fatty alcohol application to control meristematic regrowth in bottle gourd and interspecific hybrid squash rootstocks used for grafting watermelon. *HortScience* **2014**, *49*, 60–264. [CrossRef]
19. Kubota, C.; McClure, M.A.; Kokalis-Burelle, N.; Bausher, M.G.; Rosskopf, E.N. Vegetable grafting: History, use and current technology status in North America. *HortScience* **2008**, *43*, 1664–1669. [CrossRef]
20. Bie, Z.; Nawaz, M.A.; Huang, Y.; Lee, J.M.; Colla, G. Introduction of vegetable grafting. In *Vegetable Grafting: Principles and Practices*; CABI Publishing: Wallingford, UK, 2017; pp. 1–21.
21. Zhang, K.; Chu, J.; Zhang, T. Development Status and Analysis of Automatic Grafting Technology for Vegetables. *Agric. Mach. J.* **2017**, *48*, 1–13.
22. Ohkoshi, T.; Kobayashi, K. Development of automatic seedling feeding device for cucurbits grafting robot (Part 1)—Evaluation of stock feeder. *J. Jpn. Soc. Agric. Mach. Food Eng.* **2013**, *75*, 100–107.
23. Kim, H.M.; Hwang, S.J. Comparison of pepper grafting efficiency by grafting robot. *Prot. Hortic. Plant Fact.* **2015**, *24*, 57–62. [CrossRef]
24. Kang, D.H.; Lee, S.Y.; Kim, J.K.; Park, M.J.; Son, J.K.; Yun, S.W. Development of an automatic grafting robot for fruit vegetables using image recognition. *Prot. Hortic. Plant Fact.* **2019**, *28*, 322–327. [CrossRef]
25. Atlantic Group. Available online: http://tech.atlanticgroup.it (accessed on 5 September 2022).
26. Lewis, M.; Kubota, C.; Tronstad, R.; Son, Y.J. Scenario-based cost analysis for vegetable grafting nurseries of different technologies and sizes. *HortScience* **2014**, *49*, 917–930. [CrossRef]
27. Yang, L.; Liu, C.; Zhang, T. Design and experiment of vegetable grafting machine with double manipulators. *Trans. Chin. Soc. Agric. Mach.* **2009**, *40*, 175–181.
28. Jiang, K. Study on Mechanism and Experimental Device of Splice Mechanical Grafting of Cucurbit. Ph.D. Thesis, Northeast Agricultural University, Harbin, China, 2019.
29. Fu, X.; Shi, J.; Huang, Y.; Zhu, E.; Bie, Z.; Lin, W. Design and experiment of full-tray grafting device for grafted melon seedling production. *Agriculture* **2022**, *12*, 861. [CrossRef]
30. Jiang, K.; Zheng, W.G.; Zhang, Q.; Guo, R.; Feng, Q. Development and experiment of vegetable grafting robot. *Trans. Chin. Soc. Agric. Eng.* **2012**, *28*, 8–14.
31. Chen, L.; Jiang, K.; Zhang, Q.; Guo, W.; Zheng, W. Design and experiment on scion cutting mechanism of grafting robot for cucurbit. *Int. J. Agric. Biol. Eng.* **2020**, *13*, 99–106. [CrossRef]
32. Jiang, K.; Zhang, Q.; Chen, L.; Guo, W.; Zheng, W. Design and optimization on rootstock cutting mechanism of grafting robot for cucurbit. *Int. J. Agric. Biol. Eng.* **2020**, *13*, 117–124. [CrossRef]
33. Xu, P.; Zhang, T.; Chen, L.; Huang, W.; Jiang, K. Study on the Method of Matched Splice Grafting for Melon Seedlings Based on Visual Image. *Agriculture* **2022**, *12*, 929. [CrossRef]
34. Jiang, K.; Guo, W.; Chen, L.; Huang, W.; Ge, Y.; Wei, X. Design and Experiment of Automatic Clip-Feeding Mechanism for Vegetable-Grafting Robot. *Agriculture* **2022**, *12*, 346. [CrossRef]

35. Yan, G.; Feng, M.; Lin, W.; Huang, Y.; Tong, R.; Cheng, Y. Review and Prospect for Vegetable Grafting Robot and Relevant Key Technologies. *Agriculture* **2022**, *12*, 1578. [CrossRef]
36. Davis, A.R.; Perkins-Veazie, P.; Sakata, Y.; López-Galarza, S.; Maroto, J.V.; Lee, S.-G.; Huh, Y.-C.; Sun, Z.; Miguel, Z.; Stephen, R.; et al. Cucurbit Grafting. *Crit. Rev. Plant Sci.* **2008**, *27*, 50–74. [CrossRef]
37. Traka-Mavrona, E.; Koutsika-Sotiriou, M.; Pritsa, T. Response of squash (*Cucurbita* spp.) as rootstock for melon (*Cucumis melo* L.). *Sci. Hortic.* **2000**, *83*, 353–362. [CrossRef]
38. Hassell, R.L.; Memmott, F.; Liere, D.G. Grafting methods for watermelon production. *HortScience* **2008**, *43*, 1677–1679. [CrossRef]
39. Liang, H.; Jiang, K.; Shi, X.; Zhu, J.; Liu, J.; Wang, D.; Ge, M.; Zhou, M.; Shan, F. An Experimental Study on the Effect of Cutting Angle on the Growth of Grafted Watermelon Seedlings Using the One-Cotyledon Grafting Method. *Agronomy* **2023**, *13*, 250. [CrossRef]

horticulturae

MDPI

Article

IMVTS: A Detection Model for Multi-Varieties of Famous Tea Sprouts Based on Deep Learning

Runmao Zhao [1,2,3], Cong Liao [1], Taojie Yu [1], Jianneng Chen [1,2,3], Yatao Li [2,3,4], Guichao Lin [5], Xiaolong Huan [1,2,3] and Zhiming Wang [6,*]

1 School of Mechanical Engineering, Zhejiang Sci-Tech University, Hangzhou 310018, China; rmzhao@zstu.edu.cn (R.Z.); 13517960753@163.com (C.L.); 202130605347@mails.zstu.edu.cn (T.Y.); jiannengchen@zstu.edu.cn (J.C.); huanxl@zstu.edu.cn (X.H.)
2 Key Laboratory of Transplanting Equipment and Technology of Zhejiang Province, Hangzhou 310018, China; ytli@zstu.edu.cn
3 Key Laboratory of Agricultural Equipment for Hilly and Mountainous Areas in Southeastern China (Co-Construction by Ministry and Province), Ministry of Agriculture and Rural Affairs, Hangzhou 310018, China
4 School of Textile Engineering, Zhejiang Sci-Tech University, Hangzhou 310018, China
5 College of Mechanical and Electrical Engineering, Zhongkai University of Agriculture and Engineering, Guangzhou 510225, China; guichaolin@zhku.edu.cn
6 Jinhua Polytechnic Zhejiang Key Laboratory of Crop Harvesting Equipment Technology, Jinhua 321017, China
* Correspondence: jhcwzm@163.com

Abstract: The recognition of fresh tea leaf sprouts is one of the difficulties in the realization of the automated picking of fresh tea leaves. At present, the research on the detection of fresh tea leaf sprouts is based on a single variety of tea leaves for a specific period or specific place, which has no advantage for the spread, promotion, and application of the methods. To address this problem, an identification of multiple varieties of tea sprouts (IMVTS) model was proposed. First, images of three different varieties of tea (ZhongCha108 (ZC108), ZhongHuangYiHao (ZH), ZiJuan (ZJ)) were obtained, and the multiple varieties of tea (MVT) dataset for training and validating models was created. In addition, the detection effects of adding a convolutional block attention module (CBAM) or efficient channel attention (ECA) module to YOLO v7 were compared. In the detection of the MVT dataset, YOLO v7+ECA and YOLO v7+CBAM showed a higher mean average precision (mAP) than YOLO v7, with 98.82% and 98.80%, respectively. Notably, the IMVTS model had the highest AP for ZC108, ZH, and ZJ compared with the two other models, with 99.87%, 96.97%, and 99.64%, respectively. Therefore, the IMVTS model was proposed on the basic framework of the ECA and YOLO v7. To further illustrate the superiority of the model, this study also conducted a comparison test between the IMVTS model and the mainstream target detection models (YOLO v3, YOLO v5, FASTER-RCNN, and SSD) and the IMVTS model on the VOC dataset, and it is clear from the test results that the mAP of the IMVTS model is ahead of the remaining models. Concisely, the detection accuracy of the IMVTS model can meet the engineering requirements for the automatic harvesting of autumn fresh famous tea leaves, which provides a basis for the future design of detection networks for other varieties of autumn tea sprouts.

Keywords: fresh tea leaf sprouts; multiple varieties detecting; IMVTS; YOLO v7; ECA

Citation: Zhao, R.; Liao, C.; Yu, T.; Chen, J.; Li, Y.; Lin, G.; Huan, X.; Wang, Z. IMVTS: A Detection Model for Multi-Varieties of Famous Tea Sprouts Based on Deep Learning. *Horticulturae* **2023**, *9*, 819. https://doi.org/10.3390/horticulturae9070819

Academic Editors: Xinchao Wang and Jianyun Ruan

Received: 30 April 2023
Revised: 11 July 2023
Accepted: 13 July 2023
Published: 17 July 2023

1. Introduction

China is known as the home of tea. The country's planting, output, consumption, and export volume of tea are first in the world. The extensive demand for fresh tea leaves requires effective support [1]. Famous tea is a representative of high-quality tea in China. Its raw materials are one sprout and one leaf of a tea tree. Its harvesting is still mainly manual and labor intensive, and there is a shortage of tea collectors [2]. To achieve the intelligent picking of fresh tea leaf sprouts, researchers use machine vision technology to detect fresh tea leaves. Lei Zhang et al. [3] extracted the R, G, and B components of images of collected fresh tea leaves and then processed the adaptive threshold of the B components. A new component gray diagram was obtained in combination with the G components, and the thresholds were enhanced by segmental transformation, thereby improving the contrast between the tender leaves and the background, and then, the improved watershed function was used to obtain a good division effect. Liang Zhang et al. [4] used Bayesian inference to judge the principles and Bayes methods to build an identification model of the fresh leaf collection status in order to detect purple rose tea tree sprouts in April in real time. Although the above machine learning technology can accurately identify fresh tea leaf sprouts, the identified fresh leaf image background environment is relatively single, and the segmentation accuracy is greatly affected by the characteristics of the fresh tea leaf sprouts. It is often difficult to overcome this shortcoming.

With the rapid development of artificial intelligence technology, the use of deep-learning [5] target detection algorithms, such as Fast R-CNN [6], Faster R-CNN [7], You Only Look Once (YOLO) [8], and Single Shot MultiBox Detector (SSD) [9], etc., are increasingly applied in the agricultural field. In terms of fresh tea leaf sprout recognition, Yatao Li et al. [10] used the YOLO v3 network to detect and identify fresh tea leaf sprouts on an RGB-D image collected with an RGB-D camera and estimated the three-dimensional coordinates of the fresh tea leaf picking position in the corresponding depth chart of the enclosure. It reached 83.18%. Wenkai Xu et al. [11] proposed a detection and classification method for the two-level integration network of the variability domain. This method combined YOLO v3's fast detection capacity and the high-precision classification ability of DenseNet201 to achieve the accurate detection of fresh tea leaves. Yu-Ting Chen et al. [12] detected the OTTL area in a tea tree image through the Faster R-CNN model and then identified the picking point in the OTTL area with the FCN model to determine the three-dimensional coordinates of the picking point. The YOLO model is currently a widely used detection network. It adopts a one-stage algorithm. The network operates quickly, the memory occupation is small, and it has a fast detection speed [13]. Based on inheriting the advantages of the original YOLO model, YOLO v7 [14] has a higher detection accuracy and faster reasoning speed compared to the previous YOLO series models because of its more complex network structure and training techniques. However, because the YOLO v7 model uses only one main convolutional neural network to predict the category and location of different targets, it also improves the speed and makes the recognition accuracy lower. This may lead to more missed inspections and misunderstandings when using YOLO v7 to detect fresh tea leaf images. At the same time, the applicability of the models proposed in the existing literature to fresh sprouts of different varieties of tea is unknown.

Consequently, this study used the YOLO v7 model as the basic framework to research the identification of multiple varieties of fresh tea leaf sprouts and explored the impact of integrating CBAM and ECA mechanisms on YOLO v7 detection accuracy and generalization effects. According to the results of the research, an IMVTS model was proposed to solve the problem of high-precision detection of multiple tea species. In addition, to demonstrate the superiority of the proposed IMVTS model in detection, comparison tests of the IMVTS model with mainstream target detection models (YOLO v3, YOLO v5, FASTER-RCNN, and SSD) and detection tests of the IMVTS model on VOC datasets were designed.

2. Materials and Methods

2.1. Experimental Data Collection and Preparation

2.1.1. Fresh Tea Leaf Image Collection and MVT Dataset Production

At the experimental base of the Tea Research Institute of the Chinese Academy of Agricultural Sciences (Figure 1), a total of 277 fresh leaf images of 3 tea varieties were collected, of which 90 images were ZhongCha108 (ZC108), 150 were ZhongHuangYiHao (ZH), and 37 were ZiJuan (ZJ). The sampling date was 21 September 2022. The collected fresh tea leaf images were made into a dataset of fresh leaves of multiple tea varieties that could be used to train and verify the network model and were named the MVT dataset. An iQOO Neo3 mobile phone camera was used. The shooting angle was 30~60°, and the shooting distance was 30~50 cm. The images were stored in a JPG format, and the MVT dataset images are shown in Figure 2. The open-source labeling tool Labellmg was used to artificially label the sprout position on the fresh tea leaf images, and the labeled result was saved in the PASCAL VOC format.

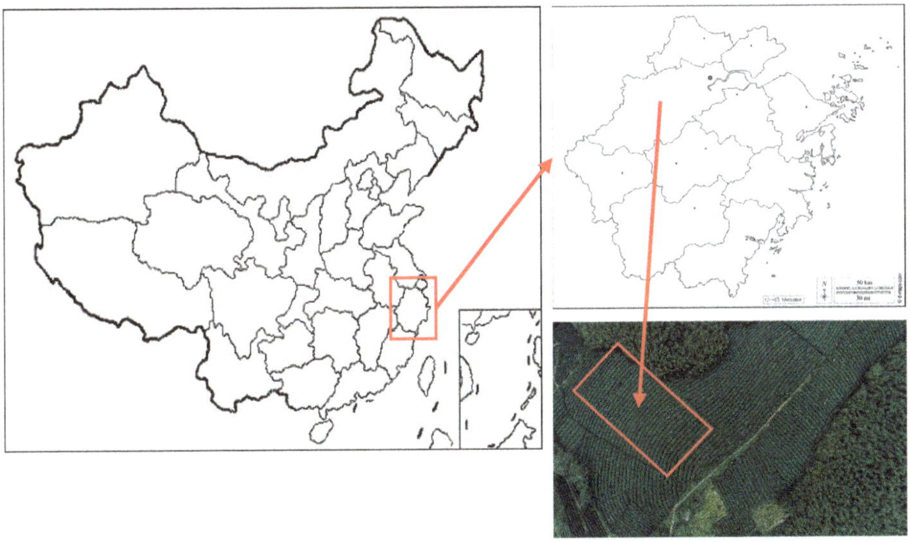

Figure 1. Experimental Base of the Tea Research Institute of the Chinese Academy of Agricultural Sciences.

(a) (b) (c)

Figure 2. MVT dataset images; (**a**) ZC108; (**b**) ZH; (**c**) ZJ.

The images of ZC108, ZH, and ZJ as the research objects are shown in Figure 2a–c. The leaves of ZC108 are long oval in shape, green in color, slightly rumbled on the leaf surface, and flat on the leaf body, and the sprouts are yellowish-green with less velvet. The leaves of ZH are oval in shape, yellowish green in color, slightly elevated on the leaf surface, slightly inflexed on the leaf body, and the sprouts are slender. ZJ is a large-leaved, medium-sprout species. The leaf blade is upwardly slanting, willow-shaped, with an acuminate tip, and purple in color. Its petiole is purple-red in color. These images can also provide data support for future studies of other tea recognition models.

2.1.2. MVT Dataset Division

The 277 obtained images and corresponding labels were divided into training sets and validation sets according to the proportion of training sets: validation sets (=9:1) so that the model training and validating in the later stage could be performed. To enrich the diversity of samples and improve the robustness of the IMVTS model, the dataset was expanded to 15,000 pieces of the original dataset, with increased or decreased brightness, increased and decreased color saturation, and horizontal flip. The application process is shown in Figure 3. After the enhancement, the number of sprouts of the fresh leaves of the three tea varieties reached 59,337, of which the number of sprouts of ZC108 was 7429, the number of sprouts of ZH was 18,902, and the number of sprouts of ZJ was 33,006.

Figure 3. *Cont.*

Figure 3. MVT dataset enhancement examples. (**a**,**e**,**i**) Image of primitive fresh tea leaves; (**b**,**f**,**j**) Image of fresh tea leaves with increased or decreased brightness; (**c**,**g**,**k**) Image of fresh tea leaves with increased or decreased color saturation; (**d**,**h**,**l**) Image of fresh tea leaves with the horizontal flip.

2.2. *Establishment and Training of the IMVTS Model*

2.2.1. Experimental Platform and Environmental Configuration

This study used a workstation equipped with the Ubuntu 18.04 operating system for model training. The CPU was Intel i7 12700F (Shenzhen, China) and the GPU was 3080Ti. The development environment was Pytorch 1.11. The development language was Python, and vs. code software was used for debugging.

2.2.2. Establishment of the IMVTS Model

This study used the YOLO v7 model as the basic framework for the recognition of multiple varieties of fresh tea leaf sprouts; the related research has shown that attention mechanisms [15] such as SENet [16], CBAM [17], and ECA [18] improved the network performance. Therefore, to improve the detection accuracy of YOLO v7's original model on the small target of fresh tea leaves, the effects of adding the CBAM or ECA attention module to YOLO v7 on the detection effect were compared; the final IMVTS network model structure obtained is shown in Figure 4. The IMVTS model mainly includes four parts: input, backbone, head, and output. First, the collected tea leaf images undergo labellmg and random data augmentation, then the tea leaf images are resized into 640 × 640 size RGB images and input to the backbone network The backbone network features the processing of fresh tea leaves with features. Second, the feature information extracted in the backbone network is combined through the characteristics of the head to obtain the characteristics of large, medium, and small sizes. Finally, the characteristics of the characteristic fusion are sent to the REP and Conv module of the head section for detection, and the final result is the output. The workflow of the algorithm is shown in Figure 5. The main network part of the YOLO v7 model is mainly composed of convolution, the extended-ELAN (E-ELAN) module, the MPConv module, and the SPPCSPC module. Among these, the E-ELAN module is based on the original ELAN, changing the calculation block while maintaining the original ELAN's transition layer structure. The network can learn more features by controlling the shortest and longest gradient path. It can resolve the relatively complicated and variable problems that can be solved for the background environment and targets appearing in the MVT dataset to extract more efficient features. The SPPCSPC module adds parallel MaxPooling operations to a convolution series to obtain different feelings so that the algorithm can adapt to fresh tea leaf images of different resolutions and can make the network speed faster; the accuracy of detecting fresh tea leaf sprouts will also be improved. The main role of the MPConv module is to sample. The experience of the current feature layer is expanded through MaxPooling and then integrated with the normal convolutional fresh tea leaf sprout feature information. It can improve the generalization of the network, suitable for the recognition scene of multiple varieties of tea leaf sprouts [19]. When introducing the attention module, the location information and detail information of the feature information are extracted from the backbone, and there is less semantic information [20]. Feature information is easily lost after processing for small targets such as fresh tea leaf sprouts. To improve the detection accuracy of the YOLO v7 original model for fresh tea leaves, an attention mechanism was added between the backbone and the head. The structure of the attention module is shown in Figure 6. CBAM combines the channel attention mechanism with the spatial attention mechanism. The CBAM module conducts global average pooling and the largest global pooling for the input feature layer, then learns the weight information of each channel through a shared full connection layer and the Sigmoid function. Then, it learns the weight information of each point in the space through a convolutional nucleus of 3 × 3 and the expansion coefficient of 2, and the Sigmoid function. ECA is a channel attention mechanism. The ECA module changes the two full connection layers to one-dimensional convolution. After the Sigmoid function is passed, the channel weight information is obtained. This module avoids the dimension of information and has good cross-channel information acquisition capabilities.

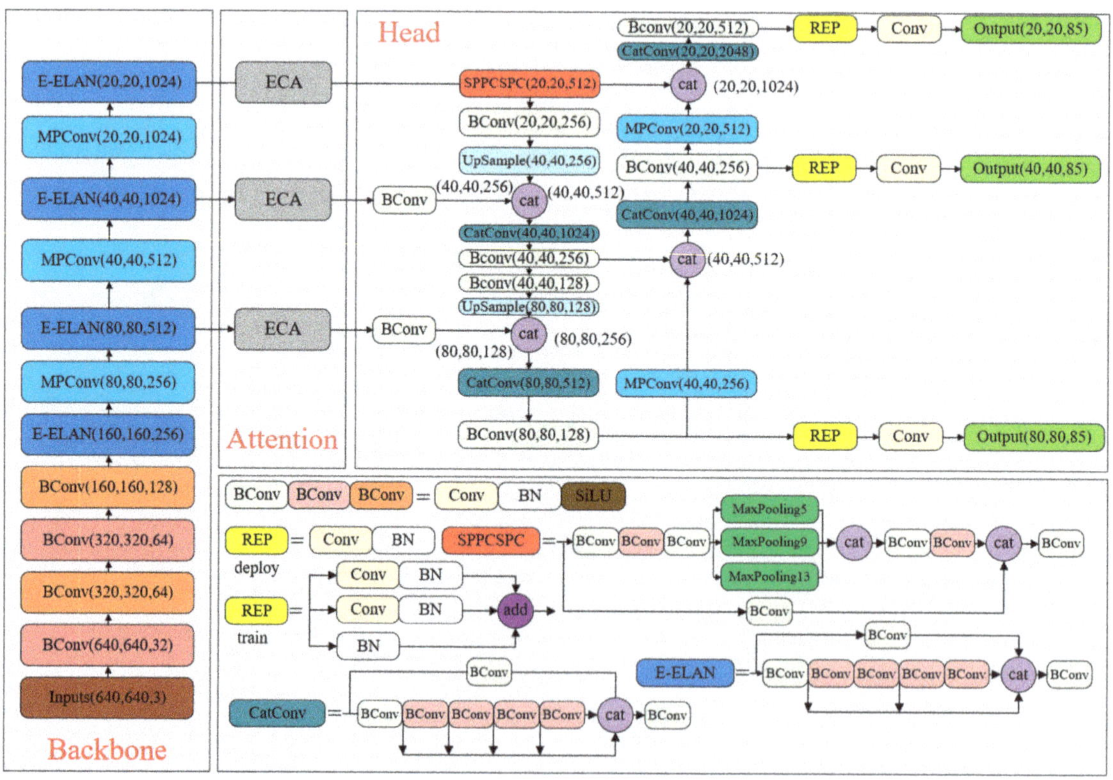

Figure 4. IMVTS model network structure diagram.

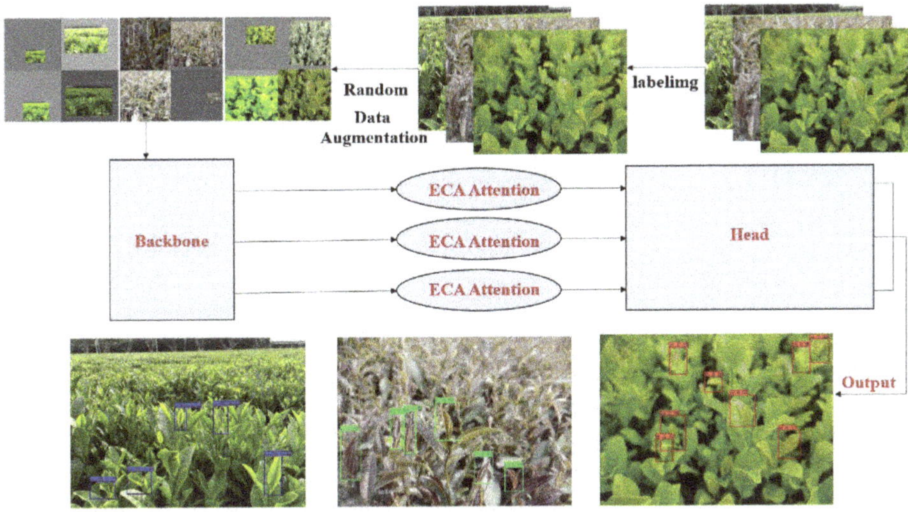

Figure 5. Algorithm workflow diagram.

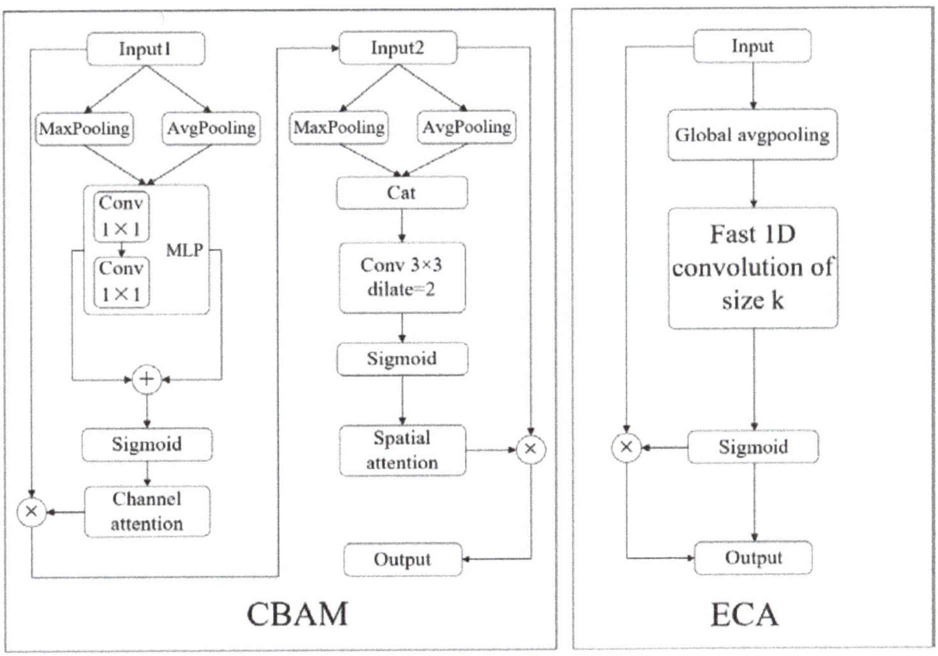

Figure 6. CBAM and ECA module structure diagram.

2.2.3. Training Parameter Settings

When training the original model of YOLO v7 and the IMVTS model, only the thawing training was performed. The thawing training set the total training generation epochs to 500; additionally, the batch size was set to 4, the initial learning rate of the model was set to 0.01, the minimum learning rate was set to 0.0001, and the SGD optimizer was used to optimize the model. The momentum parameters of the SGD optimizer were set to 0.937, and the torque annealing function was used to reduce the learning rate. In the same environment, the training loss curves of the YOLO v7 original model, YOLO v7+CBAM model, and IMVTS model are shown in Figure 7.

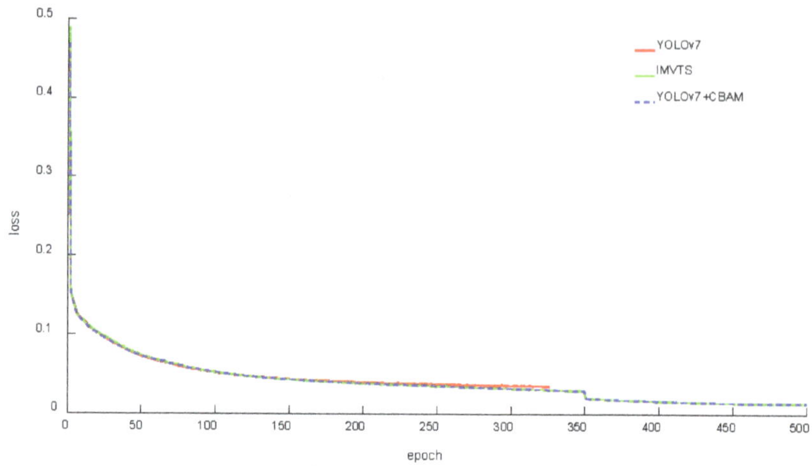

Figure 7. YOLO v7 original model, YOLO v7+CBAM model, and IMVTS model training loss curves.

2.2.4. Model Evaluation Index

To evaluate the detection effect of the IMVTS model, this study used precision (P), recall (R), F1 score, average precision (AP), and mean average precision (mAP) for measurement. Among these, P is the ratio of the number of correctly predicted sprouts to the total number of predicted sprouts in an image, R is the ratio of the number of correctly predicted sprouts to the total number of true sprouts in an image, the F1 score is the equalization of P and R, AP represents the accuracy of each kind of fresh tea leaf sprout recognition, and mAP represents the average value of three kinds of fresh tea leaf sprout recognition. They are calculated as follows (1)–(5):

$$P = \frac{TP}{TP + FP} \times 100\% \quad (1)$$

$$R = \frac{TP}{TP + FN} \times 100\% \quad (2)$$

$$F1 = 2\frac{P \times R}{P + R} \times 100\% \quad (3)$$

$$AP = \int_0^1 P(R)dR \times 100\% \quad (4)$$

$$mAP = \int_1^n \frac{AP_1 + AP_2 + \ldots + AP_n}{n} \quad (5)$$

where TP is the number of correctly predicted sprouts; FP is the number of falsely predicted sprouts; FN is the number of true sprouts that were not predicted as sprouts; and n is the number of fresh tea leaves. The values of the above parameters are obtained by counting and tallying the target detection frames in the validation sets.

3. Results and Discussion

3.1. Model Identification Results and Comparative Analysis

3.1.1. Comparison of YOLO v7 Model, YOLO v7+CBAM Model, and IMVTS Model Recognition Results

The YOLO v7 original model, YOLO v7+CBAM model, and IMVTS model were placed in the same environment for detection. The obtained model evaluation indicators are shown in Table 1. It can be seen from Table 1 that after adding an attention mechanism module to the YOLO v7 network, the P, R, F1 score, and mAP were improved. The precision value of the IMVTS model was 99.76%, 0.10% higher than the YOLO v7+CBAM and 0.24% higher than the YOLO v7 model. The recall value of the IMVTS model was 97.03%, 0.10% higher than the YOLO v7+CBAM and 2.46% higher than the YOLO v7 model. The F1 score of the IMVTS model was 0.98, as was that of the YOLO v7+CBAM model, which was 0.01 higher than the YOLO v7 model. The mAP value of the IMVTS model was 98.82%, which was 0.02% higher than the YOLO v7+CBAM and 1.36% higher than the YOLO v7 model. In summary, the IMVTS and YOLO v7+CBAM models were better than the detection effect of the YOLO v7 model, and the IMVTS model had the best detection effect. This is because the characteristic information of fresh tea leaf sprouts is easy to lose after processing. After the ECA module is introduced, it uses a dynamic convolution nucleus to perform 1*1 convolution; it can be used to extract feature information in different areas, avoiding the number of channel dimensions caused by learning channel attention information so that the model can more effectively learn the top-level information, improving the position prediction of the fresh tea sprouts, and at the same time, can reduce the model parameters, thereby increasing the model performance. For the CBAM module, although it combines the channel attention mechanism with the space attention mechanism, there is not enough space for the information-rich feature space. For small targets, such as fresh tea leaf sprouts, the actual effect is worse than the additional ECA module.

Table 1. Model performance indicator assessment comparison.

Model	P (%)	R (%)	F1-Score	mAP (%)
YOLO v7	99.52	94.57	0.97	97.46
YOLO v7+CBAM	99.66	96.93	0.98	98.80
IMVTS	99.76	97.03	0.98	98.82

3.1.2. Comparison of IMVTS Model with Mainstream Target Detection Models

To verify the advantages of the IMVTS model for multi-species tea fresh leaf shoot detection, the IMVTS model was compared with four mainstream target detection models (YOLO v3, YOLO v5, FASTER-RCNN, and SSD). The comparison experiments of the models were trained and validated under the MVT dataset. The relevant parameters were kept consistent during the experiment, and the detection effect of the model was evaluated by P, R, F1-score, and mAP, and the results are shown in Table 2. The results show that the IMVTS model proposed in this study improves in P, R, F1 values, and mAP relative to the four mainstream target detection models. Among them, IMVTS improved P by 2.38% and 1.79%, R by 7.02% and 6.31%, F1-score by 0.06 and 0.04, and mAP by 4.88% and 4.37%, respectively, relative to YOLO v3 and YOLO v5. IMVTS has higher detection accuracy than YOLO v3 and YOLO v5 because it not only inherits the advantages of the original YOLO model but also achieves higher detection speed with the same computational resources because it uses a faster convolution operation and a smaller model.

Table 2. Results of the model comparison test.

Model	P (%)	R (%)	F1-Score	mAP (%)
IMVTS	99.76	97.03	0.98	98.82
YOLO v3	97.38	90.01	0.92	93.94
YOLO v5	97.97	90.72	0.94	94.45
FASTER-RCNN	96.84	70.19	0.89	89.28
SSD	99.32	58.22	0.72	85.92

IMVTS improved P by 2.92%, R by 26.84%, F1s by 0.07, and mAP by 9.54% relative to FASTER-RCNN. The reason for the improvement in detection accuracy is that FASTER-RCNN uses resnet50 as the backbone, and its feature map only comes from the top-level features, while the MVT dataset has small targets, and there are occlusions and blurring between targets, and only the features of the top-level of the network are used to predict the targets in a single way to extract information, which is not conducive to the localization of target frames. For IMVTS compared to SSD, P improved by 0.44%, R improved by 38.81%, F1score value improved by 0.26, and mAP improved by 12.9%. This is because the SSD model adopts a deep learning network of multi-scale characteristic fusion, but the use of low-level feature information is not enough. At the same time, the resolution of the SSD model is also low, resulting in an insufficient ability to recognize the small target SSD model, which is not conducive to conducting small target detection tasks.

3.1.3. Results and Analysis of Ablation Test

In order to verify the effectiveness of the IMVTS model proposed in this study, different optimization strategies (backbone, model size, and attention) were used in this study. An ablation test was conducted, and the comparative results were shown in Table 3. As can be seen from Table 3, the mAP of the original YOLO v7 model was 97.46%, and after the introduction of the ECA attention mechanism, the mAP increased by 1.36% and reached a peak of 98.82%. Therefore, YOLO v7 and ECA are taken as the basic framework to construct the IMVTS model proposed in this study. In addition, the ablation experiments of four mainstream target detection models (YOLO v3, YOLO v5, FASTER-RCNN, and SSD) were also conducted in this study. It can be seen from Table 3 that the IMVTS model (consisting of YOLO v7 and ECA) has the best detection effect.

Table 3. Results of ablation experiments.

Model	Backbone	Model Size	Attention	mAP (%)
YOLO v7		YOLO v7	ECA	98.82
			CBAM	98.80
			NONE	97.46
		YOLO v7_x	ECA	97.86
			CBAM	97.73
			NONE	96.58
YOLO v3			ECA	94.26
			CBAM	94.20
			NONE	93.94
YOLO v5	Cspdarknet	YOLO v5_l	ECA	96.03
			CBAM	95.66
			NONE	94.45
		YOLO v5_x	ECA	95.83
			CBAM	95.72
			NONE	94.51
	Convnext_tiny	YOLO v5_l	ECA	94.28
			CBAM	94.23
			NONE	94.05
		YOLO v5_x	ECA	94.41
			CBAM	94.30
			NONE	94.12
FASTER-RCNN	Resnet50		ECA	90.24
			CBAM	89.76
			NONE	89.28
	Vgg		ECA	87.77
			CBAM	87.59
			NONE	86.33
SSD	Vgg		ECA	88.06
			CBAM	87.24
			NONE	85.92
	Mobilenetv2		ECA	85.42
			CBAM	84.11
			NONE	83.31

3.1.4. IMVTS Model for VOC Dataset Detection Test

To further illustrate the effectiveness of the IMVTS model in practical engineering applications, this study uses the IMVTS model to train and test the VOC dataset, which includes images of birds, boats, buses, etc. The relevant parameters during the experiments were all kept the same as in the previous training of the MVT dataset, and the detection effect of the model was evaluated by P, R, F1-score, and mAP. Furthermore, a comparison test with YOLO v7 was conducted, and the comparison results are shown in Table 4. From Table 4, it can be seen that the IMVTS model has better detection results compared to the original YOLO v7 model.

Table 4. Model comparison results.

Model	P (%)	R (%)	F1-Score	mAP (%)
IMVTS	87.79	72.20	0.79	83.85
YOLO v7	87.60	62.08	0.72	77.90

3.2. *Comparison of Model Recognition Effects on the MVT Dataset*

To further understand the detection effect of the models for images of fresh leaf sprouts of different varieties of tea. The YOLO v7 and the IMVTS models were used to detect the three different varieties of fresh tea leaves in the selected images. The detection results are shown in Figure 8. It can be seen that the IMVTS model detected more fresh tea leaf sprouts than the YOLO v7 model and had a better detection effect.

Figure 9 shows that when the YOLO v7 original model detects the three varieties of fresh tea leaves, there is a misunderstanding phenomenon. This is because, in the process of generating fresh tea leaf datasets, multiple fresh tea leaves and sprouts often appear in the single picture taken, and these images have problems such as blurry, chaotic positions and overlapping blocks of sprouts. As a result, the selection of the fresh tea leaf sprouts by the frame of artificial labeling is not accurate enough, resulting in error detection of the model. The IMVTS model can effectively reduce the misunderstanding phenomenon, and the specific contrast is as follows.

The results of the original model of YOLO v7, YOLO v7+CBAM, and IMVTS are given in Table 5. In Table 5, it is shown that the AP value of ZC108 can reach up to 99.87%; the AP value of ZJ can reach up to 99.64%; and the AP value of ZH is 96.97%. It can also be seen in Table 2 that compared with the ZC108 and ZJ varieties, the three models have a worse detection effect with ZH. Considering that this study used the same model for all three varieties of fresh tea leaf sprouts, the main reason should be in the images of the fresh tea leaves. Therefore, fresh leaf sprouts and old leaf images of each of the three tea varieties were selected for RGB color gamut analysis. The average and square differences of the RGB of the fresh leaf sprouts and the old leaves of the three tea varieties are shown in Figures 9 and 10. According to Figures 10 and 11, it can be seen that the differences between the three channels of the sprouts and old leaves of ZC108 and the RGB of R, G, and B are 8.33, 30.82, and 31.02, respectively; the differences between the three channels R, G, and B of the square differences of stdRGB are 2.72, 4.79, and 8.04, respectively. The differences between the three channels of the sprouts and old leaves of ZJ and the RGB of R, G, and B are 42.75, 36.52, and 33.76, respectively; the differences between the three channels R, G, and B of the square differences of stdRGB are 15.1, 13.89, and 15.35, respectively. It can be seen that there is a relatively obvious RGB gap between ZC108 and ZJ sprouts and old leaves. It is convenient for the model to distinguish the sprouts and old leaves when the model is detected. The differences between the three channels of the sprouts and old leaves of ZH and the RGB of R, G, and B are 1.34, 1.69, and 5.91, respectively; the differences between the three channels R, G, and B of the square differences of stdRGB are 2.13, 6.11, and 0.14, respectively. It can be seen that the RGB of the fresh leaf sprouts of ZH tea is very low, so the detection accuracy of the model to the fresh leaves of ZH is low.

YOLO v7 IMVTS

(**a**)

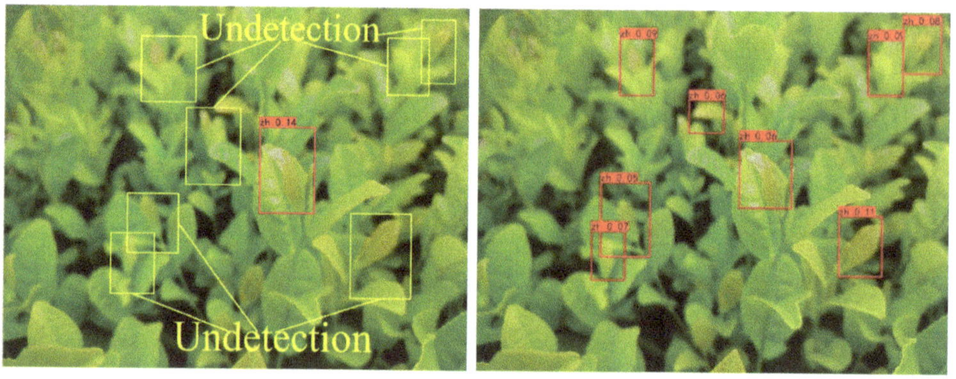

YOLO v7 IMVTS

(**b**)

YOLO v7 IMVTS

(**c**)

Figure 8. Comparison of YOLO v7 and IMVTS detection results; (**a**) ZC108; (**b**) ZH; (**c**) ZJ.

YOLO v7 IMVTS

(a)

YOLO v7 IMVTS

(b)

YOLO v7 IMVTS

(c)

Figure 9. Errors in the detection of three kinds of fresh tea leaves; (**a**) ZC108; (**b**) ZH; (**c**) ZJ.

Table 5. Comparison of the AP results of different varieties of fresh tea leaves.

Model	ZC108 (AP%)	ZH (AP%)	ZJ (AP%)
YOLO v7	99.79	93.30	99.30
YOLO v7+CBAM	99.87	96.89	99.64
IMVTS	99.87	96.97	99.64

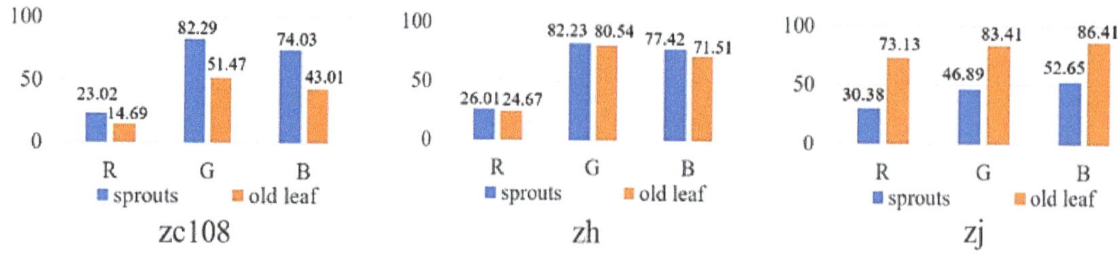

Figure 10. Average RGB of sprouts and old leaves of three varieties of tea.

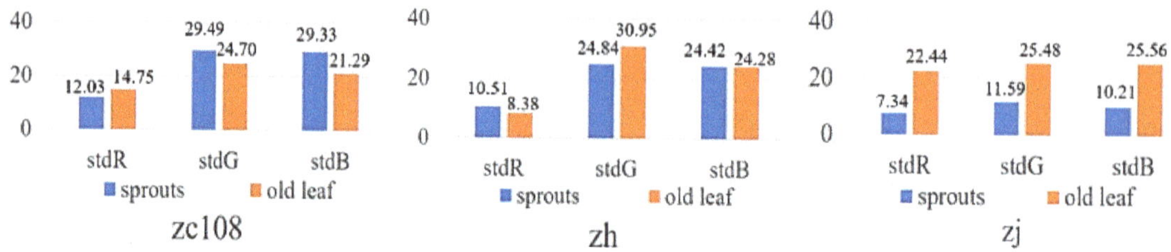

Figure 11. RGB variance stdRGB of sprouts and old leaves of three varieties of tea.

4. Conclusions

This study proposed an identification of multiple varieties of tea sprouts (IMVTS) model for the detection of fresh leaf sprouts of multiple tea species. The model training and validating dataset of multiple varieties of tea (MVT) consisted of images of three representative tea varieties (ZC108, ZH, and ZJ) in Zhejiang Province. To verify the correctness and advantages of the proposed model, IMVTS was compared with YOLO v7 and YOLO v7+CBAM. The results showed that the IMVTS model had the best detection effect, and the mean average precision (mAP) on the MVT dataset was 98.82%, which improved in comparison with YOLO v7 and YOLO v7+CBAM. In addition, this study also conducted comparison tests of the IMVTS model with mainstream target detection models (YOLO v3, YOLO v5, FASTER-RCNN, and SSD) and the IMVTS model on VOC datasets, and the results also demonstrated the superiority of the proposed IMVTS model. The average precision (AP) values of IMVTS for detecting ZC108, ZH, and ZJ tea leaves were 99.87%, 96.97%, and 99.64%, respectively. Among these, the ZH AP value was lower than those of ZJ and ZC108 because the difference in mean RGB and the difference in variance RGB between sprouts and old leaves in ZH images is smaller, and the colors of the object and background are close, making detection more difficult. In summary, the IMVTS model can improve the detection accuracy of fresh leaf sprouts of three varieties of tea, which can meet the requirements for the automatic picking of fresh leaves of the autumn famous tea and provide a basis for the future detection of fresh leaves of additional varieties of autumn tea. Future research will further focus on the improvement and design of the network structure as the IMVTS model detects objects such as ZH that are not clearly distinguished from the background.

Author Contributions: Writing—original draft preparation, R.Z.; methodology and software, C.L.; formal analysis, T.Y.; investigation, J.C.; data curation, Y.L.; supervision, G.L.; project administration, X.H.; writing—review and editing, Z.W. All authors have read and agreed to the published version of the manuscript.

Funding: This research is funded by the National Natural Science Foundation of China (Grant No. 52105284), Science and Technology Program of Meizhou, China (Grant No. 2021A0304004), Postdoctoral Science Foundation of China (Grant No. 2022M722819), and supported by China Agriculture Research System of MOF and MARA, and Key Laboratory of Crop Harvesting Equipment

Technology of Zhejiang Province (Grant No. 2021KZ04). We also thank the anonymous reviewers for their critical comments and suggestions for improving the manuscript.

Institutional Review Board Statement: Not applicable.

Informed Consent Statement: Not applicable.

Data Availability Statement: Data is available on request due to privacy.

Conflicts of Interest: The authors declare no conflict of interest.

References

1. Wang, Z. Design and Experiment of Intelligent Picking Robot for Famous Tea. Ph.D. Thesis, Shenyang University of Technology, Shenyang, China, 2020.
2. Zhou, Y.; Wu, Q.; He, L.; Zhao, R.; Jia, J.; Chen, J.; Wu, C. Design and experiment of intelligent picking robot for famous tea. *J. Mech. Eng.* **2022**, *58*, 12–23.
3. Zhang, L.; Zou, L.; Wu, C.; Jia, J.; Chen, J. Method of famous tea sprout identification and segmentation based on improved watershed algorithm. *Comput. Electron. Agric.* **2021**, *184*, 106108. [CrossRef]
4. Zhang, L.; Zhan, H.; Chen, Y.; Dai, S.; Li, X.; Kenji, I.; Liu, Z.; Li, M. Real-time monitoring of optimum timing for harvesting fresh tea leaves based on machine vision. *Int. J. Agric. Biol. Eng.* **2019**, *12*, 6–9. [CrossRef]
5. Yan, L.; Bengio, Y.; Hinton, G. Deep learning. *Nature* **2015**, *521*, 436–444.
6. Girshick, R. Fast R-CNN. In Proceedings of the 2015 IEEE International Conference on Computer Vision (ICCV), Santiago, Chile, 7–13 December 2015; IEEE: Washington, DC, USA, 2015; pp. 1440–1448.
7. Ren, S.; He, K.; Girshick, R.; Sun, J. Faster R-CNN: Towards real-time object detection with region proposal networks. In Proceedings of the 2015 IEEE Conference on Computer Vision and Pattern Recognition, Boston, MA, USA, 4 June 2015; IEEE: Washington, DC, USA, 2015; pp. 91–99.
8. Redmon, J.; Divvala, S.; Girshick, R. You only look once: Unified, real-time object detection. In Proceedings of the IEEE Conference on Computer Vision and Pattern Recognition, Las Vegas, CA, USA, 8 June 2015; IEEE: Washington, DC, USA, 2016; pp. 779–788.
9. Liu, W.; Anguelov, D.; Erhan, D.; Szegedy, C.; Reed, S.; Fu, C.-Y.; Berg, A.-C. *SSD: Single Shot Multibox Detector. European Conference on Computer Vision*; Springer: Cham, Switzerland, 2016.
10. Li, Y.; He, L.; Jia, J.; Lv, J.; Chen, J.; Qiao, X.; Wu, C. In-field tea shoot detection and 3D localization using an RGB-D camera. *Comput. Electron. Agric.* **2021**, *185*, 106149. [CrossRef]
11. Xu, W.; Zhao, L.; Li, J.; Shang, S.; Ding, X.; Wang, T. Detection and classification of tea buds based on deep learning. *Comput. Electron. Agric.* **2022**, *192*, 106547. [CrossRef]
12. Chen, Y.-T.; Chen, S.-F. Localizing plucking points of tea leaves using deep convolutional neural networks. *Comput. Electron. Agric.* **2020**, *171*, 105298. [CrossRef]
13. Liu, Y.; Cao, X.; Guo, B.; Chen, H.J.; Dai, Z.; Gong, C. Research on detection algorithm about the posture of meat goose in complex scene based on improved YOLO v5. *J. Nanjing Agric. Univ.* **2022**, *32*, 1544–1548.
14. Wang, C.-Y.; Bochkovskiy, A.; Liao, H.-Y.M. YOLOv7: Trainable bag-of-freebies sets new state-of-the-art for real-time object detectors. *arXiv* **2022**, arXiv:2207.02696.
15. Vaswani, A.; Shazier, N.; Parmar, N.; Uszkoreit, J.; Jones, L.; Gomez, A.N.; Kaiser, L.; Polosukhinl, I. Attention is all you need. In Proceedings of the NIPS'17: Proceedings of the 31st International Conference on Neural Information Processing Systems, Long Beach, CA, USA, 4–9 December 2017; pp. 6000–6010.
16. Hu, J.; Shen, L.; Albanie, S.; Sun, G.; Wu, E. Squeeze-and-excitation networks. In Proceedings of the 2018 IEEE/CVF Conference on Computer Vision and Pattern Recognition, Salt Lake City, UT, USA, 18–23 June 2018; IEEE: Washington, DC, USA, 2018; pp. 7132–7141.
17. Woo, S.; Park, J.; Lee, J.-Y. CBAM: Convolutional block attention module. In Proceedings of the 15th European Conference on Computer Vision (ECCV), Munich, Germany, 8–14 September 2018; Springer: Berlin/Heidelberg, Germany, 2018; pp. 3–19.
18. Wang, Q.; Wu, B.; Zhu, P.; Li, P.; Zuo, W.; Hu, Q. ECA-net: Efficient channel attention for deep convolutional neural networks. In Proceedings of the 2020 IEEE/CVF Conference on Computer Vision and Pattern Recognition (CVPR), Seattle, WA, USA, 13–19 June 2022; IEEE: Washington, DC, USA, 2022; pp. 11531–11539.
19. Qi, L.; Gao, J. Small Object Detection Based on Improved YOLOv7. *Comput. Eng.* **2022**, 1000–3428. [CrossRef]
20. Yu, L.; Huang, C.; Tang, J.; Huang, H.; Zhou, Y.; Huang, Y.; Sun, J. Tea Bud Recognition Method Based on Improved YOLOX Model. *Guangdong Agric. Sci.* **2022**, *49*, 49–56.

horticulturae

MDPI

Article

Simulation Model Construction of Plant Height and Leaf Area Index Based on the Overground Weight of Greenhouse Tomato: Device Development and Application

Shenbo Guo [1,2], Letian Wu [1,3,4,*], Xinwei Cao [1,3,4], Xiaoli Sun [1,3,4], Yanfei Cao [2], Yuhan Li [2] and Huifeng Shi [1,3,4,*]

1 Institute of Agricultural Machinery, Xinjiang Academy of Agricultural Sciences, Urumqi 830091, China; guoshenbo01@163.com (S.G.)
2 College of Horticulture, Northwest A & F University, Xianyang 712100, China
3 Research Center for Agricultural Engineering Facilities and Equipment Engineering Technology, Urumqi 830091, China
4 Agricultural Engineering Company, Xinjiang Academy of Agricultural Sciences, Urumqi 830091, China
* Correspondence: wuletian1982@126.com (L.W.); shihuifeng@sohu.com (H.S.)

Abstract: Plant height and leaf area index (LAI) are crucial growth indicators that reflect the growth status of tomatoes in greenhouses, enabling accurate determinations to effectively estimate crop transpiration and formulate irrigation strategies for reducing agricultural water waste. There is a need for the increased application of related models to simulate tomato growth indices in the traditional greenhouse production in China. This study proposes a nondestructive, real-time monitoring and simulation device for measuring tomato plant height and leaf area index. The weight of aboveground tomatoes was obtained by suspending tomato plants on dynamometers, while the total weight of stem and leaf organs was determined using a distribution coefficient simulation model. The R^2 value between the measurements from the electronic scale and those from the aboveground fresh weight device for tomatoes was 0.937, with an RMSE value of 0.05 kg. The monitoring device did not affect the average tomato growth during operation. The device will not affect the growth of tomatoes during monitoring. A multiple linear regression was used to compare the measured and simulated values of the plant height and leaf area index of various types of greenhouse tomatoes cultivated in different greenhouse types. The average R^2 value for simulating plant height was 0.817 with an RMSE of 10.81 cm. The average R^2 value for the leaf area index was 0.854, with an RMSE of 0.55 $m^2 \cdot m^{-2}$. The simulated values for plant height and leaf area index closely matched the measured values, indicating that the model has high accuracy and applicability in traditional Chinese greenhouses (solar greenhouses and insulated plastic greenhouses). However, further optimization is required for commercially produced, continuous plastic greenhouses equipped with greenhouse environmental control equipment.

Keywords: greenhouse; weighing device; tomato simulation model; multiple linear regression

Citation: Guo, S.; Wu, L.; Cao, X.; Sun, X.; Cao, Y.; Li, Y.; Shi, H. Simulation Model Construction of Plant Height and Leaf Area Index Based on the Overground Weight of Greenhouse Tomato: Device Development and Application. *Horticulturae* **2024**, *10*, 270. https://doi.org/10.3390/horticulturae10030270

Academic Editor: Luca Incrocci

Received: 5 December 2023
Revised: 23 February 2024
Accepted: 5 March 2024
Published: 11 March 2024

1. Introduction

In recent years, protected horticulture has experienced remarkable growth, substantially increasing the cultivated area [1,2]. China boasts over 2.8 million hectares of horticultural facilities, representing over 80% of the world's total. This area includes 810,000 hectares (29%) of solariums and 1.52 million hectares (53%) of large- and medium-sized greenhouses [3]. These figures underscore the significant and growing role of facility agriculture in meeting the world's food needs, and the supply of vegetables is crucial while promoting sustainable development in facility agriculture [4]. Nevertheless, the expansion of facility agriculture also means a rise in water usage for agricultural purposes. As water scarcity becomes a growing problem worldwide, it challenges the sustainable development of facility agriculture [5,6]. By 2050, agricultural water scarcity is expected to spike in over

80% of the world's countries [7]. This leaves only a limited scope for increasing the water supply in agriculture, underscoring the need to enhance agricultural water use efficiency. The accurate calculation of the required crop water, also known as crop evapotranspiration (ETc), can help optimize irrigation management and augment water use efficiency during the growing season [8].

Tomatoes are a vital vegetable crop grown worldwide due to their easy cultivation and high economic efficiency [9–11]. To ensure that tomatoes grow and develop optimally, they must have an adequate water supply [12]. However, in traditional tomato production facilities such as solar and plastic greenhouses, irrigation is mainly based on artificial experience, which often results in significant water wastage [13]. The leaf area (LA) is the primary organ responsible for transpiration in tomato growth, and the leaf area index (LAI) is the total tomato leaf area per unit of land area, which is directly related to the crop evapotranspiration (ETc) of tomatoes [14]. Moreover, the plant's height is an essential indicator for assessing crop evapotranspiration (ETc) and transpiration in tomatoes, which is necessary for a high and stable tomato production [15,16]. Therefore, the real-time and accurate determination of the tomato's LAI and plant height is crucial for estimating crop evapotranspiration (ETc), which, in turn, provides a theoretical basis for evaluating the tomato water demand, making irrigation decisions, and reducing water wastage. Accurate estimates of crop evapotranspiration help greenhouse managers make irrigation decisions and promote the efficient use and conservation of resources [17].

Many experts and scholars have conducted a lot of research on the accurate determination of leaf area and plant height in tomatoes. Traditional leaf area measurements usually use destructive sampling methods to collect and determine the leaves, such as the punching and weighing method [18], Image J image processing method [19], leaf area meter method [20], etc. Although these type of methods can determine the leaf area more accurately, they destroy the normal growth of tomato plants, and the workload is large and time consuming. In order to determine the leaf area nondestructively, leaf area simulation has been studied by many scholars. L. Bacci et al. [21] improved the TOMGRO model to increase the simulation accuracy of leaf area; Wang X. et al. [22] developed a leaf area simulation model for tomato processing based on the physiological development time, using the physiological development time of tomatoes as the time scale; Wang D. et al. [23] developed a simulation model of pepper leaf area applicable to solar greenhouses using auxiliary heat product as a scale. Traditional plant height measurements are usually conducted directly using a tape measure, to save labor costs. Chang Yibo et al. [24] used a binary quadratic orthogonal rotated combinatorial design to establish a logistic model for tomato plant height based on radial heat product with an irrigation limit and fertilizer application as determinants; Cheng Chen et al. [25] constructed a celery plant height simulation model based on the relationship between greenhouse celery plant height and key meteorological factors (air temperature and solar radiation) with the single-plant irradiated heat product as the independent variable; Zhai Zihe et al. [26] used the XGBoost model to establish a simulation model for the plant height growth of cucumber at five fertility periods in a solar greenhouse, and the accuracy of the model was good. The LAI and plant height can be more accurately simulated through crop growth modeling, but the model requires the input of multiple environmental parameters, and the lack of costly IoT environmental monitoring equipment in most traditional greenhouses makes these types of models less practical.

Therefore, through the real-time determination of tomato leaf area and plant height, a scientific foundation was established to assess tomato crop water requirements and make irrigation decisions in conventional greenhouses. This study's mathematical simulation model of the tomato leaf area index and plant height was constructed using multiple linear regression based on experimental data from insulated plastic greenhouse fall stubble (the leaf area index was calculated from leaf area simulation results). The model was validated with experimental data from insulated plastic greenhouse spring stubble and solar greenhouse fall stubble. Additionally, a real-time monitoring device for the aboveground weight of tomatoes was designed to analyze the distribution coefficients of aboveground stems and leaves among different stubbles. This information was used to establish a mathematical simulation model of the tomato leaf area index and plant height based on real-time fresh weight, which was further tested with experimental data from overwinter stubble in solar greenhouses and continuous plastic greenhouses. Furthermore, a low-cost and stable operation device for the real-time simulation of the leaf area index and plant height of tomatoes applicable to traditional greenhouses was developed and installed. The aim was to provide a tool for the nondestructive determination of individual or group tomato leaf areas and plant heights in traditional greenhouses while offering scientific references for estimating the evapotranspiration of tomato crops and water-saving irrigation.

2. Materials and Methods

2.1. Test Scheme

The experimental programmer involved three subtests to construct simulation models for the tomato leaf area index and plant height, a trial operation of a real-time monitoring device for tomatoes' aboveground fresh weight, and creating a simulation model for the tomato leaf area and plant height based on the real-time fresh weight data. Three different structural types of greenhouses were selected for the test greenhouses. Figure 1a presents a schematic of a large-span, asymmetric plastic greenhouse; Figure 1b presents a structural diagram of a solar greenhouse; Figure 1c presents a structural schematic of a continuous plastic greenhouse. The experimental crops involved medium- and large-fruited tomato varieties, and simulation models were conducted for the leaf area and plant height of tomatoes in the mid- and late-growth stages, specifically during the flowering and fruiting stages. The leaf area index was calculated based on the actual planted area after the leaf area simulation. The tomato plants were grown in substrate bags measuring 56 cm × 24 cm × 10 cm, with approximately 7 L of substrate per bag, and two tomato plants were planted in each substrate bag.

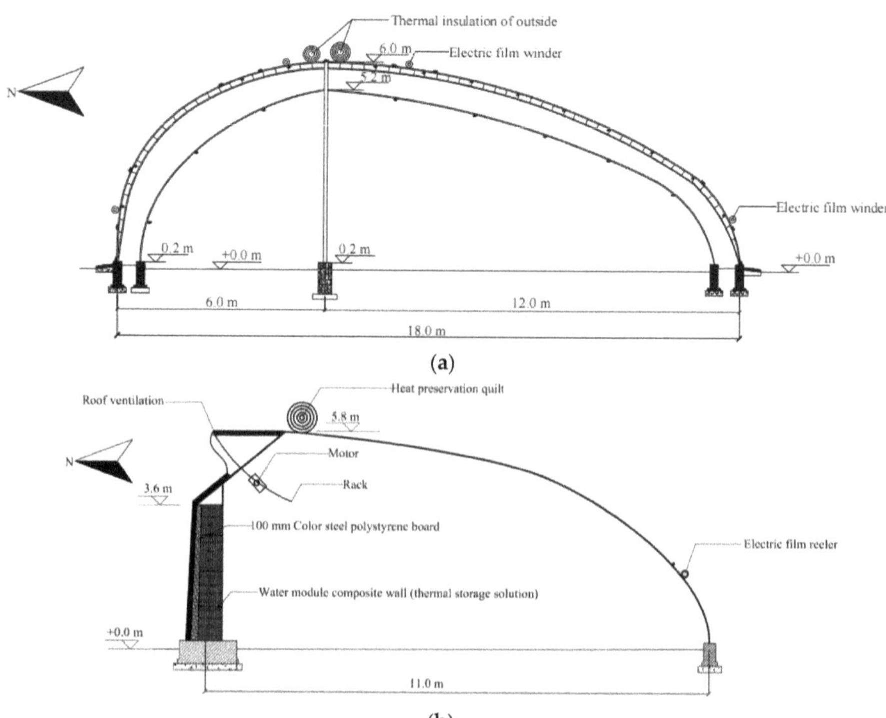

Figure 1. *Cont.*

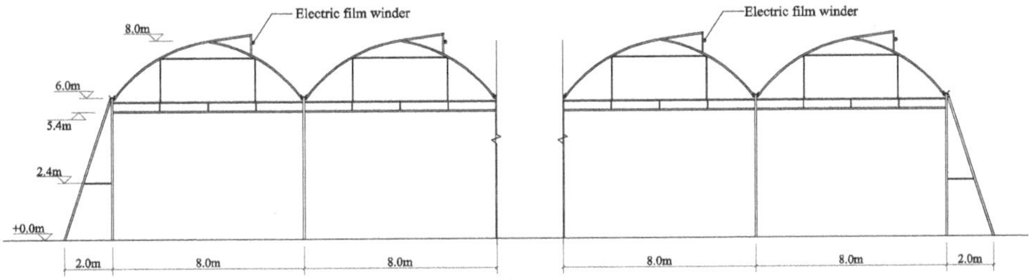

(c)

Figure 1. Three types of greenhouse structures. (**a**) Schematic of a large-span asymmetric plastic greenhouse. (**b**) Structural diagram of a solar greenhouse. (**c**) Structural schematic of continuous plastic greenhouse (partial).

2.1.1. Simulation Modeling Test for Leaf Area and Plant Height in Tomato

1. Cultivation experiment

Three stubble cultivation experiments were conducted at two locations: From January to May and from August to December 2021 in a double-arch, double-film, asymmetric, large-span insulated plastic greenhouse (Figure 1a) in the Modern Agricultural Integration and Experience Park in Yangling, Shaanxi, China (34°31′ N, 107°97′ E). The insulated plastic greenhouse was situated in a north–south direction, spanning 18.0 m with a ridge height of 6.0 m and an east–west length of 70.0 m. It had a double-layer skeleton. The inside and outside layers of the roofs on the north and south sides were covered with 0.12 mm plastic film, and the outside plastic film was covered with a 20 mm insulation quilt. Two roof vents and two bottom vents were placed on the north and south roofs for ventilation and cooling. The test tomato variety for the spring stubble was 'Provence', and the test tomato variety for the autumn stubble was 'Bao·Lu Fu Qiang', with regular field management.

The experiment was conducted from January to May 2022 in a solar greenhouse (Figure 1b) at the horticultural farm of the North Campus of Northwest Agriculture and Forestry University (34°29′ N, 108°07′ E). The greenhouse had a span of 11.0 m, a ridge height of 5.8 m, and a length of 50 m in the east–west direction, and the tomato cultivar for the test was 'Cadjarli' with conventional field management.

2. Growth indicators and plant quality data acquisition

Three tomato plants were randomly selected for height measurement every seven days during tomato cultivation, and the distance between the top of the root and the apical growing point of the tomato was measured with a tape measure. Afterward, the three tomato plants were weighed using the destructive sampling method using a TCS-300 electronic scale (Jinhua, China) for the fresh weight of roots, stems (including petioles), flowers, and fruits, respectively. The leaf area of each tomato plant was determined using a CL-202 leaf area scanner (CID BioScience, Camas, WA, USA).

3. Simulation modeling of tomato leaf area and plant height

Autumn stubble data from asymmetrically insulated plastic greenhouses from August to December 2021 were selected for the construction of simulation models of tomato plant height and leaf area based on the following considerations:

The fresh weight of tomato plant stems as the independent variable and plant height as the dependent variable; The fresh weight of tomato plant leaves as the independent variable, and leaf area as the dependent variable.

Mathematical models of the linear function, polynomial function, and power function were established using the method of least squares regression, and the accuracies of the models were verified using spring stubble data.

4. Simulation of the aboveground morphological organ allocation index of tomatoes

Based on the actual tomato planting results in the cultivation experiment, combined with the conclusions of previous studies on tomatoes' aboveground morphological organ allocation index, the method of least squares regression was utilized to fit the change curves of the allocation indices of the fresh weight of aboveground stem and leaf organs during the reproductive period of the tomatoes and to establish the calculation methods of the allocation indices of the fresh weight of aboveground stem and leaf organs of the tomatoes.

5. Model Validation

The root mean square error (RMSE) and mean absolute error (MAE) were used as indicators for the validation of the long-term monitoring accuracy of the device [26].

$$\text{RMSE} = \sqrt{\frac{1}{n}\sum_{i=1}^{n}(ac_i - sm_i)^2} \quad (1)$$

$$\text{MAE} = \frac{1}{n}\sum_{i=1}^{n}|ac_i - sm_i| \quad (2)$$

In this formula, ac_i is the measured value; sm_i is the simulated value; and n is the number of samples.

2.1.2. Operational Testing of Real-Time Tomato Aboveground Weight Monitoring Device

1. Operational test

Experiment 1 was conducted from April to June 2022 in a solar greenhouse at the horticultural farm of the North Campus of Northwest A&F University of Science and Technology (NWAFU).

2. Device design

The weighing device consists of a data acquisition module, a communication module, and a monitoring module (Figure 2). The data acquisition module adopts an S-type tension sensor (DYLY-103, range: 0–10 kg, accuracy: 0.03%, Bengbu Dayang Sensing Company based in An Hui China). One end of the sensor is secured to trailing wire, and the tomato plants are suspended on the sensor using the trailing rope at an angle to conduct real-time measurements of the fresh weights of the tomatoes on the ground. The data are transmitted analogically in real time through a 4–20 ma current to a collection cabinet, the acquisition module (DAM3232, Bei Jing China Jiuying Soaring Electronics). The communication module adopts the 485 protocol to connect the serial port of the acquisition module and the 4G IOT module wired to transmit the data to the server, and the monitoring module adopts the 485 protocol to connect the serial port of the acquisition module and the MCGS touch screen wired to realize a local real-time display of the data.

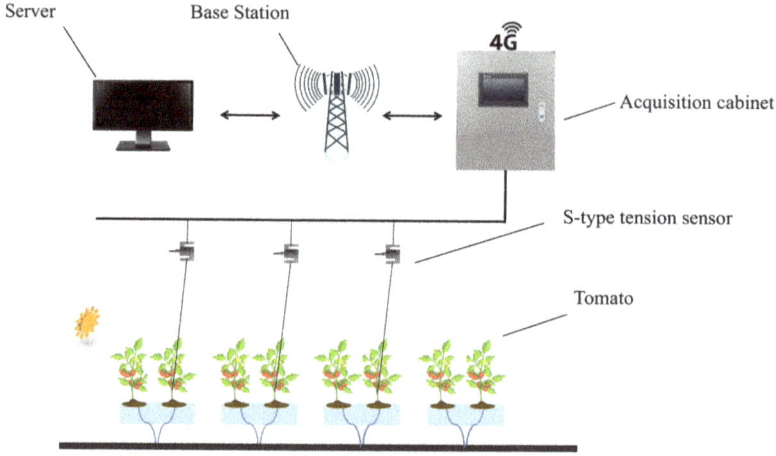

Figure 2. Schematic of real time-monitoring device for tomato aboveground weight.

3. Inspection of real-time tomato aboveground weight monitoring device

Long-term operation: From 23 May to 7 June 2022, 1.550 kg objects were suspended using three groups of S-type tension sensors. The collection interval was 1 h. Data were collected from the three sensor groups at the end of the inspection. The root mean square error (RMSE) and average absolute error (MAE) were used to verify the accuracy of the long-term monitoring device.

Measurement accuracy of weighing device: After measuring the aboveground weight of single tomato plants through the real-time tomato aboveground weight monitoring device, a destructive sampling method was used to weigh the intercepted tomato aboveground using an electronic balance, and the difference between the device measurement value and the actual measurement value was determined. A total of 45 tomato plants were selected for the accuracy test of the weighing device in the solar greenhouse on three occasions (with an interval of 20 days), and the root mean square error (RMSE) and mean absolute error (MAE) were used as indicators for verifying the long-term monitoring accuracy of the device. The root means square error (RMSE) and mean absolute error (MAE) were used to test the accuracy of the measurement.

2.1.3. Real-Time Weight-Based Simulation Model Construction Experiment for Leaf Area and Plant Height of Tomatoes

1. Cultivation experiment

The solar greenhouse in Experiment 1 and the continuous plastic greenhouse in Yangling SinoIsraeli Cooperation Smart Agriculture Demonstration Park, Shanxi Province (34°27′ N, 108°04′ E), were selected as the experimental areas for obtaining model validation data. The specifications of the solar greenhouse were the same as those of Experiment 2 and the tomato plant was planted on 18 October 2022, and the test tomato was 'Jinpeng No.1'; the continuous plastic greenhouse had a length of 71 m from north to south, 172 m from east to west, and a height of 8.4 m. The greenhouse had a total of 21 spans, each with a span of 8.0 m; the test tomato was 'Jinpeng No.1', and the tomato for the test was 'Golden Scaffold No.1'.

2. Validation method for the simulation of tomato plant height and leaf area based on real-time monitoring device.

In this experiment, 15 tomato plants were randomly selected in the solar greenhouse on 18 November 2022, and a total of 9 tomato plants were randomly selected in the continuous plastic greenhouse on 8 December 2022 and 21 December 2022, respectively, for the simulation and validation of tomato plant height and leaf area. First, a real-time monitoring device for tomatoes' aboveground fresh weight was set up in the greenhouse to tag the randomly selected tomato plants and measure their plant height. Afterward, the weighing device was employed to measure the aboveground weight of the labeled tomato plants for three consecutive occasions, and the mean value was considered the aboveground weight of the tomato plants. Finally, the labeled tomato plants were subjected to the destructive sampling method, and electronic scales were used to weigh the stem and leaf fresh weight of the tomato plants. The tomato's leaf area was determined using the CL-202 leaf area scanner, and the simulation model in Test 1 was used to calculate the height and leaf area of the tomato plant. A flowchart is shown in Figure 3.

3. Model validation

The root mean square error (RMSE) was selected as the index for validating the simulation accuracy of the tomato plant height and leaf area based on the real-time monitoring device.

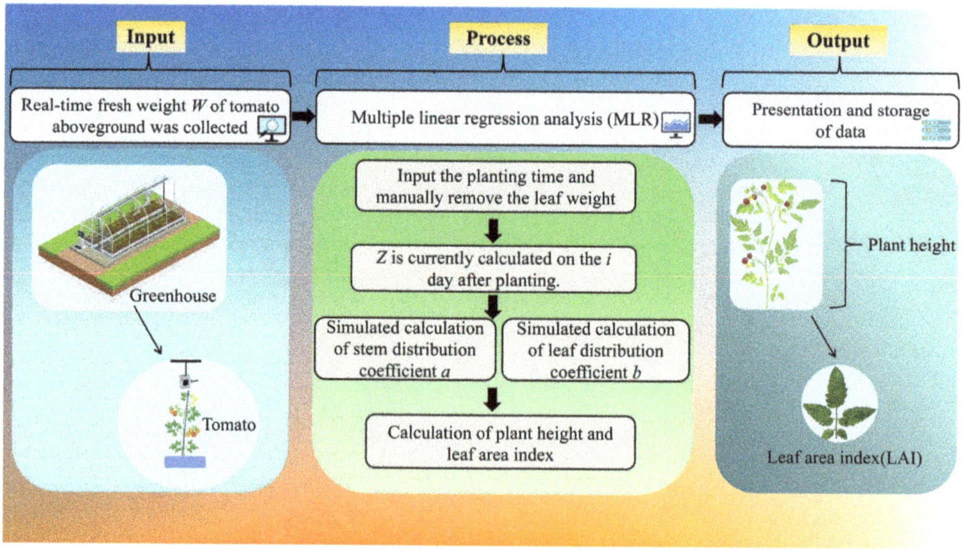

Figure 3. Simulation process of tomato growth index based on real-time monitoring device.

3. Results

3.1. Relationship of Tomatoes' Morphological Organs with Plant Height and Leaf Area

During Experiment 1's tomato reproductive period, we analyzed correlation coefficients between the stem and leaf fresh weight, and plant height and leaf area (Tables 1 and S1). The results showed a close link between stem weight and plant height in the spring and autumn. The correlation coefficients were very significant. The r values were 0.930 and 0.944. This relationship passed the significance test of F (0.01). The tomato leaves' weight was strongly linked to their size in the spring and autumn. For the spring, the correlation coefficient was 0.952, and for autumn, it was 0.984. Both coefficients passed the F (0.01) significance test.

Table 1. Correlation analysis of tomatoes' morphological organs with plant height and leaf area.

Stubble			Stem Fresh Weight			Leaf Fresh Weight
Spring stubble	Plant height	r	0.930 **	Leaf area	r	0.952 **
		p	0.001		p	0.001
Autumn stubble	Plant height	r	0.944 **	Leaf area	r	0.984 **
		p	0.001		p	0.001

Note: r is the correlation coefficient; p is significance: ** is $p < 0.01$.

3.2. Simulation Modeling of Tomato Plant Height and Leaf Area Index

3.2.1. Model Construction

Based on data from the 2021 insulated plastic shed fall stubble trial, Section 3.4.1 presents the 'Calculation method of tomato stem and leaf distribution coefficient', which integrates findings from previous studies and our own planting experience in this study. A methodology was developed to calculate the distribution coefficients (a) for stem organs and (b) for leaf organs in the aboveground parts of tomato plants. The total weight of aboveground tomatoes was measured using dynamometers, and by multiplying this weight (W) with the respective organ distribution coefficients (a and b), values W_h and W_l were obtained, respectively. Subsequently, these values were utilized along with a model to determine the height of a tomato plant (H) as well as its leaf area index (LAI). Mathematical functions were employed to construct simulation models for both the tomato plant height and leaf area index.

3.2.2. Model Validation

1. Plant height simulation model validation

The accuracy of the modeled tomato plant height simulations (Tables 2 and S2) was verified using measured data from spring stubble in insulated plastic greenhouses and overwintering stubble stem organs in solar greenhouses in Experiment 1, respectively.

Table 2. Modeling tomato morphology and appearance using different mathematical functions.

Tomato Appearance	Math Function Types	Model
Plant height (H)	Linear function	$H = 0.5286W_h + 26.304$
	Polynomial Function	$H = -0.0015{W_h}^2 + 0.8725W_h + 17.554$
	Idempotent function	$H = 9.2293{W_h}^{0.4941}$
Leaf area (LA)	Linear function	$LA = 29.182W_l + 163.75$
	Polynomial Function	$LA = -0.0917{W_l}^2 + 43.396W_l - 85.18$
	Idempotent function	$LA = 41.262{W_l}^{0.9445}$
Leaf area index (LAI)	/	$LAI = \frac{LA}{S}$

Note: W_h is the fresh weight of the tomato stems, W_l is the fresh weight of the tomato leaves, H is the height of the tomato plants, LA is the leaf area of the tomato, S is the unit footprint of the tomato plants, which was 0.24 cm × 0.28 cm in this cultivation experiment, and LAI is the leaf area index.

Through the simulation process of the tomato plant height mentioned in Figure 3, the tomato plant height was simulated based on the three formulas of plant height, as shown in Table 1, and compared with the actual values; the comparisons of Figure 4 were obtained by means of a 1:1 line. Included among these, the tomato plant height simulation for spring crops in insulated plastic greenhouses was good. The R^2 values for the linear, polynomial, and power functions were 0.886, 0.888, and 0.888; all R^2 values were above 0.8. The RMSE values were 12.96 cm, 24.05 cm, and 12.99 cm (Figure 4a). In the simulation results of the tomato plant height for overwintering stubble in solar greenhouse, the R^2 values for the linear, polynomial, and power functions were 0.905, 0.919, and 0.987, with all R^2 values above 0.9. The RMSE values were 13.37 cm, 8.69 cm, and 14.10 cm (Figure 4b).

2. Leaf area index simulation model validation

The accuracy of the modeled tomato plant leaf area index simulations (Table 2) was verified using measured data from spring stubble in insulated plastic greenhouses and overwintering stubble leave organs in solar greenhouses in Experiment 1, respectively.

Through the simulation process of the tomato plant leaf area index mentioned in Figure 3, the tomato plant leaf area index was simulated based on the three formulas of plant height, as shown in Table 1, and compared with the actual values; the comparisons of Figure 5 were obtained by means of a 1:1 line. The results showed that in the simulation of the leaf area index of spring stubble tomatoes in insulated plastic greenhouses, the R^2 values of the linear, polynomial, and power index functions were 0.881, 0.843, and 0.880. All R^2 values were above 0.8. The RMSEs were 0.98 $m^2{\cdot}m^{-2}$, 1.20 $m^2{\cdot}m^{-2}$, and 1.14 $m^2{\cdot}m^{-2}$ (Figure 5a). Meanwhile, the results show that for the simulation diagram of the leaf area index of overwintering stubble in solar greenhouse, the R^2 values of the linear, polynomial, and power index functions were 0.928, 0.902, and 0.928, with all R^2 above 0.9. The RMSEs were 0.54 $m^2{\cdot}m^{-2}$, 0.69 $m^2{\cdot}m^{-2}$, and 0.53 $m^2{\cdot}m^{-2}$ (Figure 5b).

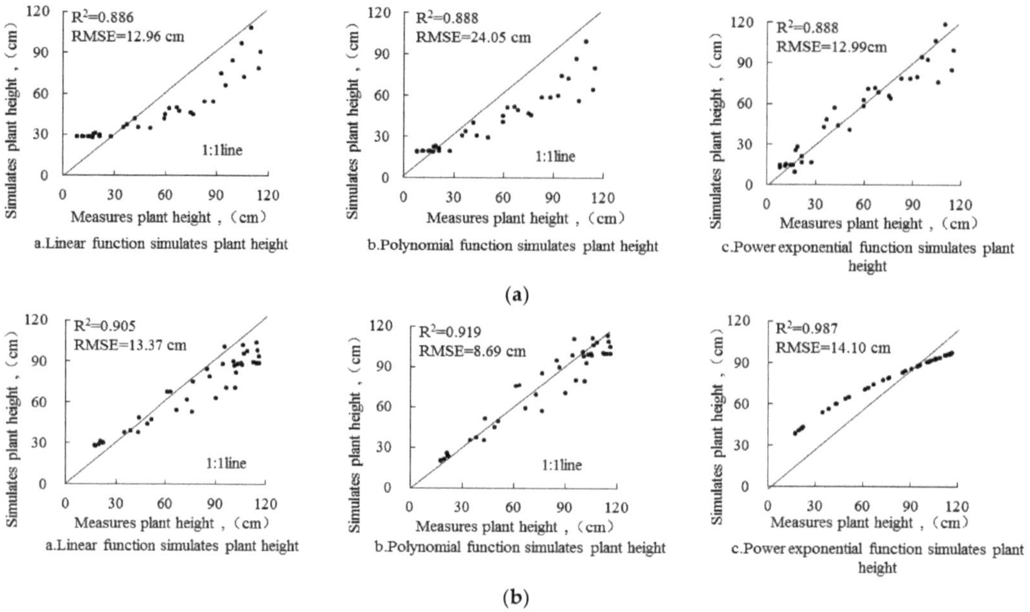

Figure 4. Comparison of simulated and measured tomato plant height values: (**a**) simulation of plant height in spring stubble under thermal insulation plastic greenhouse; (**b**) simulation of plant height during overwintering in solar greenhouse.

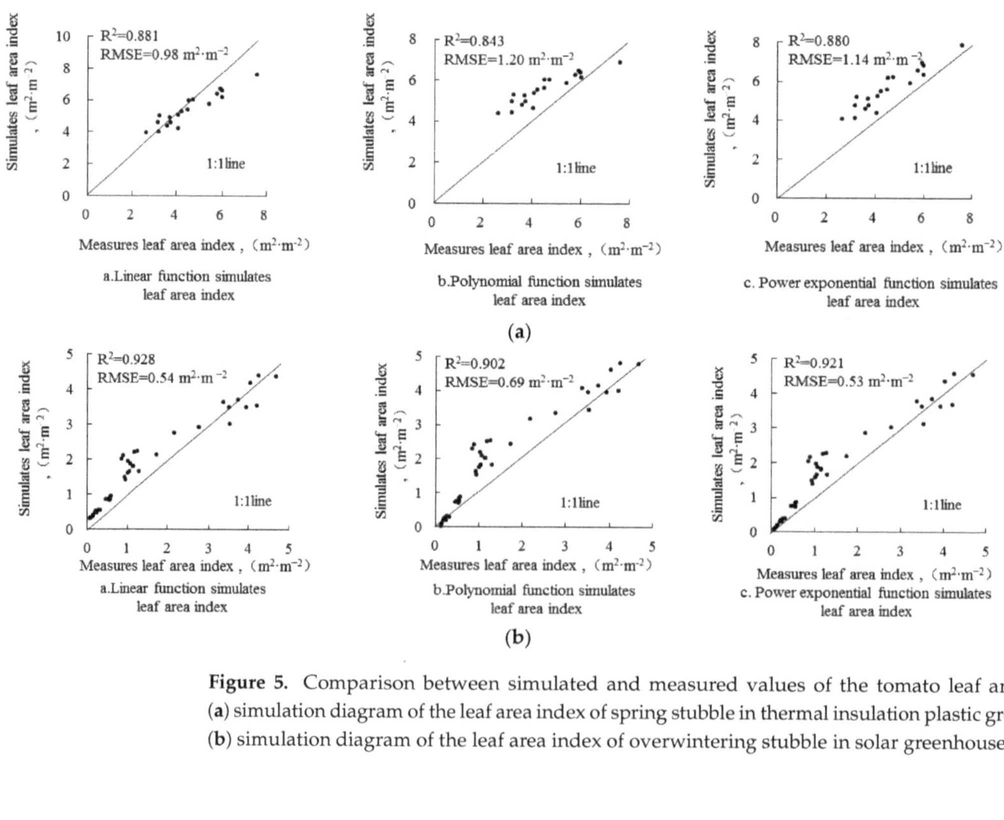

Figure 5. Comparison between simulated and measured values of the tomato leaf area index: (**a**) simulation diagram of the leaf area index of spring stubble in thermal insulation plastic greenhouse; (**b**) simulation diagram of the leaf area index of overwintering stubble in solar greenhouse.

3.3. Tomatoes' Aboveground Fresh Weight Real-Time Monitoring Device Operation Detection

3.3.1. Long-Term Monitoring of the Weight of Constant-Weight Objects

As shown in Figure 6, three sensors were used to test the stability of a tomato ground weight monitoring device. The sensors measured the weight of a constant 1.550 kg object. The S-type tension sensors collected data that showed small fluctuations, ranging from 1.55 kg up to 1.55 kg down. The monitoring weight had a maximum error of 30.81 g compared to a constant-weight object. The average error of the three groups of sensors monitoring the weight was 16.95 g. The comprehensive monitoring of the weight data of the constant-weight object was conducted for 16 consecutive d. This device can monitor the weight of suspended objects for a long time. It considers the environment, sensor voltage fluctuations, and the sensor itself. These factors affect the measurement accuracy.

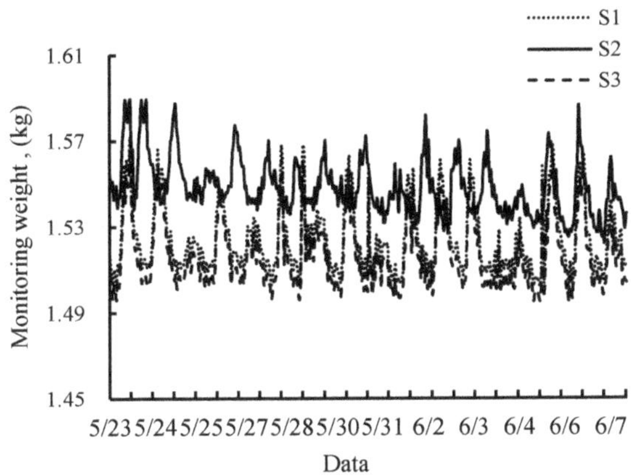

Figure 6. Real-time monitoring diagram of constant-weight object's weight.

3.3.2. Measurement Accuracy Test of Tomato Aboveground Fresh Weight Device in Different Periods

It can be seen from Figure 7 that 12 tomato plants were collected every 20 days in the solar greenhouse, for a total of 36 plants. The device measured the weight of a tomato aboveground. Then, a tool was used to cut the tomato off, and an electronic balance was used to measure its weight. Finally, the device's and electronic balance's measured values were compared. It can be seen that the fitting degree between the measured value and the measured value of the device is good; the R^2 was 0.937, RMSE was 0.05 kg, and MAE was 0.04 kg. The device can measure the weight of tomato plants aboveground in real time. It does not harm the average growth of tomatoes.

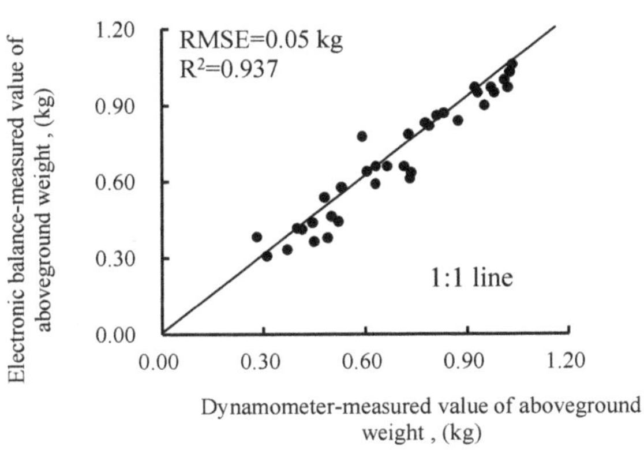

Figure 7. Measurement accuracy test of tomato aboveground fresh weight device.

3.4. Simulation of Tomato Plant Height and Leaf Area Index Model Based on Real-Time Monitoring Device

3.4.1. Calculation Method of Tomato Stem and Leaf Distribution Coefficient

In different stubbles of tomato cultivations and in tomato seedlings 10 days after planting, the slow seedling of the tomatoes ended, and they began to grow normally. The distribution index of stem organs showed a consistent and stable trend, and the distribution index of leaf organs showed a trend of slow increase at first and then decrease [27].

In previous research on the organ distribution coefficient of greenhouse tomatoes [28,29], the growth and organ distribution coefficients of the stem and leaf organs of tomato plants have been determined at different reproductive stages. Meanwhile, the allocation coefficients of stem and leaf organs at the different reproductive stages of tomato were determined by combining the actual planting of spring and fall stubbles of tomato in large-span, insulated plastic greenhouses in this study; the distribution indices of the tomato stems and leaves were fitted and analyzed, and on this basis, the change law of the distribution coefficient of both tomato stems and leaves was synthesized, and the distribution indices of the tomato stems and leaves were fitted and analyzed, and formulas for the analysis indices of the tomato stems and leaves were obtained (Figure 8).

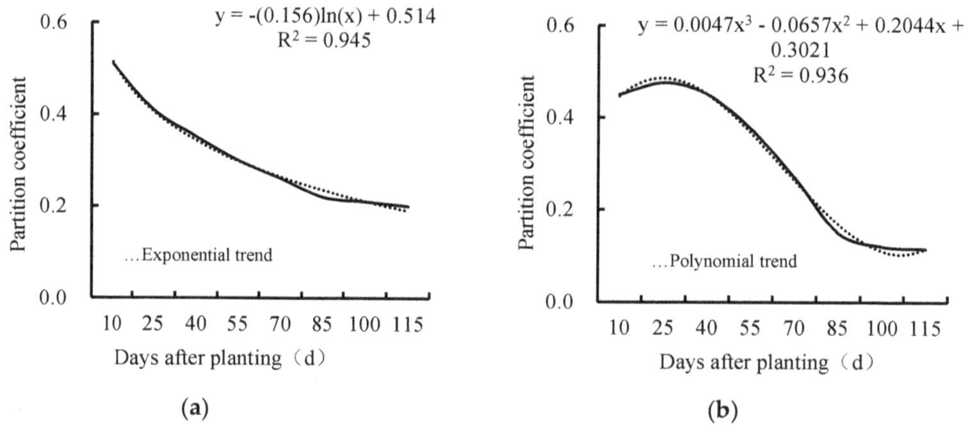

Figure 8. Changes in the distribution coefficient of tomato appearance and morphology: (**a**) distribution coefficient variation rule of stems; (**b**) distribution coefficient variation rule of stems.

3.4.2. Model Simulation Construction of Plant Height and LAI Based on Dynamometer

Based on the tomato growth curve and the results of Experiment 1 and the actual planting, a simulation model of the tomato plant height was created. The real-time monitoring device was used to calculate the height using linear, polynomial, and power functions (Tables 3, S3 and S4). Also, the leaf area index was calculated using the specific calculation formulas in Table 4.

Table 3. Plant height simulation formulas.

Name	Model
Distribution factor	$a = -0.156\ln\left(\frac{i}{15}\right) + 0.514$
Linear function	$H = 0.5286Wa + 26.304$
Polynomial function	$H = -0.0015(Wa)^2 + 0.8725(Wa) + 17.552$
Idempotent function	$H = 9.2293(Wa)^{0.4941}$

Note: i is day i after planting, a is the aboveground part of the stem allotment index, W is the aboveground weight of the tomato, and H is the height of the tomato plant.

Table 4. Simulation formulas for the leaf area index.

Name	Model
Distribution factor	$b = 0.0047\left(\frac{i}{15}\right)^3 - 0.0657\left(\frac{i}{15}\right)^2 + 0.2044\frac{i}{15} + 0.3021$
Linear function	$LA = 29.182(Wb + F) + 163.76$
Polynomial function	$LA = -0.0917(Wb + F)^2 + 43.396(Wb + F) - 85.18$
Idempotent function	$LA = 41.262(Wb + F)^{0.9445}$
Leaf area index (LAI)	$LAI = \frac{LA10^{-4}}{S}$

Note: i is the day i after planting, b is the aboveground portion of the leaf allotment index, W is the aboveground weight of the tomato, LA is the leaf area of the tomato, S is the area occupied by a single plant of tomato (0.24 m × 0.28 m in this study), and LAI is the leaf area index of the tomato.

3.4.3. Application Test of Plant Height Model

In Experiment 3, in using the simulation model of tomato plant height based on real-time fresh weight, 15 tomato plants were randomly selected in the solar greenhouse, and 12 tomato plants were randomly selected in the multispan plastic greenhouse using a random sampling method, and simulation tests of tomato plant height in the two types of greenhouses were carried out using the process presented in Figure 3 to obtain the R^2 and RMSE values of the solar greenhouses and multispan plastic greenhouses.

The results show that in the simulation of tomato plant height in autumn in the solar greenhouse, the R^2 values of the linear function, polynomial function, and power exponential function were 0.813, 0.818, and 0.835, respectively, and the RMSEs were 4.46 cm, 6.71 cm, and 4.89 cm, respectively. In the simulation of the autumn tomato plant height in the multispan plastic greenhouse, the R^2 values were 0.858, 0.711, and 0.870, respectively, and the RMSEs were 9.24 cm, 21.86 cm, and 17.71 cm, respectively. In summary, the simulation results of the solar greenhouse are better than those of the multispan plastic greenhouse. The simulation accuracy of tomato plant height in the two greenhouses is good, and different functions can accurately simulate the tomato plant height (Table 5). The significant error between the simulated value and the measured value of the multispan plastic greenhouse results from the installation of greenhouse environment control equipment (LED fill lights, circulation fans, air source heat pump) in the multispan plastic greenhouse. The microclimate in the greenhouse is evenly distributed, which is more suitable for the growth of tomatoes, resulting in the actual value of the tomato plant height being generally more significant than the simulated value.

Table 5. Tomato plant height simulation based on real-time fresh weight measurement device.

Test Greenhouse Type	Name	R^2	RMSE (cm)
Solar greenhouse	Linear function	0.813	4.46
	Polynomial function	0.818	6.71
	Idempotent function	0.835	4.89
Multispan plastic greenhouse	Linear function	0.858	9.24
	Polynomial function	0.711	21.86
	Idempotent function	0.870	17.71

3.4.4. Application Test of Leaf Area Index Model

Using the tomato leaf area index simulation model based on the real-time weights in Experiment 3, 15 and 12 tomato plants were selected in the solar greenhouse and the mul-tispan plastic greenhouse to measure the plant height. The simulated values were compared with the measured values. The simulation of the leaf area index for autumn tomatoes in the solar greenhouse had promising results. The linear, polynomial, and power exponential functions performed well. The R^2 values were 0.950, 0.955, and 0.953. The RMSE values were 0.20 $m^2 \cdot m^{-2}$, 0.18 $m^2 \cdot m^{-2}$, and 0.12 $m^2 \cdot m^{-2}$. The leaf area index of autumn tomatoes in the plastic greenhouse was simulated. The R^2 values were 0.748,

0.651, and 0.870, and the RMSE values were 1.04 $m^2 \cdot m^{-2}$, 0.52 $m^2 \cdot m^{-2}$, and 1.22 $m^2 \cdot m^{-2}$ (Table 6).

Table 6. Tomato leaf area index (LAI) simulation based on real-time fresh weight measurement device.

Test Greenhouse Type	Name	R^2	RMSE ($m^2 \cdot m^{-2}$)
Solar greenhouse	Linear function	0.950	0.20
	Polynomial function	0.955	0.18
	Idempotent function	0.953	0.12
Multispan plastic greenhouse	Linear function	0.748	1.04
	Polynomial function	0.651	0.52
	Idempotent function	0.870	1.22

4. Discussion

Facility microclimate plays a crucial role in tomato growth, characterized by intricate parameters and rapid fluctuations [30]. Transpiration is essential for assessing crop responses to microclimate changes caused by water stress, aiming to optimize the growing environment and crop breeding [31]. The growth and development of tomatoes are influenced by various environmental factors such as air temperature, relative humidity, and solar radiation [32,33]. To minimize simulation errors resulting from environmental conditions, this study utilized fresh weight measurements of different morphological organs in tomatoes to simulate plant height and leaf area index. Subsequently, the distribution indices of the stems and leaves were combined with the real-time monitoring of the above-ground fresh weight to establish a simulation model based on the real-time quantity of fresh tomatoes' plant height and leaf area index. This model was validated in different forks (varieties) and types of greenhouses. The simulated values for the plant height and leaf area index closely matched the measured values, indicating that the constructed simulation models have a high accuracy and applicability to traditional Chinese greenhouses (solar greenhouses and insulated plastic greenhouses). However, further optimization is required for commercially produced continuous plastic greenhouses equipped with greenhouse environmental control equipment.

In past studies on crop growth modeling, experts relied on the destructive sampling method to acquire crop growth index data [34–36]. However, this method needed to be improved upon in monitoring the same plant over an extended period, resulting in a sparse data collection and errors during model construction. This study utilized a weighing device to monitor plant growth continuously, while agricultural IoT equipment was used to visualize and analyze the data online. In a study by Liu et al. [22] on tomato appearance and morphology in insulated plastic greenhouses scaled through an irradiation heat product, it was found that the plant height had an RMSE of 13.66 cm. At the same time, the LAI was 1.03 $m^2 \cdot m^{-2}$. In this experiment, the average RMSE of the plant height was 10.81 cm, and the LAI was 0.55 $m^2 \cdot m^{-2}$. Overall, the findings indicate that the model developed in this study closely aligns with the simulation model based on irradiation heat accumulation. Notably, it requires monitoring only the fresh weight of fresh plants without necessitating the installation of environmental monitoring equipment. This feature makes it well suited for adoption in conventional greenhouses that lack sophisticated technology, enhancing its practicality.

The simulation results for the three methods were similar. All of the models simulated the height of tomato plants. The winter stubble in the solar greenhouse was more accurate than the spring stubble in the plastic greenhouse. The latter greenhouse did not have a back wall. This made it less accurate at preserving and storing heat. The second greenhouse's simulation may not handle cold temperatures well. This can cause slight inaccuracies. This can prevent tomato plants from growing well and make the simulated and real plant heights vary. The simulation accuracy of the leaf area index of overwintering tomatoes in the solar greenhouse is better than that of spring stubble in a thermal insulation plastic greenhouse. The reason for the error may be that the solar greenhouse's thermal insulation and heat storage capacity are more robust than those of the thermal insulation plastic greenhouse. The growth of tomato leaves is inhibited under low-temperature conditions, and the leaves are prone to disease and aging. When collecting leaf area data, the diseased or aging leaves had been removed by managers through agricultural operations, resulting in an inevitable simulation error. The superposition effect between the two led to the error between the simulated and measured tomato leaf area index values. Nevertheless, it needs to be more accurate in the multispan plastic greenhouse. The accuracy difference might be because the tomato plants were still seedlings in the solar greenhouse. The simulation is more accurate because these plants were mainly growing stems and leaves.

In the actual validation, it was found that the distribution indices of the tomato morphological organs were greatly affected by the environment, and the measured values of daylight greenhouse and continuous plastic greenhouse were generally higher than the simulated values because the environments inside their chambers were more stable and more suitable for tomato growth. When long-term low-temperature, low-light, and other growth adversities occur, the dry matter allocation of tomato stems and leaves will be inhibited to different degrees, which is consistent with the results of Gao et al. [37]. The distribution coefficient significantly affects the difference between the simulated results and the measured values of the tomato plant height and leaf area index. In addition to time factors, environmental factors, water and fertilizer measures, etc., will specifically impact the tomato distribution index [38,39]. This study considered only the linear coupling between the fresh weights of tomato organs and the plant height and leaf area index, and the influence of nonlinear coupling on the final research results remains to be further investigated.

For all experiments conducted in this study, we employed the substrate bag cultivation method. In contrast, growers in traditional Chinese greenhouses mainly use soil cultivation, which can affect tomato growth differently under different cultivation conditions [40–42]. In order to improve the applicability of the model in traditional Chinese greenhouses, the model will be validated in the future in different types of greenhouses, using different cultivation methods, and in different climatic zones, and based on this, the model will contribute to the future development of tomato-related research on water demand and irrigation decision-making devices.

5. Conclusions

The study suggests a new way for monitoring the greenhouse tomato plant height and leaf area index. This method is nondestructive and based on real-time simulation. It offers a fresh approach to building greenhouse tomato simulations. It also helps estimate tomato crop transpiration and create water-saving irrigation strategies as follows:

(1) In this study, we developed and installed a long-term, real-time monitoring device for tomato aboveground wire volume, which monitors the aboveground weight of tomato plants in real time without disrupting the average growth of tomatoes, with a coefficient of determination R^2 of 0.937, RMSE of 0.05 kg, and MAE of 0.04 kg;

(2) Simulation models of the tomato plant height and leaf area index based on real-time weight were constructed, through which the predicted values of the tomato plant height and leaf area index in different greenhouses were estimated to fit well with the measured values. The average coefficient of determination R^2 in the simulation of the plant height was 0.817, the RMSE was 10.81 cm, and the integrated simulation effect of the linear function was good; the average coefficient of determination R^2 in the simulation of the leaf area index was 0.854, the RMSE was 0.55 $m^2 \cdot m^{-2}$, and the polynomial function simulation was better. The model can be used to estimate the plant height and leaf area index in real time, which provides a new way of thinking for the construction of a simulation model of tomato growth indices in facilities and also establishes a particular foundation for the estimation of tomato evapotranspiration and the formulation of water-saving irrigation strategies.

Supplementary Materials: The following supporting information can be downloaded at https://www.mdpi.com/article/10.3390/horticulturae10030270/s1: Table S1 Correlation analysis of tomatoes' morphological organs with plant height and leaf area; Table S2 Model construction and model validation data; Table S3 Accuracy testing of dynamometer devices; Table S4 Dynamometer-based simulation model testing of plant height and leaf area index.

Author Contributions: Data collection, S.G. and L.W.; data curation, S.G.; figures, L.W.; formal analysis, Y.L. and X.S.; supervision, H.S. and Y.C.; resources, Y.C.; writing—review and editing, S.G., L.W., X.C., X.S., Y.C., Y.L. and H.S.; writing instruction, Y.C. and H.S. All authors have read and agreed to the published version of the manuscript.

Funding: This research was funded by the Xinjiang Uygur Autonomous Region Major Sciences and Technology Special Project (2022A02005-5, 2022A02005-1), Xinjiang Uygur Autonomous Region Public Welfare Scientific Research Institutes Project (KY2023036), and Key Laboratory: Xinjiang Key Laboratory of Intelligent Agricultural Facility Management and Control Technology (XJYS1703).

Data Availability Statement: The original contributions presented in the study are included in the article and supplementary material, further inquiries can be directed to the corresponding authors.

Acknowledgments: We fully appreciate the editors and all anonymous reviewers for their constructive comments on this manuscript.

Conflicts of Interest: Author Letian Wu, Xinwei Cao, Xiaoli Sun, and Huifen Shi were employed by the Agricultural Engineering Company. The remaining authors declare that the research was conducted in the absence of any commercial or financial relationships that could be construed as a potential conflict of interest.

References

1. National Greenhouse Data Device. Available online: http://data.sheshiyuanyi.com/AreaData/ (accessed on 1 November 2023).
2. Yang, S.Y.; Liu, X.; Liu, S.Y.; Chen, X.Y.; Cao, Y.F. Real Time Temperature Distribution Monitoring in Chinese Solar Greenhouse Using Virtual LAN. *Agronomy* **2022**, *12*, 1565. [CrossRef]
3. Li, T.L. Development Status of China's Facility Vegetable Industry and Outlook. *China Veg.* **2023**, *9*, 1–6.
4. Yu, X.M.; Zhang, J.W.; Zhang, Y.H.; Ma, L.L.; Jiao, X.C.; Zhao, M.F.; Li, J.M. Identification of optimal irrigation and fertilizer rates to balance yield, water and fertilizer productivity, and fruit quality in greenhouse tomatoes using TOPSIS. *Sci. Hortic.* **2023**, *311*, 11829. [CrossRef]
5. Bonotto, D.M.; Wijesiri, B.; Goonetilleke, A. Nitrate-dependent Uranium mobilisation in groundwater. *Sci. Total Environ.* **2019**, *693*, 133655. [CrossRef]
6. Wu, B.F.; Tian, F.Y.; Zhang, M.; Piao, S.L.; Zeng, H.W.; Zhu, W.W.; Liu, J.G.; Elnashar, A.; Lu, Y.M. Quantifying global agricultural water appropriation with data derived from earth observations. *J. Clean. Prod.* **2022**, *358*, 131891. [CrossRef]
7. Liu, W.F.; Liu, X.C.; Yang, H.; Ciais, P.; Wada, Y. Global Water Scarcity Assessment Incorporating Green Water in Crop Production. *Water Resour. Res.* **2022**, *58*, e2020WR028570. [CrossRef]
8. Rozenstein, O.; Fine, L.; Malachy, N.; Richard, A.; Pradalier, C.; Tanny, J. Data driven estimation of actual evapotranspiration to support irrigation management: Testing two novel methods based on an unoccupied aerial vehicle and an artificial neural network. *Agric. Water Manag.* **2023**, *283*, 108317. [CrossRef]
9. Tian, Y.Q.; Gao, L.H. Theory and Technology for Facility Cultivation of High-quality Tomato. *China Veg.* **2021**, *2*, 30–40.
10. Kehili, M.; Choura, S.; Zammel, A.; Allouche, N.; Sayadi, S. Oxidative stability of refined olive and sunflower oils supplemented with lycopene-rich oleoresin from tomato peels industrial by product, during accelerated shelf-life storage. *Food Chem.* **2018**, *246*, 296–304. [CrossRef]
11. Liu, Z.L.; Gao, J.J.; Gu, D.Y.; Zhang, Y.Y.; Jiao, J.; Liu, S.Q.; Tian, X.F. Effects of nitrogen fertilizer levels on quality, yield and nutrient absorption of substrate cultivation tomato in solar greenhouse. *J. Arid Land Resour. Environ.* **2019**, *33*, 7.
12. Nam, D.S.; Moon, T.; Lee, J.W.; Son, J.E. Estimating transpiration rates of hydroponically grown paprika via an artificial neural network using aerial and root zone environments and growth factors in greenhouses. *Hortic. Environ. Biotechnol.* **2019**, *60*, 913–923. [CrossRef]
13. Fan, S.N.; Ji, Y.H.; Liu, M.C.; Li, W.; Liang, H.; Wu, Z.H.; Wang, L.P. Physiological responses of tomato leaves to different irrigation rates. *Water Sav. Irrig.* **2023**, *1*, 19–25.
14. Ali, G.F.; Mahmoud, R.-S.; Ali, S.; Hamid, Z. Non-destructive estimation of sunflower leaf area and leaf area index under different water regime managements. *Arch. Agron. Soil Sci.* **2015**, *61*, 1357–1367. [CrossRef]
15. Liu, Z.K.; Niu, X.Y.; Wang, Y.; Hang, W.T. Estimation of plant height of winter wheat based on UAV visible image. *J. Triticeae Crops* **2019**, *39*, 859–866.
16. Li, J.; Shen, L.X.; Zhang, K.L.; Li, B.N. Effects of drip irrigation soluble organic fertilizer on tomato growth and quality. *Water Sav. Irrig.* **2022**, *12*, 37–41.

17. Elnaz, A.; Mohammad, G.; Jan, M.; Petr, K.; Marek, K. Impacts of environmental factors and nutrients management on tomato grown under controlled and open field conditions. *Agronomy* **2023**, *13*, 916.
18. Tao, H.B.; Lin, B. Comparison on disc method with copy method and length-width method for measuring leaf area of rice. *Plant Physiol. Commun.* **2006**, *42*, 496.
19. Dai, Z.C.; Du, D.L.; Si, C.C.; Lin, Y.; Hao, J.L.; Sun, F. A method to exactly measure the morphological quantity of leaf using Scanner and Image J Software. *Guihaia* **2009**, *29*, 342–347.
20. Pan, T.H.; Wang, Y.L.; Yang, J.W.; Xi, L.J.; Ding, J.J.; Zhang, J.J.; Zou, Z.R. Screening and Optimization of Rapid Measurement Method for Leaf Area of Vegetable Seedlings. *China Veg.* **2018**, *8*, 64–69.
21. Bacci, L.; Battista, P.; Rapi, B. Evaluation and adaptation of TOMGRO model to Italian tomato protected crops. *N. Z. J. Crop Hortic. Sci.* **2012**, *40*, 115–126. [CrossRef]
22. Wang, X.; Diao, M.; Ma, F.F.; Fan, H.; Cui, J.; He, H.B. Simulation of leaf area, dry matter production and accumulation of processing tomato with drip irrigation. *Agric. Mach.* **2014**, *45*, 161–168.
23. Wang, D.; Guo, S.B.; Liu, F.H.; Cao, Y.F. Adaptation and application of development stage model for greenhouse pepper in Guanzhong area. *J. China Agric. Univ.* **2023**, *28*, 74–86.
24. Chang, Y.B.; Li, J.M.; Shang, X.M.; Zhang, D.L.; Pan, T.H.; Du, Q.J. Simulation of tomato morphology growth with water-fertilizer coupling. *J. Northwest A F Univ. (Nat. Sci. Ed.)* **2015**, *43*, 126–133.
25. Cheng, C.; Dong, C.Y.; Li, Z.F.; Gong, Z.H.; Feng, L.P. Simulation model of external morphology and dry matter accumulation and distribution of celery in solar greenhouse. *Trans. Chin. Soc. Agric. Eng.* **2021**, *37*, 142–151.
26. Zhai, Z.H.; Chen, X.W.; Gao, L.P.; Zhang, T.Z. Simulation of greenhouse cucumber plant height growth based on gradient boosting algorithm. *J. China Agric. Univ.* **2022**, *27*, 134–145.
27. Xu, L.H.; Meng, F.Z.; Wei, R.H. Development and verification of tomato crop-environment interaction model in second timescale greenhouse. *Trans. Chin. Soc. Agric. Eng.* **2021**, *38*, 212–222.
28. Lee, J.H.; Goudriaan, J.; Challa, H. Using the expolinear growth equation for modelling crop growth in year-round cut chrysanthemum. *Ann. Bot.* **2003**, *92*, 697–708. [CrossRef] [PubMed]
29. Wang, D.D.; Lv, Z.N.; Li, J.; Ma, J.M.; Sun, Z.P. Comparison of dry matter partitioning model of tomato cultivated with growth-bag during different growing seasons based on product ofthermal effectiveness and photosynthesis active radiation in solar greenhouse. *Acta Agric. Boreali-Occident. Sin.* **2018**, *27*, 238–243.
30. Luo, W.H.; Li, Y.X.; Dai, J.F.; Jin, L.; Xu, G.B.; Chen, Y.S.; Chen, C.H. Simulation of greenhouse tomato drt matter partitioning and yield prediction. *Chin. J. Appl. Ecol.* **2006**, *05*, 811–816.
31. Guo, S.B.; Liu, F.H.; Wang, D.; Huang, B.; Cao, Y.F. Constrution of tomato growth rate simulation model based on climate suitability index. *Chin. J. Agrometeorol.* **2023**, *44*, 611–623.
32. Liu, H.J.; Shao, M.X.; Yang, L. Photosynthesis Characteristics of Tomato Plants and Its' Responses to Microclimate in New Solar Greenhouse in North China. *Horticulturae* **2023**, *9*, 197. [CrossRef]
33. Cao, Y.F.; Shi, M.; Liu, X.; Ding, J.J.; Li, J.M.; Sun, Y.C. Effects of thermal insulation projection on indoor light environment and tomato growth performance in plastic greenhouse. *Trans. Chin. Soc. Agric. Eng.* **2021**, *37*, 190–196.
34. Flores-Velázquez, J.; Rojano, F.; Aguilar-Rodríguez, C.E.; Villagran, E.; Villarreal-Guerrero, F. Greenhouse Thermal effectiveness to produce tomatoes assessed by a temperature-based index. *Agronomy* **2022**, *12*, 1158. [CrossRef]
35. Shi, X.H.; Cai, H.J.; Zhao, L.L.; Yang, P.; Wang, Z.S. Greenhouse tomato dry matter production and distribution model under condition of irrigation based on product of thermal effectiveness and photosynthesis active radiaton. *Trans. Chin. Soc. Agric. Eng.* **2016**, *32*, 69–77.
36. Sun, S.K.; Wang, X.T.; Xu, J.T.; Fu, H.D.; Sun, Z.P. Study of tomato growth weight distribution model based on real time plant weight in a solar greenhouse. *J. Taibah Univ. Sci.* **2021**, *15*, 1027–1037. [CrossRef]
37. Lin, D.Y.; Wei, R.H.; Xu, L.H. An Integrated Yield Prediction Model for Greenhouse Tomato. *Agronomy* **2019**, *9*, 873. [CrossRef]
38. Liu, F.H.; Guo, S.B.; Wang, D.; Huang, B.; Cao, Y.F. Construction and verification of an external morphology, substance accumulation, and distribution model of tomatoes in greenhouse. *Trans. Chin. Soc. Agric. Eng.* **2022**, *38*, 188–196.
39. Gao, H.; Wu, F.Z.; Zhou, X.G.; Gao, D.M.; Yang, W.J. Effects of sub-low temperature on dry matter accumulation, nitroge and phosphours absorption and distribution in different tomato cultivars. *Acta Agric. Boreali-Occident. Sin.* **2018**, *27*, 994–1001.
40. Li, S.L.; Li, Y.; Zhou, L.J.; Niu, X.X.; Yu, L.M. Quantitative assessment of temperature suitability of alpine summer tomato seeding in west of Henan province. *Trans. Chin. Soc. Agric. Eng.* **2019**, *35*, 194–202.
41. Doan, C.C.; Munehiro, T. Relationships between tomato cluster growth indices and Ccumulative environmental factors during greenhouse cultivation. *Sci. Hortic.* **2022**, *295*, 110803. [CrossRef]
42. Qu, Z.Y.; Sun, H.H.; Yang, B.; Gao, X.Y.; Wang, L.M.; Wang, L.P. Effects of different amendments on soil microorganisms and yield of processing tomato in saline alkali soil. *Trans. Chin. Soc. Agric. Mach.* **2021**, *52*, 311–318.

horticulturae

MDPI

Article

Human-Following Strategy for Orchard Mobile Robot Based on the KCF-YOLO Algorithm

Zhihao Huang [1], Chuhong Ou [1], Zhipeng Guo [1,2], Lei Ye [1,*] and Jin Li [1,*]

1 School of Intelligent Engineering, Shaoguan University, Shaoguan 512000, China; haurse_zh@163.com (Z.H.)
2 College of Engineering, South China Agricultural University, Guangzhou 510642, China
* Correspondence: leoye1992@sgu.edu.cn (L.Y.); sgulijin@sgu.edu.cn (J.L.)

Abstract: Autonomous mobile robots play a vital role in the mechanized production of orchards, where human-following is a crucial collaborative function. In unstructured orchard environments, obstacles often obscure the path, and personnel may overlap, leading to significant disruptions to human-following. This paper introduces the KCF-YOLO fusion visual tracking method to ensure stable tracking in interference environments. The YOLO algorithm provides the main framework, and the KCF algorithm intervenes in assistant tracking. A three-dimensional binocular-vision reconstruction method was used to acquire personnel positions, achieving stabilized visual tracking in disturbed environments. The robot was guided by fitting the personnel's trajectory using an unscented Kalman filter algorithm. The experimental results show that, with 30 trials in multi-person scenarios, the average tracking success rate is 96.66%, with an average frame rate of 8 FPS. Additionally, the mobile robot is capable of maintaining a stable following speed with the target individuals. Across three human-following experiments, the horizontal offset Error Y does not exceed 1.03 m. The proposed KCF-YOLO tracking method significantly bolsters the stability and robustness of the mobile robot for human-following in intricate orchard scenarios, offering an effective solution for tracking tasks.

Keywords: orchard scene; mobile robot; KCF-YOLO; human-following; visual tracking

Citation: Huang, Z.; Ou, C.; Guo, Z.; Ye, L.; Li, J. Human-Following Strategy for Orchard Mobile Robot Based on the KCF-YOLO Algorithm. *Horticulturae* **2024**, *10*, 348. https://doi.org/10.3390/horticulturae10040348

Academic Editor: Stefano Poni

Received: 18 February 2024
Revised: 23 March 2024
Accepted: 29 March 2024
Published: 31 March 2024

1. Introduction

The increasing use of autonomous mobile robots in collaborative transportation tasks has positioned this field as a rapidly advancing research area [1–7]. Over the past two decades, significant advancements in sensor technology, robotic hardware, and software have driven the widespread adoption of mobile robots in diverse industries [8–13]. In agriculture, collaborative mobile robots have emerged as key replacements for labor-intensive tasks, highlighting their potential as essential servo technologies. Human-following technology has attracted considerable attention in structured environments, such as factory logistics, transportation, and airport operations, and has seen partial commercialization [14–16]. However, this technology is still experimental in complex orchard environments. The complex nature of an orchard environment, coupled with numerous obstacles and occlusions, makes the human tracking process susceptible to recognition failure, target loss, and other problems. These challenges lead to the failure of the mobile tracking function of robots, which is a significant obstacle to the successful operation of mobile robots.

To handle the aforementioned challenges, this paper proposes a human-following strategy for an orchard mobile robot based on KCF-YOLO. The method includes two main components: personnel detection and tracking, and human-following. The KCF algorithm, as a traditional visual tracking method, mainly achieves target tracking by adopting a histogram of oriented gradients (HOG) features, which results in higher tracking accuracy compared to grayscale or color feature methods. The YOLO v5s algorithm, as an efficient object detection algorithm, boasts excellent detection speed and accuracy. Therefore, this method combines the strengths of both the KCF and YOLO v5s algorithms, achieving continuous and stable tracking of target individuals in orchard environments (Section 3.3). Building upon stable visual tracking, spatial information of individuals is obtained through binocular stereo vision methods and transformed into coordinates in the vehicle's frame. Due to the potential positional oscillations in the three-dimensional spatial information acquired through cameras, dynamic modeling of target trajectories is implemented. The unscented Kalman filter (UKF) is introduced to predict the trajectories of individuals (Section 3.4), thereby enhancing the stability of following. Simultaneously, multi-person interference experiments are conducted on the KCF-YOLO algorithm to evaluate its robustness. Furthermore, the integration of the KCF-YOLO algorithm onto a mobile robot is performed to assess its performance in real orchard environments (Section 4.2).

This study aimed to devise a method for accurately recognizing and tracking a target within a complex scene, thereby achieving stable mobile robots in such intricate environments. The main contributions of this paper are summarized as follows:

1. A visual tracking algorithm is proposed in the paper, with the YOLO algorithm as the main framework and the KCF algorithm introduced for auxiliary tracking, aiming to achieve continuous and stable tracking of targets in orchard environments.
2. A KCF-YOLO human-following framework has been constructed in the paper, which can be employed for human-following based on visual mobile robots in real orchard scenarios.

The remainder of this paper is organized as follows: In Section 2, related works are discussed. Section 3 provides a detailed description of the proposed KCF-YOLO visual tracking algorithm, along with the human-following method. Section 4 describes the experimental validation of the proposed algorithm and discusses the experimental results. Finally, Section 5 concludes the paper with a summary and outlook on the proposed methodology.

2. Related Works

With the rapid development of computer vision technology, human-following methods based on machine vision have received widespread attention [17–19]. These methods involve two crucial steps: the first is target detection and tracking, and the second is human-following. For target detection and visual tracking, Bolme et al. [20] proposed a minimum output sum of squares of error filter that generates a stable correlation filter via single-frame initialization to enhance the tracking robustness against rotation, scale variation, and partial occlusion. However, this method is sensitive to changes in the target color and brightness, making it prone to tracking errors when the target moves rapidly or closely resembles the background. Henriques et al. [21] proposed a high-speed kernel correlation factor (KCF) algorithm that uses a cyclically shifted ridge regression method to reduce memory and computation significantly, thereby improving the execution speed of the algorithm. When encountering changes in the appearance of non-rigid targets, such as the human body, adaptation through the online updating of the tracking model is essential.

Nonetheless, online updating can lead to drifts in tracking. To solve the drift problem, Liu et al. proposed a real-time target response-adaptive and scale-adaptive KCF tracker that can detect and recover from drifts [22]. Despite this, drift errors persist in long-term target tracking, impacting the system robustness owing to frequent changes in the target attitude and appearance. To mitigate the error caused by tracking drift and enhance system robustness, Huan et al. proposed a tracking method using a structured support vector machine and the KCF algorithm [23]. This approach optimizes the search strategy for tracking the motion characteristics of a target, thereby reducing the search time for dense sampling. Consequently, it improves the search efficiency and classifier accuracy compared with the traditional KCF algorithm in the setting of a dense sample. Search efficiency and classifier accuracy in dense sampling improve computational efficiency and target tracking performance in complex environments, solving the problem of failing to accurately track the target owing to the drift caused by changes in target size and rapid movement. Nevertheless, the KCF algorithm faces the significant drawback of poor tracking robustness owing to obstacle occlusion. Bai et al. introduced the KCF-AO algorithm to solve the tracking failure problem caused by occlusion in the KCF algorithm [24]. This algorithm employs the confidence level of the response map to assess the tracking result of each frame. In cases where the target disappearance is detected, it employs the A-KAZE feature point matching algorithm and the normalized correlation coefficient matching algorithm to complete the redetection of the target. The position information is then fed back to KCF to resume tracking, which improves the robustness of the tracking performance of the KCF algorithm. Mbelwa et al. proposed a tracker based on object proposals and co-kernelized correlation filters (Co-KCF) [25]. This tracker utilizes object proposals and global predictions estimated using a kernelized correlation filter scheme. Through a spatial weight strategy, it selects the optimal proposal as prior information to enhance tracking performance in scenarios involving fast motion and motion blur. Moreover, it effectively handles target occlusions, overcoming issues such as drift caused by illumination variations and deformations. The studies mentioned underscore researchers' substantial contributions to overcome challenges related to obstacle occlusion, target size variation, and the rapid movement faced by the KCF algorithm in mobile robot tracking. These findings pave the way for novel advancements in visually guided human-following techniques. However, the KCF target-tracking algorithm is computationally complex, and detection accuracy based on artificially designed features is unsatisfactory for partially occluded human bodies.

In contrast, deep learning methods offer innovative approaches in the realm of people detection and following. Gupta et al. proposed the use of the mask region-based convolutional neural network (mask RCNN) and YOLO v2-based CNN architectures for personnel localization, along with speed-controlled tracking algorithms [26]. Boudjit et al. introduced a target-detecting unmanned aerial vehicle (UAV) following a method based on the YOLO-v2 architecture and achieved UAV target tracking by combining detection algorithms with proportional–integral–derivative control [27]. Additionally, the single-shot multi-box detector (SSD) target detection algorithm proposed by Liu et al. is even faster than the so-called faster RCNN detection method. It offers a significant advantage in the mean average precision achieved compared with YOLO [28]. Algabri et al. presented a framework combining an SSD detection algorithm and state-machine control to identify a target person by extracting color features from video sequences using H-S histograms [29]. This framework enables a mobile robot to effectively identify and track a target person. The aforementioned methods involve integrating deep learning with traditional computer vision techniques for fine-grained target detection and tracking.

Stably following a target person in complex scenarios poses a significant challenge in human-following. In the following problem, effectively avoiding obstacles while continuously following a target and keeping the target within the robot's field of view is an important part of realizing a human-following task. Han et al. utilized the correlation filter tracking algorithm to track the target individual [30]. In instances of tracking failure, they introduced facial matching technology for re-tracking, achieving continuous tracking of the target person in indoor environments and improving the stability of the tracking process. Cheng-An et al. obtained obstacle and human features using an RGB-D camera and estimated the next moment state of pedestrians using the extended Kalman filter (EKF) algorithm to achieve stable human-following in indoor environments [31]. However, outdoor environments are more complex than indoor environments. Human stability is affected by various factors, including diverse terrains, unstructured obstacles, and dynamic pedestrian movements. Gong et al. proposed a point cloud-based algorithm, employing a particle filter to continuously track the target's position. This enables the robot to detect and track the target individual in outdoor environments [32]. Tsai et al. achieved human-following in outdoor scenes using depth sensors to determine the distance between the tracking target and obstacles [33]. A Kalman filter predicts the target person's position based on the relative distance between the mobile robot and the target person. However, the applicability of this method is limited to relatively simple scenes, making it unstable in complex orchard environments.

Human-following techniques for orchard environments face a lack of effective solutions. The traditional KCF algorithm exhibits instability when confronted with challenges, such as obstacle occlusion, variations in target size, and rapid movements. Consequently, they lack persistent tracking capabilities for specific individuals or targets, which restricts their effectiveness in orchards. Although the YOLO target detection algorithm demonstrated high accuracy in orchards, its role as a detection tool limited its ability to track specific individuals, limiting its applicability in complex scenarios. Ensuring stability and devising effective strategies for robots are paramount in an unstructured orchard environment. The absence of state estimation and prediction capabilities in robots operating in uncertain and complex environments results in a lack of stability in target following. Therefore, a novel human-following strategy is proposed to enhance the robustness of target tracking and improve the adaptability and stability of robot movements in orchard scenarios.

3. Algorithm

Based on the KCF and the YOLO v5s algorithms, this paper proposes a comprehensive human-following system framework. This framework utilizes camera sensors to acquire three-dimensional spatial information of individuals, which is then transformed into the coordinate system of a mobile robot. The unscented Kalman filter (UKF) algorithm is employed to predict the trajectory of individuals based on their three-dimensional information. When the target individual moves out of the robot's field of view, the KCF-YOLO algorithm is used to retrack the target individual upon re-entry into the field of view, enabling continuous tracking of individuals. The overall framework of the human-following system is illustrated in Figure 1.

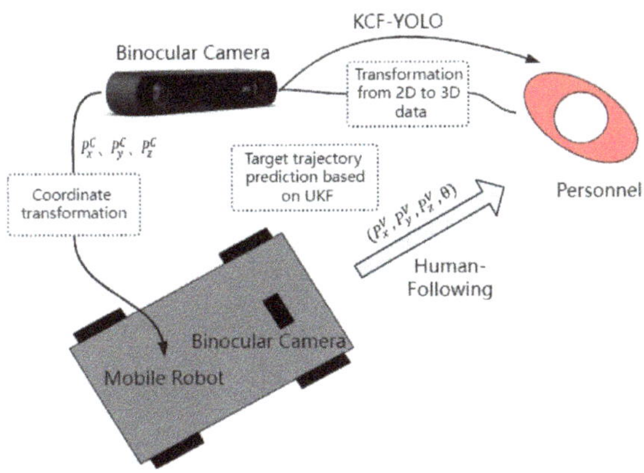

Figure 1. Human-following framework. P_x^C, P_y^C, and P_z^C represent the three-dimensional coordinates of the target personnel in the camera coordinate system. $(P_x^V, P_y^V, P_z^V, \theta)$ represents the posture of the target personnel at a specific moment in time.

3.1. *Kernel Correlation Filter (KCF)*

The KCF is a target-tracking algorithm based on online learning that encompasses three key steps: feature extraction, online learning, and template updating. Initially, the algorithm extracts HOG features from the target, generating a Fourier response. Subsequently, the correlation of the Fourier response is computed to estimate the target location. Following that, the classifier is trained by cyclically shifting image blocks around the target location and adjusting the weights of the KCFs through a ridge regression formulation. Continuous target tracking is achieved through online learning and updating, leveraging the real-time detected target position and adjusted filter weights. The flow of the KCF algorithm is illustrated in Figure 2.

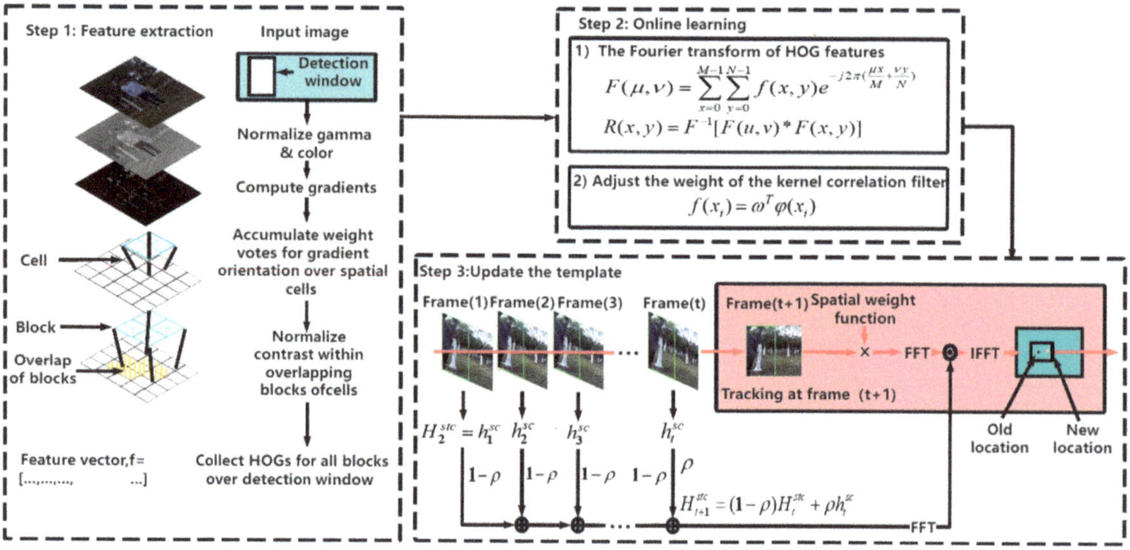

Figure 2. KCF algorithm.

In Step 1, the HOG feature extraction process is illustrated. In Step 2, $F\ (u,\ v)$, a complex-valued spectrum in the frequency domain, represents information regarding the frequency component (u, v). $f\ (x, y)$ denotes the pixel intensity value at coordinates (x, y) in the image, and $R\ (x, y)$ signifies the response at position (x, y) in the image. In Step 3, the template update process in the KCF algorithm is elucidated, where FFT represents the fast Fourier transform, and IFFT represents the inverse Fourier transform. The Gaussian kernel function in the KCF algorithm plays a crucial role in modeling the similarity between the target and candidate regions. The *sigma* value, an essential parameter of the Gaussian kernel function, determines the bandwidth of the Gaussian kernel function, thereby directly affecting the stability of the KCF algorithm. The Gaussian kernel function in the frequency domain is typically represented as shown in Equation (1).

$$K(u,v) = e^{-2\pi^2\sigma^2(u^2+v^2)} \tag{1}$$

The KCF algorithm demonstrates robust real-time capabilities and accuracy in practical applications, particularly in real-time video target tracking, and effectively addresses the challenges associated with target deformation and scale changes. However, in complex orchard environments, the robustness of the tracking performance of the KCF algorithm diminishes because of factors such as occlusions between trees. Therefore, the integration of additional techniques is essential to enhance the tracking performance in specific application scenarios.

3.2. YOLO

The YOLO series of algorithms is a fast and efficient object detection algorithm that can perform object detection and classification directly in the entire image. It provides both the position and category probability for each detected object box [34]. The network structure of the YOLO v5s algorithm was categorized into four modules: input, backbone, neck, and prediction. The network architecture is shown in Figure 3.

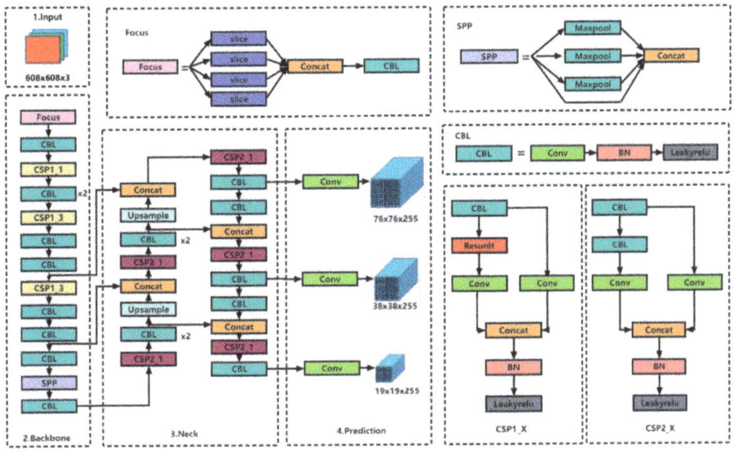

Figure 3. YOLO v5s framework.

Initially, the preprocessed image undergoes characterization and a series of convolutional processes in the backbone layer. Subsequently, the neck layer integrates the feature pyramid network and path aggregation network to construct multi-scale feature information. Finally, the prediction layer utilizes three feature maps to predict the target class and generate information regarding the target box location.

The algorithm proposed in this paper serves as a framework, not limited to a specific version of YOLO. The YOLOv5s algorithm strikes a balance between performance and speed, catering to specific scenarios and the requirements of existing hardware. Simultaneously, its deployment is more straightforward on mobile robots. While higher versions may offer additional features, they demand increased computational resources. In the context of mobile robot tracking tasks, real-time information is crucial for system stability. Therefore, this study harnesses the advantages of YOLO v5s regarding target detection accuracy, coupled with its real-time capabilities and adaptability to the continuous tracking of KCF. A target detection algorithm that integrates the strengths of both approaches to enhance the stability and reliability of target visual tracking is proposed.

3.3. *Proposed KCF-YOLO Algorithm*

The KCF algorithm is known for its high efficiency and accuracy in real-time tracking, excelling when the target size remains relatively constant and there are no occlusions. However, in intricate orchard environments characterized by occlusions, such as trees and overlapping pedestrians, the KCF algorithm faces challenges that lead to failures in target tracking. Conversely, the YOLO v5s algorithm demonstrates a rapid and accurate response in target recognition yet encounters difficulties in distinguishing and localizing specific objects among similar targets. To address the limitations of both algorithms in specific target tracking, this section introduces the KCF-YOLO fusion visual tracking algorithm. The implementation of this algorithm is illustrated in Figure 4.

The KCF-YOLO algorithm leverages the YOLO v5s algorithm for target detection. The algorithm identifies the target detection box and determines the position of the target center in the image. Through the continuous calculation of the range between the real-time target center and the edges of the field of view, the algorithm evaluates whether the detected target is on the verge of leaving the field of view. Based on this evaluation, the algorithm determines whether the KCF algorithm should intervene to provide auxiliary tracking. Suppose the algorithm determines that the tracked target is positioned at the edge of the field-of-view window. In that case, the KCF-YOLO algorithm utilizes the target detection frame obtained by the YOLO v5s algorithm as the region of interest to initialize the KCF algorithm for auxiliary tracking. Notably, the parameters distance serves as the trigger region for the KCF algorithm, denoted as *dis*, which represents a certain distance from the left or right boundary of the image to the image center, as shown in Figure 5. It is primarily used to determine whether the target individual is about to leave the frame and initiate the KCF algorithm for auxiliary tracking, directly impacting the detection efficiency of the KCF-YOLO algorithm. This study experimentally validated the KCF-YOLO algorithm using different *dis* and *sigma* values as test variables.

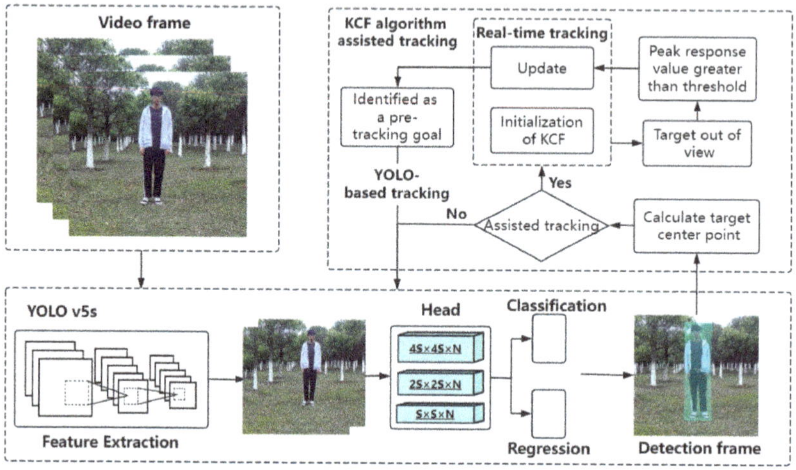

Figure 4. KCF-YOLO algorithm.

Figure 5. The trigger region of the KCF algorithm. The distance is denoted as *dis* in this paper.

There are two common target-loss scenarios during personnel visual tracking. Tracking loss occurs when the target is positioned at the edge of the image. To address this problem, the KCF algorithm calculates the response value by performing correlation operations in the region surrounding the region of interest. If the peak response value exceeds a preset threshold, the KCF-YOLO algorithm considers the position as a new tracking position. It continuously updates the tracking region of the target until the personnel exit the edge of the image. Second, in the complex environment of an orchard, the target is prone to tracking loss owing to obscuration and other circumstances. The system captures the frame of the image before the target leaves its field of view and performs a response value calculation. When the target-tracking frame is at the edge of the image, waiting for the target to re-enter the tracking area, the system waits for and monitors the target. If the target re-enters the image field of view area and the peak response value exceeds the set threshold, the system recognizes it as a specific target that has previously lost its field of view. The KCF is used to accomplish assisted visual tracking. Once a specific target returns to the field of view and retracing is confirmed, the system turns to the YOLO algorithm to complete the detection and tracking of that target. The process is illustrated in Figure 6.

Figure 6. Visual tracking of personnel. (**a**) Out of sight. (**b**) Re-tracking.

In complex orchard environments, the KCF algorithm may encounter difficulties in tracking owing to obstructions, such as bushes and fruit trees, or issues with pedestrian overlap. In contrast, the YOLO v5s algorithm struggles to determine whether the target re-entering the field of view after being lost is the original target. However, the KCF algorithm supports target tracking under specific conditions. When the target is lost and re-enters the field of view, YOLO v5s, as the primary tracker, can identify and continue tracking the previously lost target by combining it with the KCF. This addresses the deficiency of YOLO v5, which cannot confirm the original target when it re-enters the field of view. Consequently, this integration improves the stability and robustness of visual tracking in a complex orchard environment.

3.4. Human-Following Control Strategy

To maintain personnel within the field of view of the robot's camera, appropriate control commands must be generated based on the personnel's positional information. The human-following control strategy includes personnel moving trajectory acquisition, offset calculation of the personnel positions, and subsequent control. The process is described below.

3.4.1. Obtaining Personnel Trajectory

Human-following requires acquiring the position information of the target person and executing the corresponding behavior based on that person's spatial information. This study utilizes the software development kit (SDK) provided by the camera, version ZED 3.8.2, to extract 3D spatial information from two-dimensional (2D) image data, as illustrated in Figure 7a. This process stabilizes the tracking of the target person through the KCF-YOLO fusion algorithm, producing a target-tracking detection box, as shown in Figure 7b. The geometric center of the detection box is calculated based on its four corners. The coordinates of this geometric center in the 2D plane image are defined as the mapped 3D spatial information of the person in the camera coordinate system. After obtaining the geometric center of the person in the 2D plane image, a stereo-matching algorithm is used to find the corresponding feature points in the image, establishing the relationship between the feature points in the 2D image and the actual positions in the 3D space. Finally, the spatial position of the tracking target in the camera coordinate system is determined based on the binocular disparity parameters. The spatial position of the person is shown in Figure 7c.

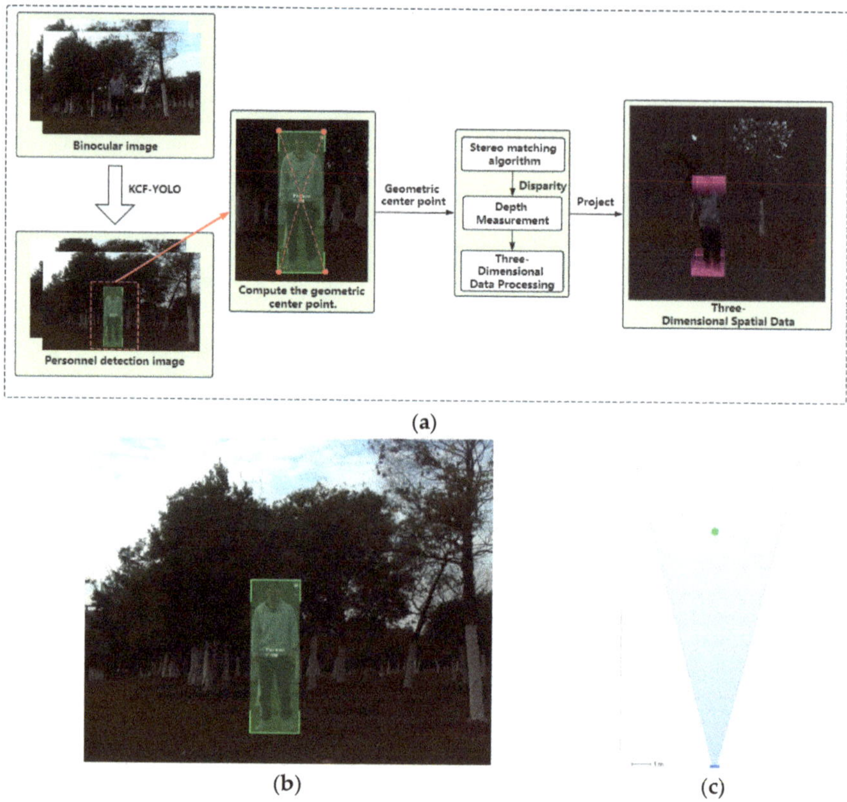

Figure 7. Personnel spatial data. (**a**) Transformation of 2D to 3D data. (**b**) Detection of personnel. (**c**) Location in space.

3.4.2. Personnel Trajectory Calculation

The task of human-following involves the processing of multiple coordinate systems. These encompass the world coordinate system W, the vehicle coordinate system V of the mobile robot, and the camera coordinate system C of the binocular camera, as illustrated in Figure 8.

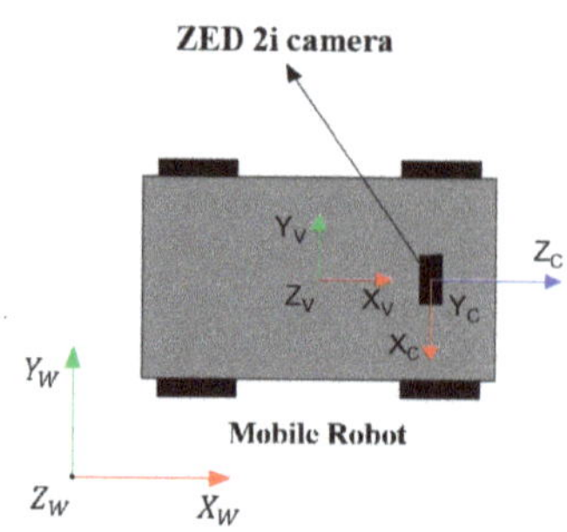

Figure 8. Coordinate relationship of mobile robot.

The origins of the camera and vehicle coordinate systems are defined as the geometric centers of the stereoscopic camera and mobile robot structure, respectively. The rotation relationship R and translation relationship T of the two coordinate systems are known. The spatial information of the individuals in the camera coordinate system can be transformed into the vehicle coordinate system through coordinate transformation, as expressed in Equation (2). The mobile robot can then acquire the spatial information of the dynamic personnel.

$$P_V = \begin{bmatrix} R & T \\ \mathbf{0}^{\mathrm{T}} & \mathbf{1} \end{bmatrix} P_C \tag{2}$$

Owing to the uncertainty in the personnel trajectory, relying solely on the spatial position obtained from the stereo camera is prone to result in positional oscillations, making real-time human-following tasks challenging. The UKF algorithm proves to be effective in predicting the trajectory of a target person. Even if the target exits the field of view of the camera, the mobile robot can track the target based on the predicted trajectory, guiding it back into the field of view. This significantly enhances the stability and robustness of human-following in uncertain environments. The trajectory of a target person's movement can be conceptualized as a combination of multiple curves and straight-line trajectories. For simplicity, the target person is assumed to travel along a straight line and move at a fixed turning rate. This motion model is defined as a constant turn rate and velocity (CTRV) motion model, as shown in Figure 9.

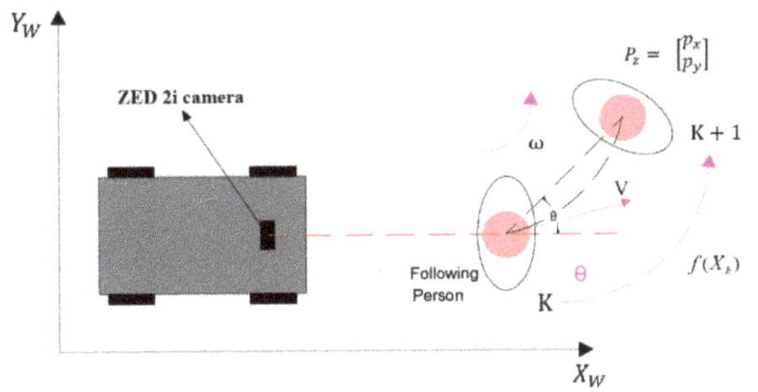

Figure 9. Human-following constant steering and velocity motion model. The personnel exhibit velocity in both the x- and y-directions in the robot coordinate system. In the CTRV model, it is assumed that the target moves with constant velocity and turn rate in the plane. Therefore, 'v' represents the velocity in the multi-vector space.

In the CTRV model, the state of the target person can be defined using Equation (3),

$$X = [p_x \; p_y \; v \; \theta \; \omega \,]^T \tag{3}$$

where p_x and p_y are the coordinates of the target person in the vehicle coordinate system, v is the linear velocity of the target, θ is the heading angle of the target moving in the vehicle coordinate system, and ω is the angular velocity of the target heading.

In real-world scenarios, achieving a uniform speed state for the target person is challenging. Therefore, it becomes necessary to introduce perturbations in the target person's motion model through noise simulation. The line acceleration α and angular acceleration $\dot{\omega}$ are considered to be process noise. The two are assumed to follow Gaussian distributions with a mean of 0 and variances of σ_a^2 and $\sigma_{\dot{\omega}}^2$, respectively. In other words, there exist $a \sim N(0, \sigma_a^2)$ and $a \sim N(0, \sigma_{\dot{\omega}}^2)$ such that the state transfer process noise is denoted as $W = \begin{bmatrix} a & \dot{\omega} \end{bmatrix}^T$, and the covariance of W can be expressed as shown in Equation (4).

$$Q = \begin{bmatrix} \sigma_a^2 \; 0 \\ 0 \; \sigma_{\dot{\omega}}^2 \end{bmatrix} \tag{4}$$

The human-following motion model can be expressed as $X_k = f(X_{k-1}, W_k)$, as shown in Equation (5).

$$X_k = \begin{cases} X_{k-1} + \begin{bmatrix} \frac{v}{\omega}[\sin(\omega_{k-1} \cdot \Delta t + \theta_{k-1})] - \sin(\theta_{k-1}) \\ -\frac{v}{\omega}[\cos(\omega_{k-1} \cdot \Delta t + \theta_{k-1})] - \cos(\theta_{k-1}) \\ 0 \\ \omega_{k-1} \cdot \Delta t \\ 0 \end{bmatrix} + \begin{bmatrix} \frac{1}{2}\Delta t^2 \cos(\theta_{k-1}) a_k \\ \frac{1}{2}\Delta t^2 \sin(\theta_{k-1}) a_k \\ \Delta t \cdot a_k \\ \frac{1}{2}\Delta t^2 \cdot \dot{\omega}_k \\ \Delta t \cdot \dot{\omega}_k \end{bmatrix} & \omega \neq 0 \\ X_{k-1} + \begin{bmatrix} v\cos(\theta_{k-1}) \\ v\sin(\theta_{k-1}) \\ 0 \\ \omega_{k-1} \cdot \Delta t \\ 0 \end{bmatrix} + \begin{bmatrix} \frac{1}{2}\Delta t^2 \cos(\theta_{k-1}) a_k \\ \frac{1}{2}\Delta t^2 \sin(\theta_{k-1}) a_k \\ \Delta t \cdot a_k \\ \frac{1}{2}\Delta t^2 \cdot \dot{\omega}_k \\ \Delta t \cdot \dot{\omega}_k \end{bmatrix} & \omega = 0 \end{cases} \tag{5}$$

The observation equation for the stereo camera of the target pedestrian is given by Equation (6):

$$Z_k = \begin{bmatrix} p_x \\ p_y \end{bmatrix} = HX_k + V_k = \begin{bmatrix} 1 & 0 & 0 & 0 & 0 \\ 0 & 1 & 0 & 0 & 0 \end{bmatrix} \begin{bmatrix} p_x \\ p_y \\ v \\ \theta \\ \omega \end{bmatrix} + \begin{bmatrix} r_k \\ r_y \end{bmatrix}_k \tag{6}$$

where $V_k = \begin{bmatrix} r_x & r_y \end{bmatrix}^T$ is the observation noise satisfying $r_x \sim N(0, \sigma_{r_x}^2)$ and $r_y \sim N(0, \sigma_{r_y}^2)$. Therefore, $V_k \sim N(0, R), R = \begin{bmatrix} \sigma_{r_x}^2 & 0 \\ 0 & \sigma_{r_y}^2 \end{bmatrix}$.

In this paper, the Q and R matrices are set according to default parameters, with a value of 0.8 for σ_a, 0.55 for $\sigma_{\dot{\omega}}$, and 0.15 for both σ_{r_x} and σ_{r_y}. In this process, the system state X_0 is first initialized, and a set of sigma points χ_i is generated based on the personnel $k-1$ moment states. The corresponding weights W_i for χ_i are constructed, and a nonlinear state function is used to predict the k-moment sigma points. The mean and covariance of the state at moment k are calculated. Subsequently, the means and covariances of the measurements are predicted. Finally, the Kalman filter gain K_k is derived from the measurements at moment k to estimate the state and variance at moment k. A flowchart of the process is shown in Figure 10.

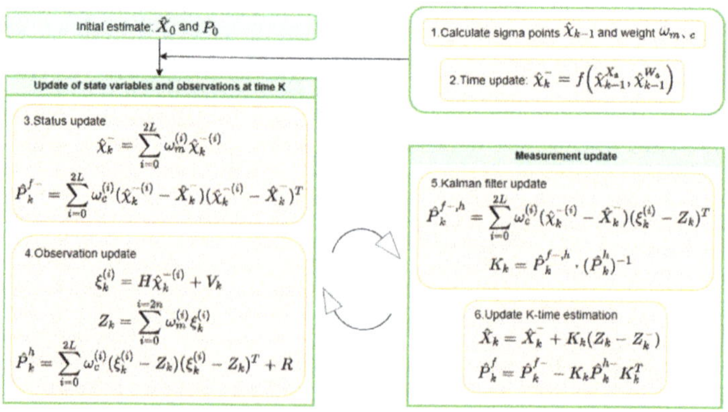

Figure 10. Human trajectory prediction based on the CTRV model.

3.4.3. Human-Following Control

After obtaining the predicted value of the personnel space, the x-axis of the robot is defined as the positive forward direction of the mobile robot. The angular offset between the personnel position and the robot is then calculated using Equation (7). p_x^v and p_y^v represent the personnel distances along the x- and y-axes, respectively, in the mobile-robot coordinate system.

$$\theta_{\mathrm{p}} = \arctan(\frac{p_y^v}{p_x^v}) \tag{7}$$

The mobile robot controls the steering angle according to the offset, thus keeping the target person in the field of view and following it at a safe distance. The strategy for the mobile robot is shown in Figure 11.

During the human-following process, the deviation angle from the robot when the target person is on one side of the field of view of the camera is θ_{p}, where there exists a heading-angle error threshold θ between the target and the robot. The robot does not need to execute steering commands within this threshold range. When the deviation angle θ_{p} exceeds a certain heading-angle error threshold, the robot controls the steering angle based on the magnitude of the deviation angle, which is denoted as θ_{p}. Simultaneously, the safe distance between the target person and the robot is defined as X_{safe}. When the distance p_z^v between the person being followed and the robot is larger than the predefined X_{safe}, the mobile robot continues to follow. Steering commands are executed to adjust the body position, ensuring the personnel returns to the center of the field of view of the camera. When the personnel stops moving and the distance p_z^v is less than or equal to X_{safe}, the mobile robot brakes slowly until it stops. By this process, a mobile robot can automatically track the target person.

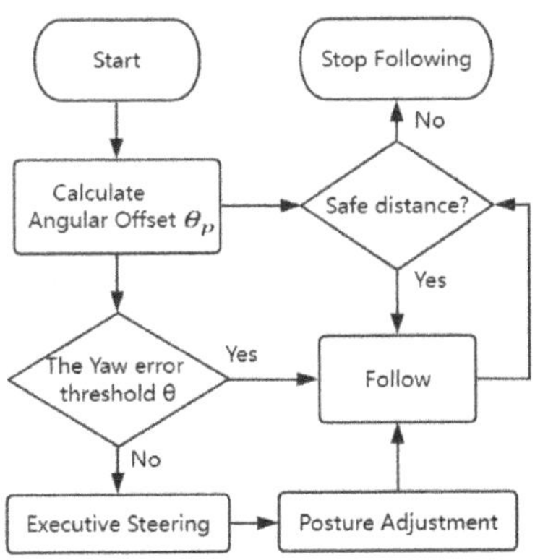

Figure 11. Human-following strategy.

4. Experiments and Discussions

4.1. Experimental Platform and Equipment

This study is based on an experimental design executed by a self-developed mobile robot in an orchard. The robot achieves intelligent autonomous following by integrating environmental sensors and an underlying control module. The overall test platform is illustrated in Figure 12. The key parameters are listed in Table 1. In addition, the mobile robot is outfitted with a binocular stereo camera with a resolution of 1920 × 1080 pixels and a frame rate of 30 FPS. Table 2 shows the main parameters of the binocular stereo camera.

Figure 12. Test platform.

Table 1. Main parameters of the test platform.

Parameter	Value
Length	160 mm
Width	80 mm
Height	120 mm
Maximum grade	20°
Running speed	0–3.6 km/h

Table 2. Main parameters of the binocular stereo camera.

Parameter	Value
Resolution	1920 × 1080 px
Frame rate	30 FPS
Baseline	12 cm (4.72 in)
Field of view (H × V × D)	Max. 72° (H) × 44° (V) × 81° (D)

To assess the robustness and stability of the proposed method, an experiment was conducted in a complex orchard environment. The ground in the orchard exhibits a firm texture, and simultaneously, it is covered with a dense layer of grass, comprising both herbaceous species and other low vegetation, as illustrated in Figure 13. The robot operating system platform was used for data processing. The evaluation of the tracking robustness of the KCF-YOLO algorithm and the stability of the mobile robot are discussed in the following subsections.

Figure 13. Test scenarios. The scene is mainly presented from an aerial view.

4.2. Experimental Results

To assess the recognition effectiveness of the KCF-YOLO algorithm in an orchard setting, experiments were conducted on two major modules: visual tracking and human-following. In the visual tracking section, we conducted three sets of experiments: tracking experiments under multi-person interference, investigations of the effects of different *dis* and *sigma* values on the efficiency of the KCF-YOLO algorithm, and comparisons between the KCF-YOLO algorithm and other algorithms. The evaluation indices for the algorithm performance included the average frame rate and recognition success rate. The reason for choosing the average frame rate as a quality metric is because it directly reflects the real-time performance evaluation of the system and its responsiveness. The average frame rate represents the average number of image frames displayed per second during the KCF-assisted tracking. When the algorithm fails to track a person, the FPS is recorded as 0 and is not included in the calculation of the average frame rate. This metric serves as a crucial indicator for measuring system real-time performance. A higher frame rate implies a more timely system response, consequently enhancing the robot's tracking performance of personnel. The recognition success rate is the ratio of the successful tracking of the target person when entering and leaving the field of view of the camera to the total number of entries and exits. The human-following experiment evaluates the stability of following through two different paths.

4.2.1. Visual Tracking Experiments under Multiple Interferences

In practical applications of human-following in orchard environments, encountering situations with multiple people simultaneously is common. To evaluate the impact of personnel interference on the recognition accuracy of the KCF-YOLO algorithm, tests were conducted in orchard scenarios with different numbers of individuals. Four sets of experiments were carried out with varying numbers of individuals: 1, 2, and 3, respectively. One of the experiments focused on visual tracking under different environmental conditions, potentially affecting detection stability and other factors. Tests were conducted in overcast weather conditions. The *dis* of the KCF-YOLO algorithm is set to 200 pixels, along with a *sigma* value of 0.2 for the Gaussian kernel function. Each set of experiments comprised 30 tests.

At the onset of the experiment, the personnel are positioned at the center of the field of view of the camera. Subsequently, the personnel gradually shift from the center to the edge of the field of view. As the target person approaches the edge of the frame, the KCF algorithm assists in tracking. The target person continues to move leftward, eventually exiting the field of view of the image. Once out of sight, the system enters a waiting state for the target person's re-entry into the field of view. Upon re-entry, the system re-identifies the target person. Finally, the system reverts to the YOLO algorithm to resume tracking the target person. The entire process is deemed as successful tracking by the KCF-YOLO algorithm. The detection effect of the algorithm on the target person is shown in Figure 14.

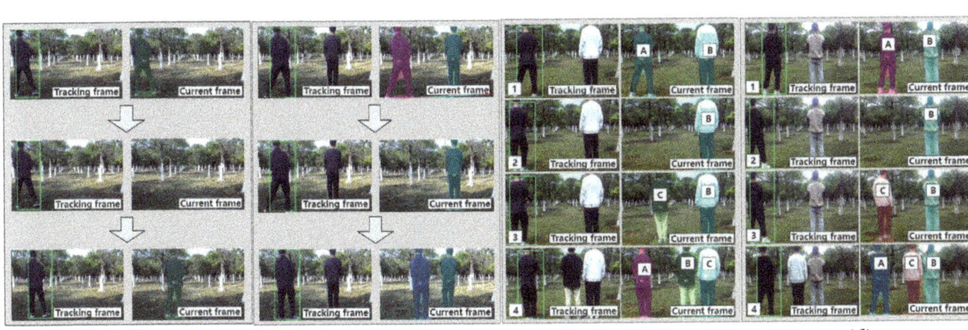

(a) (b) (c) (d)

Figure 14. Detection effect for different numbers of people. (**a**) An individual positioned within the field of view. (**b**) Two individuals positioned within the field of view. (**c**) Three individuals positioned within the field of view. (**d**) Visual tracking in overcast weather conditions. Person A is the tracking target. At the beginning, only person A and person B are in the field of view. When person A leaves the field of view, person C enters the field of view and will not be misidentified as person A by the system. When person A re-enters the field of view, it is considered that the tracking is successful if it is re-tracked by the system.

As shown in Table 3, the average frame rate for the four sets of trials reaches 9 FPS, 9 FPS, 8 FPS, and 6 FPS, respectively. The recognition success rates are 100%, 96.667%, 96.667%, and 93.333%, respectively. These data indicate that the KCF-YOLO algorithm can reliably and accurately recognize a target despite personnel interference. Additionally, it exhibits good recognition speed with minimal impact from the interfering individuals. Meanwhile, the tracking performance of the algorithm is minimally affected under overcast conditions.

Table 3. Performance of the algorithm with different numbers of persons.

Number of Persons	FPS	Success Rate/%
1	9	100
2	9	96.667
3	8	96.667
3 (overcast)	6	93.333

4.2.2. Effects of Different *dis* and *sigma* Values on the Efficiency of the KCF-YOLO Algorithm

In the KCF-YOLO algorithm, the *dis* and *sigma* values are crucial parameters that substantially influence the detection efficiency and stability of the algorithm. To validate the efficiency and stability of the KCF-YOLO algorithm, experiments were conducted using various *dis* and *sigma* values. The experiment includes a total of three participants, with each group conducting 30 trials.

During the experiment, setting the *dis* value to 175 pixels while maintaining the *sigma* value at 0.1 resulted in the relatively low effectiveness of the KCF-YOLO algorithm, with a recognition success rate of only 86.667%. The small *dis* value prompted the early intervention of the KCF algorithm for tracking assistance, causing the peak response value in the region of interest (ROI) calculation correlation to fall below the set threshold, resulting in the failure of the mobile robot to track the target. In addition, with a *sigma* value of 0.1, the Gaussian kernel function proved to be excessively sensitive to the input samples, affecting the stability of the algorithm. The recognition success rate of the KCF-YOLO algorithm reached 100% when the *dis* value was set to 200 pixels, and the *sigma* values were taken as 0.1 and 0.2, respectively. A higher recognition speed was observed when the *sigma* value was set to 0.2. However, at a *sigma* value of 0.3, the recognition success rate of the KCF-YOLO algorithm is only 86.667%. Owing to the large *sigma* value, the discriminative ability of the KCF-YOLO algorithm toward target details decreased, thereby affecting the recognition accuracy. Efficient recognition efficiency was achieved by setting the *dis* value to 225 pixels and using *sigma* values of 0.1, 0.2, and 0.3 for optimal performance of the KCF-YOLO algorithm. The experimental results are presented in Table 4.

Table 4. Algorithm performance for different *dis* and *sigma* values.

dis/Pixel	*sigma*	FPS	Success Rate/%
175	0.1	9	86.667
	0.2	9	90
	0.3	11	93.33
200	0.1	11	100
	0.2	9	100
	0.3	8	86.667
225	0.1	8	96.667
	0.2	9	96.667
	0.3	12	96.667

The experimental results demonstrate that the KCF-YOLO algorithm performs well in recognition and tracking. Specifically, when the *dis* is set to 200 pixels and *sigma* is set to 0.1, the tracking performance is optimal, with an average frame rate of 11 FPS and a recognition success rate of 100%.

4.2.3. A Comparative Experiment with Other Algorithms

To compare the performance of different algorithms, experiments were conducted to compare the YOLO v5s algorithm, the KCF algorithm, and the KCF-YOLO algorithm in similar scenarios. In this study, the YOLO v5s algorithm utilizes the official provided code. The KCF algorithm employed the official version provided by OpenCV. Simultaneously, the KCF-YOLO algorithm utilized the optimal parameters obtained from the aforementioned experiments, with a *dis* set to 200 pixels and *sigma* set to 0.1.

In the YOLO experiment, targets were detected using a pre-trained YOLO v5s model. Subsequently, attempts were made to track the targets based on the detected target position information in consecutive frames. In the KCF experiment, person A was set as the target for tracking and designated as the initial target. Afterwards, the scenario was simulated where individual A moved out of the camera's field of view, representing the target leaving the view. After a certain period, individual A re-entered the camera's field of view, simulating the target reappearing. The tracking performance of the KCF algorithm was recorded and analyzed. The KCF-YOLO experiment process was similar to that under multiple person interference. The experimental process is illustrated in Figure 15.

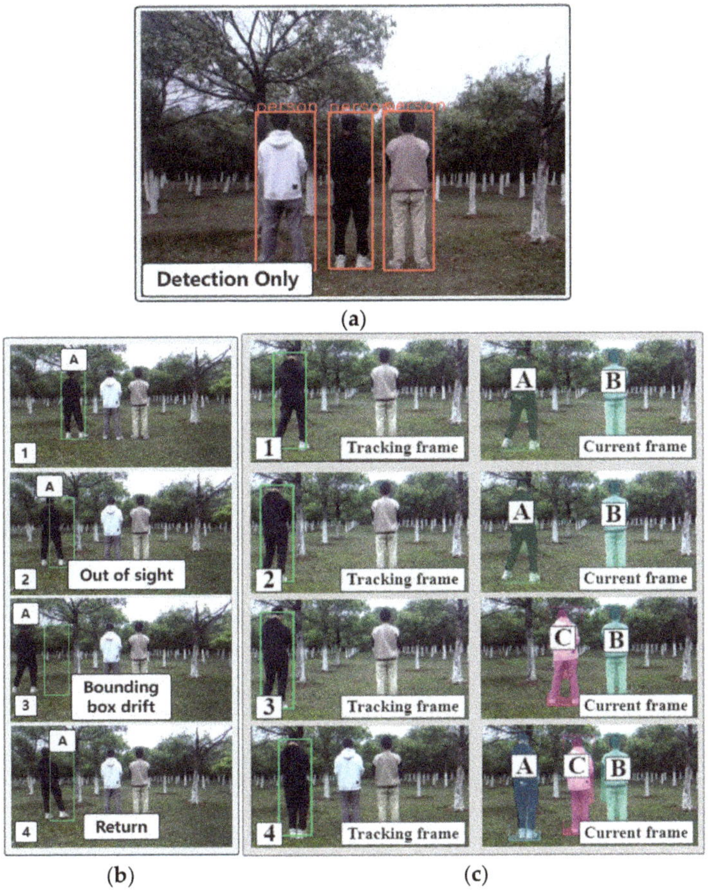

Figure 15. The process of comparative experiments among the three algorithms. (**a**) The YOLOv5s algorithm. (**b**) The KCF algorithm. (**c**) The KCF-YOLO algorithm.

The experimental results indicate that the YOLOv5s algorithm only performs target detection on individuals. However, due to its inherent design characteristics, it lacks the capability for continuous target tracking. While the KCF algorithm is capable of tracking target individuals, it suffers from issues such as target bounding box loss and drifting, resulting in poor stability. In contrast, the KCF-YOLO algorithm combines the strengths of the YOLOv5s and KCF algorithms. It not only leverages the precise target detection capability of YOLOv5s but also effectively tracks targets using the KCF algorithm. It demonstrates excellent performance in terms of tracking stability and resistance to interference.

4.2.4. Human-Following Experiment

In intricate orchard scenarios, challenges arise because of factors such as the varying heights of fruit trees, narrow passages, and randomly distributed obstacles, which intensify the complexity of human-following tasks. Consequently, this study integrates the human-following function into a test scenario by synergizing the KCF-YOLO algorithm with a mobile robot control chassis. This study explored human-following performance when personnel were in a dynamic state. In the experiment, the safe following distance between personnel and the robot was set to 3 m. When the mobile robot detected a distance less than 3 m from the personnel, it would cease following. The human-following experiment involved two participants and included following scenarios under target loss, occlusion, and overcast weather conditions.

In the first scenario, when the target person A exits the robot's field of view, the robot temporarily halts following target A. Concurrently, the robot activates the KCF-YOLO algorithm for assisted localization of the target person A. Upon the target person A's re-entry into the robot's field of view and recognition as the previously tracked target, the robot resumes its tracking program to continue following the movement trajectory of the target person A, as shown in Figure 16.

Figure 16. Repositioned human-following. (**a**) Following from the camera's perspective. (**b**) Following from the real-world perspective.

In the second scenario, even when the target individual is occluded by fruit trees, the system can still identify the person using the YOLO v5s algorithm, as shown in Figure 17a. When the distance between the target individuals falls below the safety threshold, the mobile robot stops following them. When a person appears from the side, the mobile robot resumes following, as illustrated in Figure 17b.

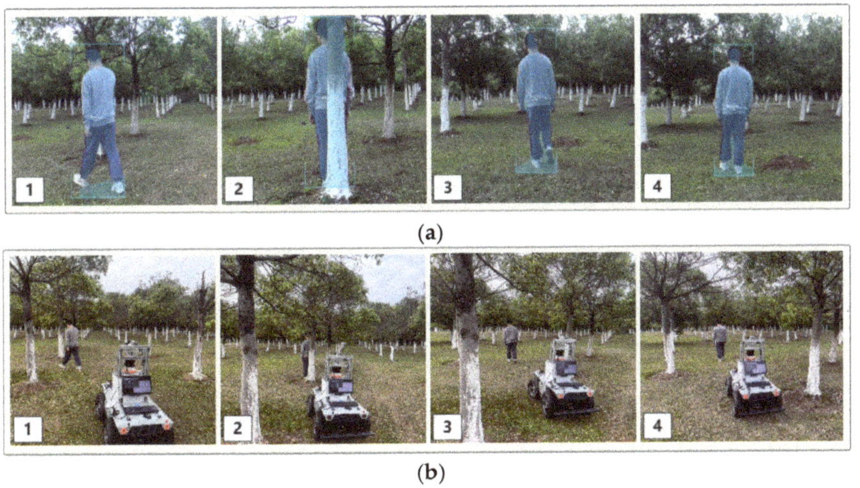

Figure 17. Human-following under fruit tree occlusion. (a) Following from the camera's perspective. (b) Following from the real-world perspective.

Furthermore, in addressing the issue of human-following effectiveness in different environments, the stability of mobile robot following was further evaluated under overcast weather conditions. The experimental process is illustrated in Figure 18.

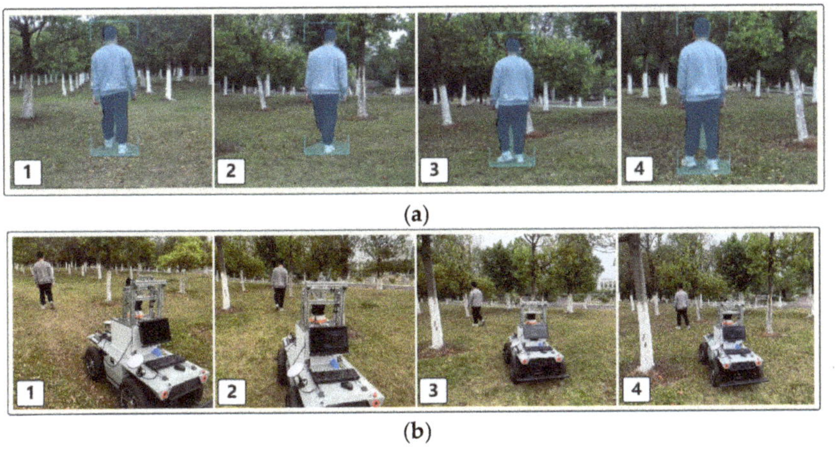

Figure 18. Human-following under overcast conditions. (a) Following from the camera's perspective. (b) Following from the real-world perspective.

The mobile robot is equipped with a SLAM module to record its own position. Based on this module, after the stereo camera captures the spatial coordinates of the personnel, the personnel's spatial information is transformed into the world coordinate system through the static spatial relationship between the stereo camera and the robot, as well as the rotation matrix of the robot in the world coordinate system. Consequently, the complete trajectory of the personnel in the spatial environment is obtained. Figure 19 shows the motion trajectories of the target and robot in the human-following experiment. The red line segment denotes the following trajectory of the mobile robot, and the blue line segment illustrates the actual movement trajectory of the personnel.

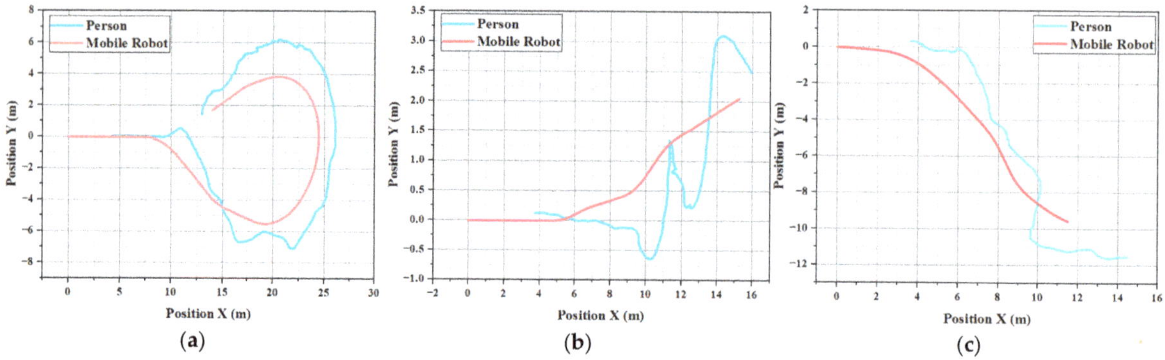

Figure 19. Mobile robots and target personnel movement trajectories. (**a**) Under the condition of target loss. (**b**) Under the condition of target occlusion. (**c**) Under the condition of overcast conditions.

Figure 20 illustrates the distribution of the spatial positional offsets for the X- and Y-components in the motion trajectories of the follower and mobile robot. Error X represents the variance in the vertical distance between the personnel and the mobile robot, whereas Error Y indicates the degree of deviation in the horizontal distance during the mobile human-following process.

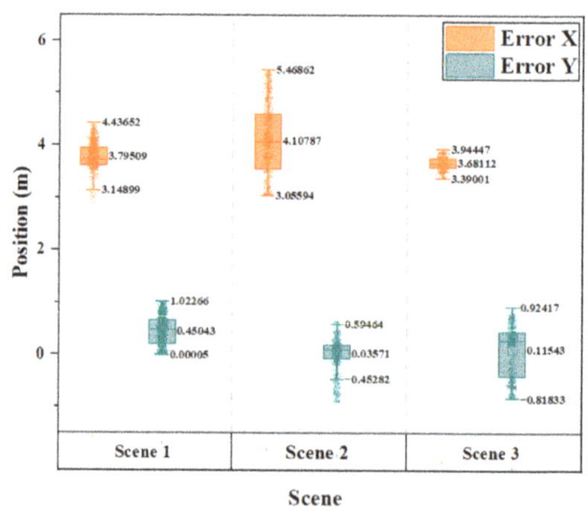

Figure 20. Distribution of offsets. Scene 1: Under target loss. Scene 2: Under target occlusion. Scene 3: Under overcast conditions.

To ensure the stability of human-following in an orchard environment, the UKF algorithm predicts the trajectories of personnel. This prediction allows the mobile robot to stabilize the movement trend of the personnel during the following process, thereby minimizing the lateral swaying of the robot. The average angular velocity of the steering joint of the robot is shown in Figure 21. The dense part of the illustration represents the robot executing steering commands. However, owing to the complexity of the orchard terrain, the angular velocity of the steering joint is affected by the landscape during the following phase, resulting in some noise. Overall, the steering joint of the robot maintains a relatively stable angular acceleration during the following process. Figure 22 shows the linear acceleration situations during the following processes of the mobile robot in three groups. The illustration indicates that the linear acceleration of the robot remains within a certain range, suggesting stable following at a consistent average speed.

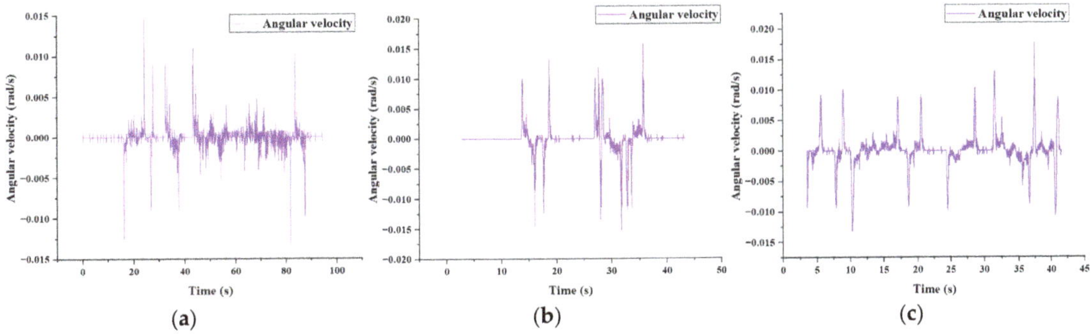

Figure 21. Angular velocity of the steering joint. (**a**) Under the condition of target loss. (**b**) Under the condition of target occlusion. (**c**) Under the condition of overcast conditions.

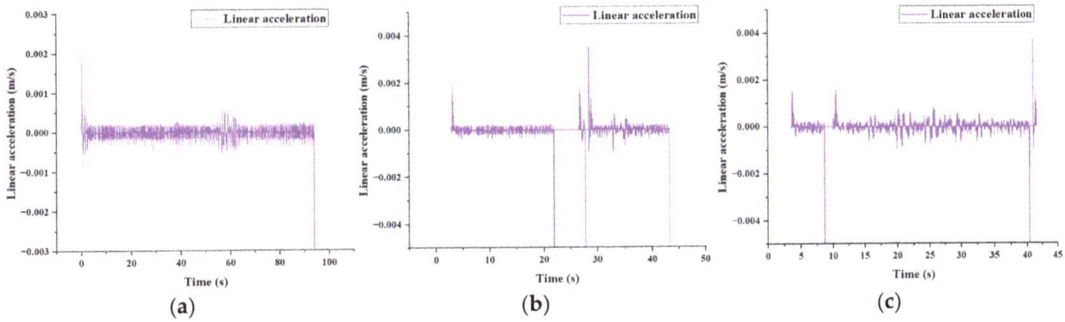

Figure 22. Line acceleration of the mobile robot. (**a**) Under the condition of target loss. (**b**) Under the condition of target occlusion. (**c**) Under the condition of overcast conditions.

In all three different human-following experiments under varying conditions, the maximum horizontal distance Error Y does not exceed 1.03 m. The robot's following trajectory is relatively smooth, allowing it to follow the target at a steady pace until the end of the follow-up task, demonstrating good reliability and stability.

4.3. Results Analysis and Discussion

In the above-mentioned experiments, this study primarily validates the stability and feasibility of the KCF-YOLO algorithm from experiments conducted in two major modules: visual tracking and human-following. In the multi-person interference experiments, when the number of subjects is two or three, during the recognition process, the fluctuation in the scale of the target individuals and the similarity between the features identified from interfering persons and those of the target individuals affect the algorithm's decision mechanism. Consequently, there is a decrease in the recognition success rate. The KCF-YOLO algorithm can still stably track target individuals, demonstrating good robustness in detection performance. Under overcast conditions, both the success rate and the frame rate have decreased, possibly due to the dim lighting in overcast scenes, leading to a blurred background. This increases the computational load for processing the background in the KCF-YOLO algorithm. The algorithm requires more computational resources to identify and filter targets, thereby reducing tracking efficiency and frame rate. However, the ability to track target individuals has not been significantly affected, with a success rate reaching 93.33%. Thus, the KCF-YOLO algorithm still maintains good stability.

In terms of the impact of different *dis* and *sigma* values on the algorithm, this study experimented with various values of *dis* and *sigma* for validation. When the *dis* value was set to 200 pixels and the *sigma* value to 0.1, the KCF-YOLO algorithm demonstrated the best performance. For *sigma* values of 0.2 and 0.3, the Gaussian kernel function performs poorly in processing target feature information. The reason might be that when the *sigma* value is large, the blurring effect of the Gaussian kernel function leads to loss of target feature information, thereby reducing the expressive power of the algorithm. By combining the optimal solutions for *dis* and *sigma* values, the algorithm in this study effectively captured target individual features, leading to the best overall performance. By adjusting the *dis* and *sigma* values according to specific scenes, the algorithm can meet diverse scene requirements, thereby enhancing its robustness and accuracy. In order to better evaluate the tracking stability and anti-interference ability of the algorithm, this paper conducted comparative experiments between the YOLO v5s algorithm and the KCF algorithm. The experimental results indicate that the YOLO v5s algorithm can only recognize target individuals but lacks the ability to track them, while the KCF algorithm exhibits instability and poor tracking performance. In contrast, the proposed KCF-YOLO algorithm not only achieves accurate tracking of target individuals but also possesses real-time detection capabilities, demonstrating excellent performance in both tracking and detection.

Building upon the visual experiments, this paper conducted experiments focused on human-following to assess the practicality and reliability of the KCF-YOLO algorithm in real-world scenarios. Due to the slightly faster movement speed of the target individual compared to the constant speed maintained by the mobile robot, there may be differences in the Error X between the mobile robot and the target individual. However, this does not affect the stability of the KCF-YOLO algorithm during the following process. Meanwhile, it is evident that the prediction of personnel trajectories through UKF reduces the lateral swaying of the robot, resulting in smoother following paths for the robot. The horizontal distance Error Y between the three trajectories does not exceed 1.03 m. However, due to terrain effects, there is relative jitter between the personnel and the mobile robot, leading to some errors in the obtained personnel trajectories. In particular, Figure 22b illustrates the linear acceleration of the mobile robot when personnel following are obstructed by fruit trees. During the period from 20 s to 30 s, there is a segment where the acceleration value is 0, indicating that the mobile robot has stopped following. The linear accelerations of the three axes remain within a certain range without significant fluctuations, indicating that the speed of the mobile robot following is smooth. Compared to sunny conditions, the tracking error under overcast conditions may slightly increase. This could be due to changes in lighting conditions, resulting in less clear visual features of the target person and causing some delay in the steering of the mobile robot. Although lighting conditions under overcast skies may affect the tracking accuracy of the mobile robot to some extent, the experimental results demonstrate that the mobile robot system is still capable of effectively tracking the target person.

In summary, the algorithm combines the strengths of the KCF and YOLO v5s algorithm. By integrating the YOLO v5s algorithm into the target detection module, precise identification of target individuals is achieved. When the target individual is about to leave the field of view, the KCF algorithm is introduced to accurately track the target individual, ensuring that when the target person returns to the center of the field of view, YOLO v5s can identify and continue tracking the lost target. The algorithm effectively utilizes the advantages of both algorithms, thereby enhancing the stability of following within complex orchard environments.

5. Conclusions and Future Work

5.1. Conclusions

To address the stability challenges faced by autonomous mobile robots in orchard scenarios during human-following functions, this study proposes a vision-tracking method based on KCF-YOLO fusion. This method aimed to overcome issues, such as obstacle occlusion and overlapping personnel in unstructured orchard environments. First, target localization and specific person identification were achieved by leveraging the target detection accuracy advantage of YOLO v5s, combined with the real-time capabilities of KCF and its adaptability to continuous tracking. Second, fusion with the unscented Kalman filter algorithm to fit the personnel movement trajectory achieved human-following in a complex orchard scenario. Experiments were conducted in a complex orchard environment to verify the effectiveness of the method in practical applications. The experimental results showed that the KCF-YOLO algorithm performed well in tracking in orchard scenarios, with an average tracking success rate of 96.66% and an average frame rate of 8 FPS. The mobile robot maintained a constant speed and followed the target person steadily, maintaining the horizontal offset from the followed person within a certain range. These results suggest that the KCF-YOLO vision fusion human-following method proposed in this paper offers an innovative solution for achieving human-following tasks in complex orchard environments, providing more efficient and reliable assistance tools for fruit farmers.

5.2. Future Work

The proposed KCF-YOLO vision fusion method can effectively address the problems of target tracking loss and repositioning failure of mobile robots in complex environments. However, setting different *dis* values in different scenarios to ensure the robustness of the algorithm is a crucial task. The reason lies in the difficulty of determining the optimal value of this parameter across different scenes. Meanwhile, the *dis* value has a certain impact on the stability of the subsequent human-following. In future work, further research and adoption of adaptive algorithms are needed to enable the system to dynamically adjust parameters based on real-time scenarios, thereby improving the robustness and adaptability of the algorithm. Although the KCF-YOLO vision fusion method performs well in target tracking, it lacks autonomous obstacle avoidance functionality. To enhance the practicality and safety of mobile robots in orchard environments, future work will focus on researching the autonomous navigation capabilities of robots to achieve obstacle avoidance during the human-following. By delving into advanced navigation algorithms and sensor technologies, and leveraging the characteristics of orchard terrain, robots will be able to accurately identify and evade obstacles, ensuring smooth progress during the human-following.

Author Contributions: Methodology, software, writing—original draft, Z.H.; data curation, C.O.; data curation, Z.G.; supervision, methodology, writing—review and editing, L.Y.; supervision, writing—review, J.L. All authors have read and agreed to the published version of the manuscript.

Funding: This work was supported by grants from the Guangdong University Characteristic Innovation Project (grant number 2018KTSCX207), the Key research project of Shaoguan University (grant number SZ2023KJ14), the National College Student Innovation Training Project (grant number 202310576008) and the Shaoguan Science and Technology Project (grant numbers 210725144530830 and 230403118030785).

Data Availability Statement: Data are contained within the article.

Conflicts of Interest: This study was conducted without any commercial or financial relationships that could be construed as potential conflicts of interest.

References

1. Lei, Y.; Jieli, D.; Zhou, Y.; Xiangjun, Z.; Mingyou, C.; Sheng, Z. Collision-free motion planning for the litchi-picking robot. *Comput. Electron. Agric.* **2021**, *185*, 106151.
2. Wang, H.; Li, H. Design of Intelligent Ground Air Multi Robot Collaborative Transportation System. *J. Adv. Artif. Life Robot.* **2023**, *4*, 94–97.
3. Hichri, B.; Adouane, L.; Fauroux, J.-C.; Mezouar, Y.; Doroftei, I. Flexible co-manipulation and transportation with mobile multi-robot system. *Assem. Autom.* **2019**, *39*, 422–431. [CrossRef]
4. Sirintuna, D.; Giammarino, A.; Ajoudani, A. Human-robot collaborative carrying of objects with unknown deformation characteristics. In Proceedings of the 2022 IEEE/RSJ International Conference on Intelligent Robots and Systems (IROS), Kyoto, Japan, 23–27 October 2022; IEEE: New York, NY, USA, 2022; pp. 10681–10687.
5. Daegyun, C.; Donghoon, K. Intelligent Multi-Robot System for Collaborative Object Transportation Tasks in Rough Terrains. *Electronics* **2021**, *10*, 1499. [CrossRef]

6. Ramasubramanian, A.K.; Papakostas, N. Operator-mobile robot collaboration for synchronized part movement. *Procedia CIRP* **2021**, *97*, 217–223. [CrossRef]
7. Lei, Y.; Fengyun, W.; Xiangjun, Z.; Jin, L. Path planning for mobile robots in unstructured orchard environments: An improved kinematically constrained bi-directional RRT approach. *Comput. Electron. Agric.* **2023**, *215*, 108453.
8. Van Toan, N.; Do Hoang, M.; Khoi, P.B.; Yi, S.-Y. The human-following strategy for mobile robots in mixed environments. *Robot. Auton. Syst.* **2023**, *160*, 104317. [CrossRef]
9. Wang, C.; Zou, X.; Tang, Y.; Luo, L.; Feng, W. Localisation of litchi in an unstructured environment using binocular stereo vision. *Biosyst. Eng.* **2016**, *145*, 39–51. [CrossRef]
10. Wang, C.; Li, C.; Han, Q.; Wu, F.; Zou, X. A Performance Analysis of a Litchi Picking Robot System for Actively Removing Obstructions, Using an Artificial Intelligence Algorithm. *Agronomy* **2023**, *13*, 2795. [CrossRef]
11. Tang, Y.; Chen, M.; Wang, C.; Luo, L.; Li, J.; Lian, G.; Zou, X. Recognition and Localization Methods for Vision-Based Fruit Picking Robots: A Review. *Front. Plant Sci.* **2020**, *11*, 510. [CrossRef] [PubMed]
12. Luo, L.; Tang, Y.; Lu, Q.; Chen, X.; Zhang, P.; Zou, X. A vision methodology for harvesting robot to detect cutting points on peduncles of double overlapping grape clusters in a vineyard. *Comput. Ind.* **2018**, *99*, 130–139. [CrossRef]
13. Luo, L.; Tang, Y.; Zou, X.; Ye, M.; Feng, W.; Li, G. Vision-based extraction of spatial information in grape clusters for harvesting robots. *Biosyst. Eng.* **2016**, *151*, 90–104. [CrossRef]
14. Li, G.; Li, Z.; Su, C.-Y.; Xu, T. Active human-following control of an exoskeleton robot with body weight support. *IEEE Trans. Cybern.* **2023**, *53*, 7367–7379. [CrossRef]
15. Li, S.; Milligan, K.; Blythe, P.; Zhang, Y.; Edwards, S.; Palmarini, N.; Corner, L.; Ji, Y.; Zhang, F.; Namdeo, A. Exploring the role of human-following robots in supporting the mobility and wellbeing of older people. *Sci. Rep.* **2023**, *13*, 6512. [CrossRef] [PubMed]
16. Kästner, L.; Fatloun, B.; Shen, Z.; Gawrisch, D.; Lambrecht, J. Human-following and-guiding in crowded environments using semantic deep-reinforcement-learning for mobile service robots. In Proceedings of the 2022 International Conference on Robotics and Automation (ICRA), Philadelphia, PA, USA, 23–27 May 2022; IEEE: New York, NY, USA, 2022; pp. 833–839.
17. Kapgate, S.; Sahu, P.; Das, M.; Gupta, D. Human following robot using kinect in embedded platform. In Proceedings of the 2022 1st International Conference on the Paradigm Shifts in Communication, Embedded Systems, Machine Learning and Signal Processing (PCEMS), Nagpur, India, 6–7 May 2022; IEEE: New York, NY, USA, 2022; pp. 119–123.
18. Zhu, Y.; Wang, T.; Zhu, S. A novel tracking system for human following robots with fusion of MMW radar and monocular vision. *Ind. Robot Int. J. Robot. Res. Appl.* **2022**, *49*, 120–131. [CrossRef]
19. Thakran, A.; Agarwal, A.; Mahajan, P.; Kumar, S. Vision-Based Human-Following Robot. In *Advances in Data Computing, Communication and Security: Proceedings of I3CS2021*; Springer: Berlin/Heidelberg, Germany, 2022; pp. 443–449.
20. Bolme, D.S.; Beveridge, J.R.; Draper, B.A.; Lui, Y.M. Visual object tracking using adaptive correlation filters. In Proceedings of the 2010 IEEE Computer Society Conference on Computer Vision and Pattern Recognition, San Francisco, CA, USA, 13–18 June 2010; IEEE: New York, NY, USA, 2010; pp. 2544–2550.
21. Henriques, J.F.; Rui, C.; Pedro, M.; Jorge, B. High-Speed Tracking with Kernelized Correlation Filters. In *IEEE Transactions on Pattern Analysis and Machine Intelligence*; IEEE: New York, NY, USA, 2014; pp. 583–596.
22. Liu, Z.; Lian, Z.; Li, Y. A novel adaptive kernel correlation filter tracker with multiple feature integration. In Proceedings of the 2017 IEEE International Conference on Image Processing (ICIP), Beijing, China, 17–20 September 2017; IEEE: New York, NY, USA, 2017; pp. 2572–2576.
23. Huan, L.; Jingqi, M.; Xingjian, L. Target tracking method based on the fusion of structured SVM and KCF algorithm. In Proceedings of the 2020 Chinese Control and Decision Conference (CCDC), Hefei, China, 22–24 August 2020; IEEE: New York, NY, USA, 2020; pp. 1174–1178.
24. Bai, S.; Tang, X.; Zhang, J. Research on Object Tracking Algorithm Based on KCF. In Proceedings of the 2020 International Conference on Culture-Oriented Science & Technology (ICCST), Beijing, China, 28–31 October 2020; pp. 255–259.
25. Mbelwa, J.T.; Zhao, Q.J.; Wang, F.S. Visual tracking tracker via object proposals and co-trained kernelized correlation filters. *Vis. Comput.* **2020**, *36*, 1173–1187. [CrossRef]
26. Gupta, S.C.; Majumdar, J. Convolutional neural network based tracking for human following mobile robot with LQG based control system. In Proceedings of the Third International Conference on Advanced Informatics for Computing Research, Shimla, India, 15–16 June 2019; pp. 1–7.
27. Kamel, B.; Naeem, R. Human detection based on deep learning YOLO-v2 for real-time UAV applications. *J. Exp. Theor. Artif. Intell.* **2022**, *34*, 527–544.
28. Liu, W.; Anguelov, D.; Erhan, D.; Szegedy, C.; Reed, S.; Fu, C.-Y.; Berg, A.C. SSD: Single Shot MultiBox Detector. In *Computer Vision, Proceedings of the ECCV 2016, Amsterdam, The Netherlands, 11–14 October 2016*; Springer: Cham, Switzerland, 2016; pp. 21–37.
29. Algabri, R.; Choi, M.-T. Deep-Learning-Based Indoor Human Following of Mobile Robot Using Color Feature. *Sensors* **2020**, *20*, 2699. [CrossRef]
30. Han, D.; Peng, Y. Human-following of mobile robots based on object tracking and depth vision. In Proceedings of the 2020 3rd International Conference on Mechatronics, Robotics and Automation (ICMRA), Shanghai, China, 16–18 October 2020; IEEE: New York, NY, USA, 2020; pp. 105–109.

31. Yang, C.A.; Song, K.T. Control Design for Robotic Human-Following and Obstacle Avoidance Using an RGB-D Camera. In Proceedings of the International Conference of the Society for Control Robot Systems (ICCAS), Jeju, Republic of Korea, 15–18 October 2019; IEEE: New York, NY, USA, 2019; pp. 934–939.
32. Linxi, G.; Yunfei, C. Human Following for Outdoor Mobile Robots Based on Point-Cloud's Appearance Model. *Chin. J. Electron.* **2021**, *30*, 1087–1095. [CrossRef]
33. TsungHan, T.; ChiaHsiang, Y. A robust tracking algorithm for a human-following mobile robot. *IET Image Process.* **2021**, *15*, 786–796.
34. Redmon, J.; Divvala, S.K.; Girshick, R.B.; Farhadi, A. You Only Look Once: Unified, Real-Time Object Detection. In Proceedings of the IEEE Conference on Computer Vision and Pattern Recognition, Las Vegas, Nevada, 26 June–1 July 2016; IEEE: New York, NY, USA, 2016; pp. 779–788.

horticulturae

Article

An Original UV Adhesive Watermelon Grafting Method, the Grafting Device, and Experimental Verification

Xin Zhang [1,2], Linghao Kong [1], Hanwei Lu [1], Qingchun Feng [2], Tao Li [2], Qian Zhang [3,*] and Kai Jiang [2,*]

[1] Mechanical and Electrical Engineering Institute, Xinjiang Agricultural University, Urumqi 830052, China; zhangx@nercita.org.cn (X.Z.); kolinghao@163.com (L.K.); kingluhanwei@outlook.com (H.L.)
[2] Research Center of Intelligent Equipment, Beijing Academy of Agriculture and Forestry Sciences, Beijing 100097, China; fengqc@nercita.org.cn (Q.F.); lit@nercita.org.cn (T.L.)
[3] Research Center of Information Technology, Beijing Academy of Agriculture and Forestry Sciences, Beijing 100097, China
* Correspondence: zhangq@nercita.org.cn (Q.Z.); jiangk@nercita.org.cn (K.J.); Tel.: +86-10-5150-3504 (K.J.)

Abstract: This study is aimed at traditional vegetable grafting using a large number of plastic clips, which cannot be recycled in time and cause serious pollution within the planting environment. This paper proposes a new grafting method based on a UV adhesive instead of plastic clips. First of all, a UV adhesive spray grafting device was designed. The structure includes seedling adsorption positioning mechanisms, a butt joint mechanism, a handling mechanism, a spray valve, a UV curing lamp, etc., to facilitate the adhesive spraying. For the rootstock and scion, a horizontal, lateral seedling and negative pressure adsorption and positioning method is adopted, with fluid dynamics simulation of the diameter and quantity of the adsorption holes in the rootstock adsorption mechanism carried out using Fluent 2022 R1 software and completion of the optimization of the parameters of the adsorption and positioning mechanism. The fluid volume method is used to simulate the adsorption and positioning mechanism. For optimization, the volume of fluid method (VOF) and the discrete particle method (DPM) are used in a coupled simulation of the UV adhesive spraying process, and the value range of the spraying influencing factors is determined: the selected glue pressure, atomization pressure, and spraying height for three-factor, three-level orthogonal simulation. A grafting test is also verification, deriving the significance ranking of their impact on the success rate of the grafting: atomization pressure > spraying height > glue pressure. Under the condition of a 0.25 Mpa atomization pressure, a 0.15 Mpa glue supply pressure, and a 10 mm spraying height, the grafting success rate for watermelon was 100%, the effective spraying rate was 83.03%, the healing success rate was 94.5%, and the length of the film was 7.86 mm. The results of the study can provide a research basis for the research and development of new types of spraying and grafting robot technology.

Keywords: watermelon grafting; UV adhesive; fluent; VOF-DPM numerical simulation; grafting device; test

Citation: Zhang, X.; Kong, L.; Lu, H.; Feng, Q.; Li, T.; Zhang, Q.; Jiang, K. An Original UV Adhesive Watermelon Grafting Method, the Grafting Device, and Experimental Verification. *Horticulturae* **2024**, *10*, 365. https://doi.org/10.3390/horticulturae10040365

Academic Editor: Darius Kviklys

Received: 27 February 2024
Revised: 28 March 2024
Accepted: 2 April 2024
Published: 5 April 2024

1. Introduction

According to the Food and Agriculture Organization of the United Nations (FAO), as of 2019, the global watermelon production was 100,414,900 t, and China's watermelon production was 60,681,200 t. This accounted for approximately 60.42% of global watermelon production, making China the world's largest producer and consumer of watermelon [1]. The continuous planting of watermelon on a piece of land is called continuous cropping, which causes the production of plant root pathogens due to heavy cropping, leading to plant wilt, leaf blight, virus diseases, and other hazards, which seriously affect the growth and yield of the crops [2–4]. The application of pesticides can alleviate disease, but it results in high drug residues in agricultural products and pollution of the ecological environment. Breeding new disease-resistant varieties is a long and complicated process, and large-scale crop rotation is even more impractical [5]. Grafting technology is an effective measure for solving the problems of continuous cropping and pests and diseases [6–8]. The selection of disease-resistant rootstocks can significantly reduce or avoid yellow wilt, wilt, root rot, green blight, and root-knot nematodes and other soil-borne pests and diseases in vegetable crops, can effectively overcome the barriers to continuous cultivation, and can increase the number of planting stubble in the same plot of land to improve the land utilization rate. At present, watermelon seedlings are grafted artificially [9–11], fewer people work in grafting as China's rural population ages [12], and the quality and efficiency of manual grafting can no longer meet the production requirements of factory nurseries. Mechanical grafting has the advantages of standardization, high quality, and high efficiency [13,14], which is the development direction for factory-grafting seedlings [15].

In the early 1980s, Japan became the first to start researching grafting machines, many agricultural machinery enterprises have been involved in research work, with the formation of dozens of commercialized products. From the late 1990s onward, Korean and Chinese scholars conducted a lot of research on grafting methods, executive mechanisms, and devices, while obtaining many valuable research results. Subsequently, the Netherlands developed grafting machines suitable for tomato seedlings. In 1994, ISEKI & Co., Ltd. (700 Umaki-cho, Matsuyama-shi, Ehime-ken, 799-2692 Japan) launched the GR800 series of vegetable grafting robots [16]. This machine is used to feed seedlings on a gap holder and is suitable for fixing rootstock and scion seedlings with a productivity of 800 plants/h and a success rate of 95%. It also launched the GRF800-U fully automatic grafting machine for melons in 2010 [17,18], which requires only one person for hole tray supplying. It employs an automatic feeding mechanism, replacing manually feeding the seedlings, to achieve automatic support for and cutting and clamping of the seedlings in the hole tray. Plastic clips are used to secure the grafting seedlings, which has a production efficiency of 800 plants/h and a grafting success rate of 95%. In 1998, Zhang Tiezhong et al. at China Agricultural University in Beijing, China developed the 2JSZ-600 semi-automatic grafting machine [19], which employed plastic clips for securing the grafting seedlings and featured a one-piece rotary cutter frame. The cutter frame rotation allowed for simultaneous cutting operations for both the rootstock and scion. However, the adjustment of the cutting angle was intricate, reaching a production efficiency of 600 plants/h and a grafting success rate of 90%. In 2009, Gu Song et al. at South China Agricultural University in Guangdong, China developed a 2JC-600-type oblique insertion semi-automatic grafting machine [20], designed for the upper surface of the rootstock clamp adsorption holes. The use of negative pressure fixes the rootstock cotyledons for horizontal adsorption so that the two cotyledons unfold in order to facilitate punching and scion insertion, but the rootstock petiole is easy to damage, with a production efficiency of 600 plants/h and a grafting success rate of only 90%. In 2011, ISO Group Co., Ltd. (Middelkampseweg 9, 5311 PC, Gameren, Netherlands) developed Graft 1000, 1100, and 1200, three kinds of silicone sleeve grafting machines [21,22] which are suitable for grafting tomato seedlings of a young seedling age. A synchronized rootstock cutting mechanism was developed to achieve the same cutting angle (0° or 45°) for the rootstock. The fixation clamp was a silicone sleeve, and a special sleeve supply mechanism was designed to enable the sleeve to open up, wrap, and fix the cuts of the rootstock and scion into the butts. The production efficiency was up to 1200 plants/h, with a grafting success rate of 98%, but the requirement for seedling–stem matching of the rootstock seedlings was high, and frequent sleeve replacement was required. In 2016, Helper Robotech Co., Ltd. (82, Yuha-ro 226beon-gil, Gimhae-si, Gyeongsangnam-do 621-834 Korea) developed the AFGR-800CS grafting machine for Cucurbitaceae and Solanaceae [23–25], which fixes the grafted seedlings with plastic clips, and developed a visual information recognition system for the seedling incision, utilized for calculating the seedling incision centerline position and achieving an overlap between the rootstock and scion incision centerlines during the butt joint session, with a production efficiency of 800 plants/h, and the grafting success rate was 98%. In 2020, Jiang Kai et al. at Beijing Academy of Agriculture and Forestry Sciences in Beijing, China developed the

2TJGQ-1000 four-station grafting machine [26,27], also using plastic clips to fix the grafted seedlings, determined the cutting angle for the rootstock (white-seeded pumpkin) and scion (watermelon) to match the grafting, and realized the synchronous operation of cutting the seedlings and joining the butts on four stations, which for the first time exceeded the efficiency to 1052 plants/h for cucurbits, with a grafting success rate of 96.67%. In 2021, AiGRAFT Co., Ltd. (the second floor of Shengya Office Building, No. 789 Mingchuan Road, High-tech Zone, Hefei City, Anhui Province) in China launched the JFT-A1200T grafting machine [28], which adopts special plastic clips to fix the grafted seedlings, and the main structure is a three-dimensional rotary four-station structure, including the seedling supply, cutting, butt joint, and lowering, with a production efficiency of 1200 plants/h. Due to the use of a horizontal direction for the seedling supply, seedling supply personnel are easily fatigued by the prolonged operation.

The above grafting machines all graft using plastic clips or silicone clips, and it is difficult to recycle them during the healing and planting periods of the grafted seedlings. China's annual demand for grafted seedlings reaches 50 billion plants, and each year, due to grafting, the use of plastic clip products amounts to hundreds of thousands of tons. Year after year, the use of plastic clips faces the plight of recycling difficulties. Therefore, this paper proposes a UV adhesive grafting method instead of traditional plastic clips; determines the influencing factors and parameters for adhesive spray atomization by researching a simulation model of the UV glue spray grafting parameters; designs a UV adhesive spray grafting device; analyzes the significance of the influence of each factor on the success rate of grafting using an orthogonal test; and determines the optimal conditions for the success of the adhesive spray grafting. The survival rate of the grafted seedlings will also be verified. The successful development of this technology is expected to improve the traditional mechanical grafting methods and seedling nursery mode and to provide a research basis for the development of a new adhesive spray grafting robot.

2. Materials and Methods

2.1. Theory, Design, and Modeling of the UV Adhesive

2.1.1. Grafting Theory

Splice grafting generally uses plastic clips to fix the rootstock and scion seedlings, as shown in Figure 1a. A certain impact force is exerted on the grafted seedling at the moment of the plastic clamp's closure, and the clamping force of the plastic clamp is constant during the healing period, which leads to a lack of flexible treatment methods for the fixation of the grafting seedlings. UV adhesive grafting refers to the use of spray valves to atomize the adhesive, sprayed onto the periphery of the rootstock and scion incision butt joints, under UV light source irradiation, triggering the adhesive to quickly solidify to form a thin adhesive film coating, instead of plastic clips to complete the fixation of the grafted seedling, in the red circle as shown in Figure 1b. The comparison found that with UV adhesive grafting of a lightweight parcel layer, the formation of the parcel force is more flexible than that of a plastic clip. The parcel layer has the characteristics of automatic shedding. The UV adhesive spraying process necessitates precise control over parameters such as the atomization pressure, glue supply pressure, and spraying height. Establishing a nozzle model for numerical simulation of the adhesive atomization process is imperative, aiming to ascertain the optimal parameters for grafting through spraying.

The UV adhesive used in grafting should be green and harmless to ensure the safety of the operators and seedlings. LOCTITE AA 3321 (Henkel, Düsseldorf, Germany, 40589) was selected as a medium-viscosity transparent liquid with a density of 1.08×10^3 kg/m^3 and a viscosity of 5.5 kg/(m·s); its dyne value was measured to be 34 using a dyne pen, which means that the surface tension was 0.034 N/m, and its chemical composition is polyurethane acrylate (PUA). The cured wrapping layer has a high resistance to abrasion, adhesion, flexibility, etc. and passed the ISO-10993 biocompatibility test to reach a medical grade [29]. Due to the limited space for grafting operations, the NordsonEFD 781mini spray valve was selected, with a nozzle aperture of 0.01 in, which is a low-flow, low-pressure spray valve, as shown in Figure A1. In a small space, to achieve micro-precision spraying, a working frequency of up to 400 times/min was used, suitable for high-frequency cyclical work scenarios. Spray valve work requires access to a 3-way air source (glue supply pressure, driving pressure, atomization pressure), driving pressure to force the liquid into the nozzle. Under the joint action of the glue supply pressure and atomization pressure, inside the nozzle, we see the formation of a pressure drop so that the liquid atomizes into tiny droplets.

(a) (b)

Figure 1. Comparison of grafting methods. (**a**) Plastic clip grafting; (**b**) UV adhesive grafting.

The control circuit for the spray valve is shown in Figure 2. First of all, the gas is compressed using a pressure-regulating filter (pressure range of 0~1.0 Mpa). The pressure-regulating filter is divided into three gas paths: the first for the spray valve for the glue pressure inlet, with pressure gauge 1 to regulate the amount of spray (pressure range of 0.15~0.25 Mpa); the second for the spray valve drive pressure inlet, with solenoid valve 1 to move the spray valve needle up and down (pressure range of 0.4~0.5 Mpa); and the third for the spray valve atomization pressure inlet, with solenoid valve 2 to control the intervention of atomized gas (pressure range of 0.20~0.30 Mpa). Solenoid valve 1 and solenoid valve 2 are controlled using the controller.

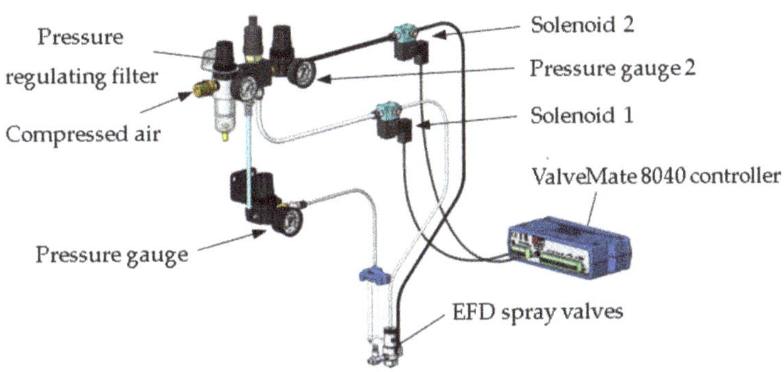

Figure 2. Spray valve control wiring diagram.

The ValveMate 8040 controller is shown in Figure A2. It includes a programmable dispensing time, a digital display for time reading, and four independent solenoid valve drives and can communicate inputs and outputs with the host PLC, with microprocessor circuitry for precise spray control. The external solenoid provides low-volume, low-pressure (LVLP) air to the nozzle by connecting the nozzle air pressure regulator for a high transfer efficiency and does not result in over-spraying. In addition, it has the characteristics of high-frequency cycle control, so it can adjust the time of a single working cycle with an accuracy of 0.001 s. When the work cycle ends, the air in the nozzle continues for a few milliseconds to ensure a clean, clog-free cut-off, and then the controller signals that the process is complete.

The UV adhesive curing speed is affected by the light intensity, spectral distribution, irradiation time, substrate light transmission, and other factors. The spray valve is a micro-spray. The point light source curing machine model is HY-UV0003UV (Zhuhai Hao Yun Optoelectronics Technology Co., Ltd., No. 222, Cuizian North Third Street, Zhuhai, Guangdong, China), as shown in Figure A3. Through the conversion of electro-optical energy, the LED high-power ultraviolet diode chip produces high-purity 365 nm monochromatic ultraviolet light at an irradiation distance of 10 mm and with a light intensity of 6000 MW/cm^2. We estimated using a preliminary test that the adhesive dose of a single grafted seedling was about 0.01 g, and the weight of a single plastic clip is about 1 g. Watermelon grafting requires a lower dosage of UV adhesive, forming an adhesive film around the grafted seedling with a thickness of 0.1 mm to 0.3 mm, and rapid curing is accomplished within a few seconds. Grafted seedlings are sprayed as shown in Figure 3. The cross-section of the seedling stem was approximately elliptical, and the UV adhesive was sprayed from the spray valve in the form of a solid cone, downward and perpendicular to the long axis of the seedling incision at the butt joint.

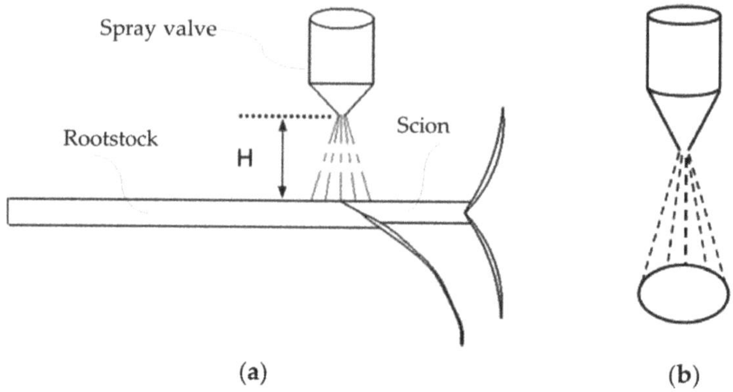

Figure 3. Grafted seedling spraying schematic. (**a**) Main view of spraying; (**b**) side view of spraying.

The coverage area of the seedling spraying depends on the spray cone angle and spray height; the larger the spray cone angle and spray height, the larger the film-forming area, forming a semi-encapsulated adhesive film on the outside of the grafted seedling's incision butt joints, an ellipsoidal film with a thickness of 0.1 mm to 0.3 mm. The shaded portion is the film-forming range of the adhesive under the conditions of different spraying parameters, as shown in Figure 4.

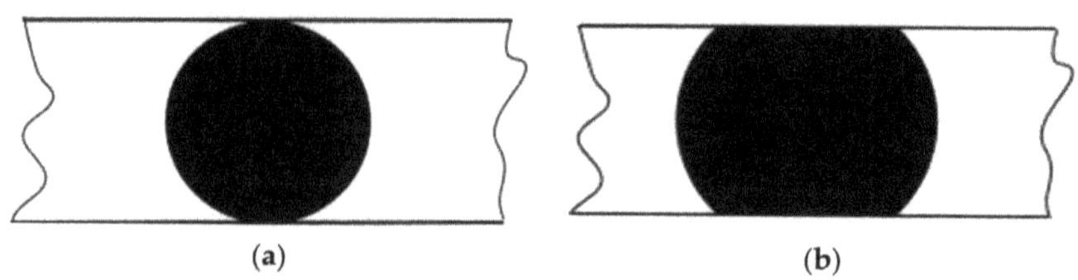

Figure 4. Adhesive film-forming range. (**a**) Small-area film formation; (**b**) large-area film formation.

The spray valve's spraying flow rate function is recorded as f, indicating the spray valve, the unit of time, and the stereo angle of the liquid volume sprayed, that is

$$f = \frac{dV}{d\Omega dt} \tag{1}$$

The cumulative rate of the spray thickness d_S at any point S on the sprayed surface in the conical spraying area is:

$$d_S = \frac{f \cdot \cos\gamma}{r^2} \tag{2}$$

The three-dimensional spatial distribution of the spray is shown in Figure 5. γ is the angle between the unit vector and the XOY plane from point P of the nozzle outlet to point S of the spray surface, and r is the length of PS.

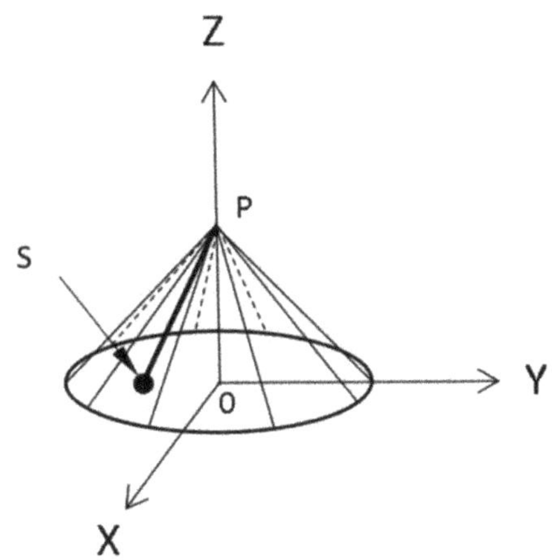

Figure 5. Spraying schematic.

The cutting angle is 30° (rootstock on the left, scion on the right), b_1 and b_2 are the short-axis dimensions of the rootstock and scion stems, respectively, and the adhesive film-forming length needs to be greater than L if the sprayed adhesive is to completely encapsulate the cut, as shown in Figure 6.

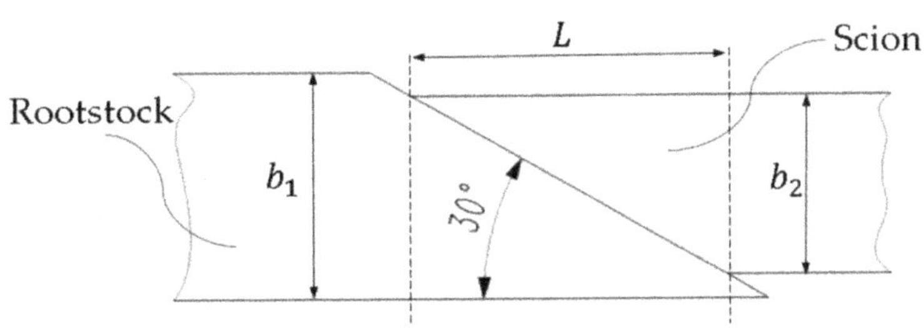

Figure 6. Schematic diagram of incision length.

The film-forming length is:

$$L = \frac{b_2}{\tan 30^{\circ}} \tag{3}$$

2.1.2. Grafting Device Design

The grafting device consists of a spray valve, a rootstock adsorption mechanism, a scion positioning plate, a rotary cylinder, a sliding plate, a frame, a Y-axis linear slide, a scion adsorption mechanism, an X-axis linear slide, a UV curing lamp, etc., as shown in Figure 7. It is divided into 5 processes, namely the seedling supply, butt joining, spraying, curing, and seedling dismounting.

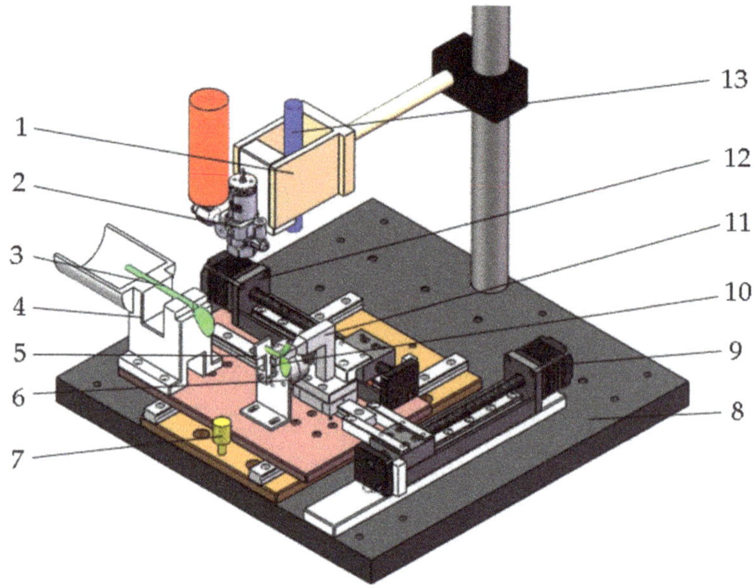

Figure 7. Spray grafting device. 1. Fixture. 2. Spray valve. 3. Rootstock seedling. 4. Rootstock adsorption mechanism. 5. Scion positioning plate. 6. Rotary cylinder. 7. Sliding plate. 8. Frame. 9. Y-axis linear slide. 10. Scion seedling. 11. Scion adsorption mechanism. 12. X-axis linear slide. 13. UV curing lamp.

The installation layout is as follows:

① The spray valve (2) and UV curing lamp (13) in the vertical direction are fixed and mounted onto fixture (1), with fixture (1) through the column mounted onto frame (8);

② The rootstock adsorption mechanism (4), rotary cylinder (6), and X-axis linear slide (12) are installed on the sliding plate (7). The sliding plate (7) is installed on the frame (8) using a linear slide block;

③ The scion adsorption mechanism (11) is mounted onto the X-axis linear slide (12), and the X-axis linear slide (12) drives the scion adsorption mechanism (11) to complete the incision butt joint movement for the rootstock seedling;

④ The rootstock adsorption mechanism (4), the rotary cylinder (6), and the scion adsorption mechanism (11) are arranged from left to right, and the centers of the adsorption grooves of the rootstock adsorption mechanism (4) and the scion adsorption mechanism (11) are coaxial, which facilitates the realization of accurate butt joining of the scion seedling incision;

⑤ The scion positioning plate (5) is mounted onto the rotary cylinder (6), used for the scion on the seedling limit;

⑥ The Y-axis linear slide (9) is also installed on frame (8); the Y-axis linear slide (9) and sliding plate (7) are connected and fixed, for the rootstock and scion, and transported to the spray valve (2) and UV curing lamp (13) below, respectively, for glue spraying and curing operations.

The principle of spray grafting is shown in Figure 8. The working process is as follows:

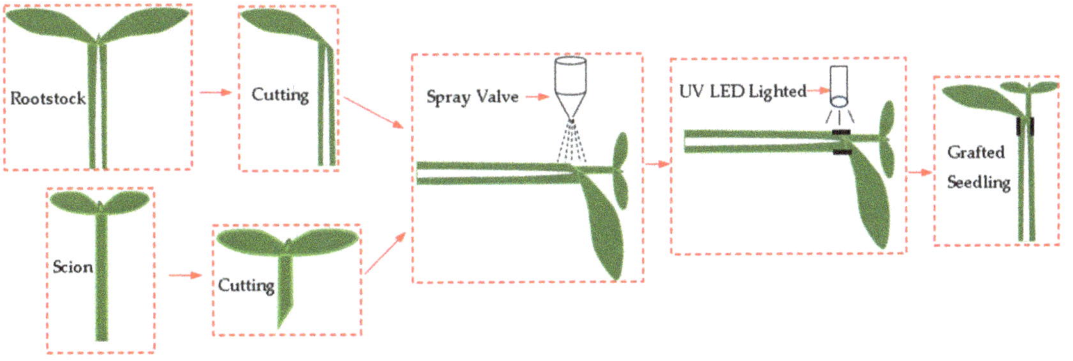

Figure 8. Spray grafting principle.

① Supplying: The cutting of the rootstock seedling is placed horizontally upward in the adsorption slot of the rootstock adsorption mechanism, the foot switch is stepped on once, and the adsorption slot generates negative pressure adsorption and fixes the stem of the rootstock seedling; the cutting of the scion seedling is placed horizontally and downward into the adsorption slot of the scion adsorption mechanism, and the location of the scion on the seedling is determined using the scion locating plate as the reference. The foot switch is stepped on twice, and the adsorption slot generates negative pressure adsorption and fixes the stem of the scion seedling;

② Butt joining: The scion positioning plate is rotated counterclockwise 90° to a horizontal state, and the scion seedlings on the X-axis linear slide are driven to move to the butt joint position so that the scion seedling incision and rootstock seedling incision fit each other;

③ Spraying: The rootstock and scion seedlings move under the drive of the Y-axis linear slide to the underside of the spray valve, which atomizes the UV adhesive and sprays it out to spray the periphery of the scion incision fitting section;

④ Curing: The rootstock seedlings and scion seedlings continue to move under the drive of the Y-axis linear slide underneath the UV curing lamp, and the UV curing lamp irradiates the UV adhesive sprayed onto the grafted seedlings to cure them rapidly;

⑤ Dismounting: The Y-axis linear slide is reset, the grafted seedling is carried to the butt joint station, the foot switch is stepped on 3 times, the negative pressure air source of the rootstock and scion adsorption mechanism is disconnected, the grafted seedling is manually taken out, and the other mechanisms are reset in turn.

A. Seedlings Adsorption Mechanism

Most of the grafting machines use the clamping method to position the seedlings, and EVA cushioning material is pasted inside the clamping hand to realize flexible clamping. Due to the differences in the stems of the seedlings, if the strength of the clamping method is not adjustable, the clamping force is not enough for smaller seedlings, resulting in unstable clamping, and the clamping force is too large for larger seedlings, thus damaging the seedlings [30]. The negative pressure adsorption method can be applied to seedlings of different sizes to meet positioning stability and safety.

Adhesive spraying requires the spray valve to be installed in the vertical direction; therefore, the horizontal direction is selected for seedling placement, adsorption, and positioning, and the seedlings adsorption mechanism is equipped with a positioning groove and adsorption holes in the groove. To determine the structural dimensions of the seedlings adsorption mechanism, it is necessary to measure the external characteristic parameters of the seedling. The rootstock and scion selected were Jingxin rootstock No. 2 of white-seeded pumpkin and Jingxin No. 4 of watermelon, respectively, and both the rootstock and scion seedling stems were oval, defining the seedling stems in the direction of the cotyledon unfolding as the short axes and the seedling stems in the perpendicular direction to the short axis as the long axes. The rootstock plant height H was measured, as shown in Figure 9a, as were the seedling stem long axis a_1 and short axis b_1, as shown in Figure 9b; the length of the hypocotyl after scion cutting was 10 mm to 15 mm, so the plant height of the scions did not need to be measured, only the seedling stems long axis a_2 and short axis b_2.

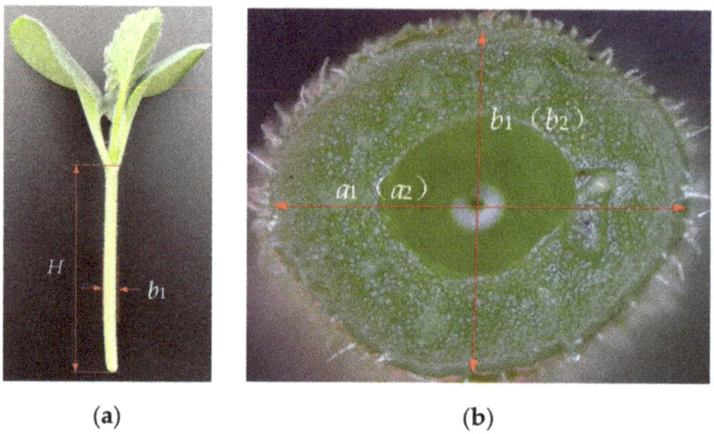

Figure 9. Seedling characterization parameters. (**a**) Main view; (**b**) sectional view of stem.

The external morphological characteristics of the white-seeded pumpkins and watermelons were measured after 7 days and 10 days of planting, respectively. The sample size was 50 plants each, and the results are shown in Table 1. For the seedlings to be placed smoothly in the positioning groove, the width of the groove should be slightly larger than the size of the long axis of the seedling stems, the depth of the groove should be 3/4 of the size of the short axis of the seedling stems, and the length of the groove should be smaller than the height of the plant.

Table 1. External characteristic dimensions of seedlings.

Parameters	Stem of Rootstock a_1 (mm)	Stem of Rootstock b_1 (mm)	Rootstock Height H (mm)	Stem of Scion a_2 (mm)	Stem of Scion b_2 (mm)
Average value	3.66	3.19	66.38	2.34	1.86
Maximum value	4.06	3.71	74.11	2.68	2.05
Minimum value	3.12	2.59	53.19	2.07	1.45

According to Table 1, the width, depth, and length of the locating groove for the rootstock adsorption mechanism are set to 4.5 mm, 2.5 mm, and 40 mm, respectively, as shown in Figure 10a. Additionally, the rootstock adsorption mechanism is equipped with a soil lump support, making it suitable for both rooted and rootless grafting of rootstocks. The locating groove dimensions for the scion adsorption mechanism are set at 3.5 mm, 1.5 mm, and 5 mm, as depicted in Figure 10b.

The negative pressure inlet of the rootstock adsorption mechanism is located in the lower part, and there are two adsorption holes in the positioning slot, as shown in Figure 11a; the negative pressure inlet of the scion adsorption mechanism is located in the rear part, and there is one adsorption hole in the positioning slot, as shown in Figure 11b. The arrow direction is the direction of the airflow, and the seedlings are adsorbed and fixed into the positioning slot due to the pressure difference.

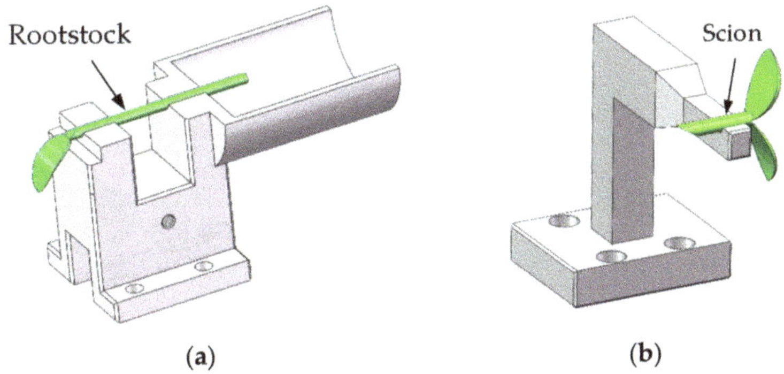

Figure 10. Seedlings adsorption mechanism. (**a**) Rootstock adsorption; (**b**) scion adsorption.

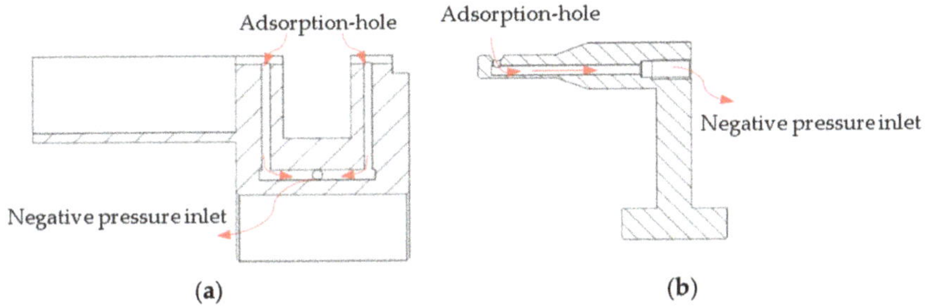

Figure 11. Seedlings adsorption mechanism airflow distribution. (**a**) Rootstock adsorption airflow distribution; (**b**) scion adsorption airflow distribution.

The force for the adsorption and positioning of the seedlings is shown in Figure 12 (along the stem direction of seedlings). The larger the adsorption force, the better the adsorption performance, provided that the adsorption is stable and the seedlings are not damaged [31].

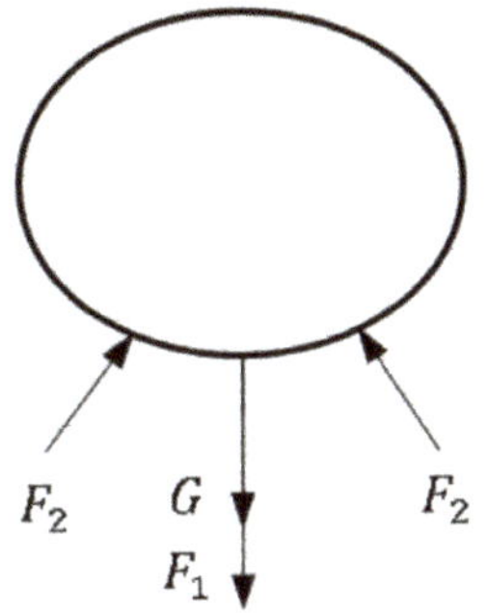

Figure 12. Schematic diagram of seedling adsorption force. G represents the gravity of the seedling itself (N); F_1 represents the adsorption force on the seedling (N); F_2 represents the support force on the seedling (N).

The equilibrium equation for the adsorption forces on the seedlings is as follows:

$$F_1 + G - F_2 = 0 \quad (4)$$

To ensure that seedlings are not destroyed by excessive adsorption,

$$F_1 < F_\sigma \tag{5}$$

$$F_\sigma = P_\sigma S \tag{6}$$

$$S = \frac{\pi d^2}{4} \tag{7}$$

In the formula, S is the cross-sectional area of the adsorption hole (mm^2); F_1 is the adsorption force (N); F_σ is the seedling stem's breaking pressure (N); P_σ is the seedling diameter breaking pressure (Pa); d is the adsorption hole diameter (mm).

In terms of the adsorption size:

$$F_1 = \frac{\pi \rho C_d d^2 v^2}{8} \tag{8}$$

In the equation, C_d represents the drag coefficient, set to 1.0; ρ is the air density, taken as 1.29kg/m^3; v represents the average velocity of the force equilibrium airflow field.

B. Butt Joining and Handling Mechanisms

The butt joint and handling of the rootstock and scion seedlings are completed under the drive of the X- and Y-axis linear slides; the linear slides have a lead of 12 mm, an effective stroke of 100 mm, and a maximum moving speed of 0.1 m/s. A butt joint and handling diagram is shown in Figure 13, and the processes are as follows:

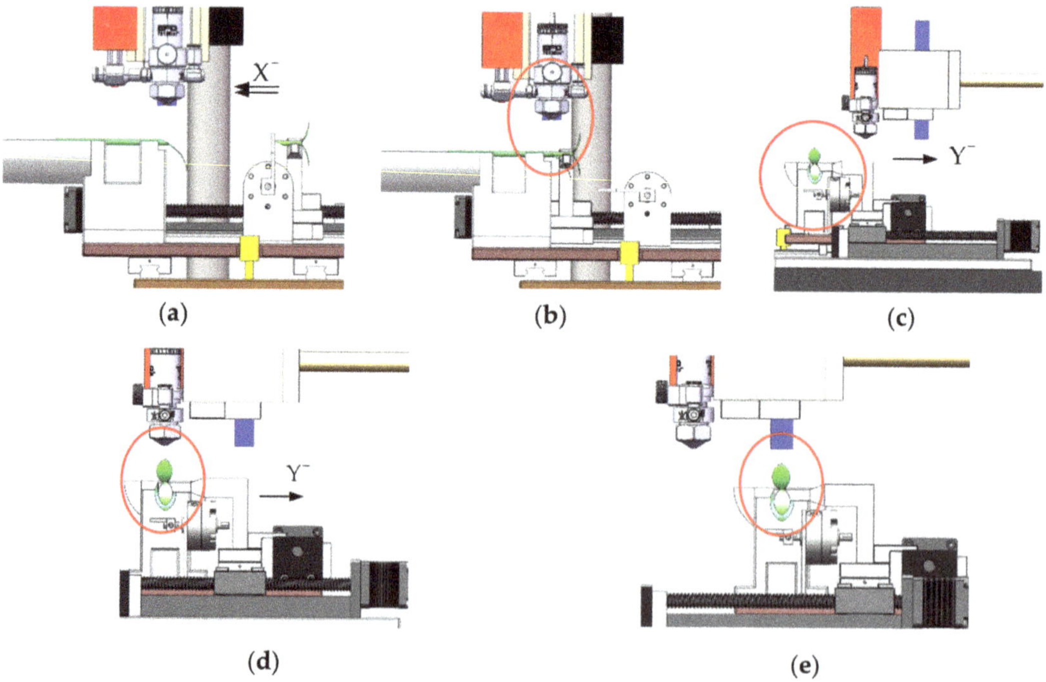

Figure 13. Schematic diagram of the butt joint handling process. (**a**) Seedling supply; (**b**) butt joint; (**c**) butt joint side view; (**d**) spraying; (**e**) curing.

① Completing the adsorption positioning of the rootstock and scion seedlings using the adsorption mechanism and scion positioning plate.

② The X-axis linear sliding table drives the scion to move 50 mm in the X negative direction to complete the butt joint.

③ The Y-axis stepper slide drives the handling mechanism to the spraying and curing stations.

④ The Y-axis linear slide drives the rootstock and scion seedling as a whole to move 50 mm in the negative direction of the Y-axis to the spraying station; the spraying height is 5 mm to 15 mm.

⑤ After the spraying is completed by the spraying valve, it will continue to move 42 mm in the negative direction of the Y-axis to reach the curing station. The UV curing lamp works for 1~2 S.

Finally, each mechanism is reset in turn.

2.1.3. Simulation Modeling

① Adsorption Modeling

To determine the optimum parameters for the adsorption mechanism, hydrodynamic simulation analysis is carried out for different pore diameters and numbers of pores. For the calculation, the fluid inside the air chamber is considered a continuous phase, and the air fluid is considered an incompressible viscous fluid in the negative pressure adsorption process, which must follow the laws of the conservation of mass, momentum, and energy [32], with the Navier–Stokes equation satisfying these conditions being:

$$\frac{\partial \rho}{\partial t} + \frac{\partial}{\partial x_i}(\rho u_i) = 0 \tag{9}$$

To keep the set of equations closed, the fluid inside the gas chamber of the adsorption mechanism was numerically simulated using the standard $k - \varepsilon$ turbulence model, viz:

$$\frac{\partial}{\partial t}(\rho k) + \frac{\partial}{\partial x_i}(\rho k u_i) = \frac{\partial}{\partial x_j}\left[\left(\mu + \frac{\mu_t}{\sigma_k}\right) + \frac{\partial k}{\partial x_j}\right] + G_k + G_b - \rho\varepsilon - Y_M + S_k \tag{10}$$

$$\frac{\partial}{\partial t}(\rho \varepsilon) + \frac{\partial}{\partial x_i}(\rho \varepsilon u_i) = \frac{\partial}{\partial x_j}\left[\left(\mu + \frac{\mu_t}{\sigma_\varepsilon}\right)\frac{\partial \varepsilon}{\partial x_j}\right] + C_{1\varepsilon}\frac{\varepsilon}{k}(G_k + C_{3\varepsilon}G_b) - C_{2\varepsilon}\rho\frac{\varepsilon^2}{k} + S_\varepsilon \tag{11}$$

Among this,

$$\mu_t = \rho C_\mu \frac{k^2}{\varepsilon} \tag{12}$$

$$G_k = \mu_t\left(\frac{\partial \mu_i}{\partial x_i} + \frac{\partial \mu_j}{\partial x_i}\right)\frac{\partial \mu_i}{\partial x_i} \tag{13}$$

$$G_b = \beta g_i \frac{\mu_t}{Pr}\frac{\partial T}{\partial x_i} \tag{14}$$

$$Y_M = 2\rho\varepsilon Ma^2 \tag{15}$$

where G_k is the production term for the turbulent kinetic energy k due to the mean velocity gradient; G_b is the generation term for the turbulent kinetic energy k from buoyancy; μ_t is the turbulent viscosity; β is the thermal expansion coefficient; Y_M is the contribution of pulsation expansion in compressible turbulence; Pr is the turbulent Prandtl number; ε is the dissipation rate; Ma is the Mach number; S_k, S_ε are source terms; the empirical constants $C_{1\varepsilon}$, $C_{2\varepsilon}$, $C_{3\varepsilon}$ are taken as 1.44, 1.92, and 1.0, respectively, and σ_ε is set to 1.2.

The size of the diameter of the adsorption hole in the positioning groove affects the contact area between the airflow and the seedlings. The maximum hole diameter should be smaller than the size of the long axis of the seedlings, and the hole diameter should not be too small; otherwise, it will make the airflow and the contact area of the seedlings too small and thus lead to adsorption instability. Three sizes of 1.5 mm, 2.0 mm, and 2.5 mm were selected as the pore size variables. The number of adsorption holes affects the force points of the seedlings. The more adsorption holes there are, the more points of force the seedlings are subject to, so the seedlings become more stable, but because of the dispersion of forces, the adsorption force is smaller. Conversely, with fewer adsorption holes, the adhesion force of the seedlings is greater but less stable. The selected number of adsorption holes is 1 to 3 in this research.

The standard $k - \varepsilon$ turbulence model is selected, the inlet boundary condition is the pressure inlet, the pressure magnitude is −40 kPa, the outlet boundary condition is standard atmospheric pressure, the wall is set as a no-slip wall boundary condition, and the adsorption hole size is small, so it is necessary to take into account the influence of the boundary layer, using the standard wall functions to this end [33].

② Spraying Modeling

Firstly, the fluid domain model of the spray valve is established using SolidWorks 2022 3D software, and the model is meshed using Fluent Meshing. Due to the large change in the velocity gradient at the nozzle exit, the mesh encryption process is carried out at the nozzle exit, and the mesh type is a tetrahedral mesh, with a total number of meshes of 939,340 and a minimum orthogonal mass of 0.46, as shown in Figure A4. The flow field distribution of the gas–liquid two-phase flow in the nozzle is simulated using the VOF method [34], and the liquid-phase velocity, diffusion angle, and mass flow rate at the nozzle outlet are monitored to obtain them and use them as boundary conditions for the DPM injection model [35].

In the VOF model calculations, the realizable k-ε turbulence model was used, with Rosin–Rammler distribution for the droplet size and a diffusion factor of 3.5. The sprayed wall was set as a wall film model, and the rest were set as exit boundary conditions. The pressure–velocity coupled PISO algorithm was used in second-order upwind format.

The spray valve's spray adhesive mist droplets move in the flow field mainly on the basis of their own gravity and air impact [36]. The droplet force balance equation is:

$$m\frac{du_p}{dt} + F_D + G = 0 \tag{16}$$

where

$$F_D = \frac{1}{2}C_D\rho_f A_f\left(u - u_p\right)^2 \tag{17}$$

where F_D is the trailing force on the droplet, m is the droplet mass, G is the droplet weight, u_p is the droplet velocity, C_D is the coefficient of the trailing force, ρ_f is the fluid density, A_f is the reference area, and u is the fluid velocity.

The droplet–wall impact model can be categorized into four modes: adhesion, rebound, wall-attached jet, and splashing [37,38]. The Weber number can be used as a basis for judging the impact model, and the Weber number is calculated as:

$$We = \frac{\rho v^2 l}{\sigma} \tag{18}$$

where ρ is the fluid density, v is the characteristic flow rate, l is the characteristic length, and σ is the fluid surface tension coefficient.

(1) When $We < 5$, the collision of the droplet with the wall is modeled as adhesion;
(2) When $5 \le We \le 10$, the collision between the droplet and the wall is modeled as a rebound;
(3) When $We > 10$, the droplet–wall collision is modeled as a splash/wall-attached jet.

The droplet's Weber number should not be too large if more droplets are made to adhere to the wall on first collision.

The following equation is used for the conservation of the droplet's mass:

$$\frac{\partial h}{\partial t} + \nabla_s \cdot \left[h\overline{V}_l\right] = \frac{\dot{m}_s}{\rho_l} \tag{19}$$

where ρ_l is the density of the liquid film $\left(\mathrm{kg/m^3}\right)$, h is the thickness of the liquid film (m), $\overline{V}_l$ is the average velocity of the liquid film flow $(\mathrm{m/s})$, and $\dot{m}_s$ is the mass source of the liquid film per unit wall area $\left(\mathrm{kg/\left(m^2{\cdot}s\right)}\right)$.

We use the following conservation of momentum equation:

$$\frac{\partial h\overline{V}_l}{\partial t} + \nabla_s \cdot \left[h\overline{V}_l\overline{V}_l\right] = \frac{h\nabla_s P_L}{\rho_l} + (\overline{g}_\tau)h + \frac{3}{2\rho_l}\overline{\tau}_{fs} - \frac{3v_l}{h}\overline{V}_l \tag{20}$$

2.2. *Tests*

2.2.1. Test Materials

The seedling nursery was located in the solar greenhouse of the Beijing Academy of Agricultural and Forestry Sciences (Haidian District, Beijing, China, 39°56′28″ N, 116°16′53″ E). Jingxin rootstock No. 2, a white-seeded pumpkin variety, was chosen as the rootstock, while Jingxin No. 4, of watermelon, was selected as the scion. The rootstock and scion were sown separately using standard hole trays of 5 × 10 and 6 × 12, respectively. For optimal rootstock grafting, the seedling age was set at 8 to 10 days, with the external morphology of the rootstock characterized by two spreading cotyledons and a stem thickness of about 3.5 mm, as depicted in Figure 14a. In contrast, the scion was sown three days earlier than the rootstock, displaying 2 to 3 leaves in its growth [39,40], as shown in Figure 14b.

Figure 14. Test materials. (**a**) Rootstock seedlings; (**b**) scion seedlings.

The designed spray grafting device was subjected to drawing, machining, module assembly, and control software development to integrate the spray grafting device as shown in Figure A5, which only requires one person to operate it.

2.2.2. Test Contents and Methods

To determine the optimal grafting process parameters for the spray valve, with atomization pressure, glue supply pressure, and spraying height as the independent variables and grafting success rate as the dependent variable, we analyzed the success rate of watermelon grafting using an orthogonal test, verified the grafting survival rate with the optimal combination of factors, and analyzed the effect of the adhesive film-forming length on the effective spraying rate.

The atomization pressure, glue supply pressure, and spraying height of the spray valve were selected for a 3-factor, 3-level orthogonal experimental design using the L_9 (3^3) table, and the coding of the factor levels is shown in Table 2. A total of 9 groups of experiments were arranged, and 30 grafted seedlings were completed in each group of experiments.

The grafting process is, sequentially, seedling supply, butt joining, spraying, curing, and seedling dismounting. First, the operator uses a blade to cut the rootstock and scion seedlings at a 30° cutting angle; then, the rootstock and scion seedlings are put into the seedlings adsorption mechanism, respectively, as seen in Figure 15a,b; and then butt joining and spraying, as in Figure 15c, and curing and seedling dismounting are completed automatically, as in Figure 15d.

Table 2. Factor level coding table.

Levels	Factors		
	Atomizing Pressure (A)/Mpa	**Glue Supply Pressure (B)/Mpa**	**Spraying Height (C)/mm**
1	0.20	0.15	5
2	0.25	0.20	10
3	0.30	0.25	15

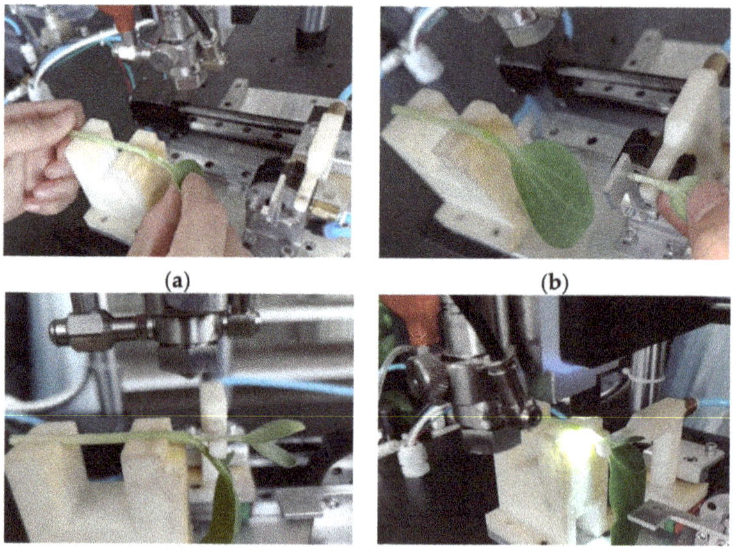

Figure 15. Grafting procedure. (**a**) Rootstock seedling supply; (**b**) scion seedling supply; (**c**) butt joining and spraying; (**d**) curing process.

The test indicators are as follows:

(1) The grafting success rate for each group of tests was counted, and the grafting was considered successful if the cut surfaces of the rootstock and scion were closely adhered to form a single unit after spray grafting. Otherwise, it was regarded as a grafting failure, as shown in Figure A6.

(2) Since the adhesive cannot be completely sprayed onto the grafted seedling, the amount of adhesive sprayed onto the grafted seedling is called the effective spray volume. The mass of the rootstock and scion before spraying, the mass of the adhesive applied in a single spray by the spray valve, and the mass of the grafted seedling after spraying were weighed separately. The formula for calculating the effective spray rate R_{es} for the adhesive is as follows:

$$Effective spray rate R_{es} = \frac{Mass\ of\ grafted\ seedlings\ after\ spraying\ (g)}{Mass\ of\ rootstock\ and\ scion\ before\ spraying\ (g) + Mass\ of\ adhesive\ applied\ in\ a\ single\ spray\ (g)} \times 100\%$$

(3) The adhesive film length along the axis of the grafted seedling was measured.
(4) The grafted seedlings were completed under the conditions of the optimal process parameters for adhesive spray grafting 200 plants and put into an incubator for healing treatment. Environmental conditions: 80% to 90% humidity, daytime temperature 28 °C, nighttime temperature 25 °C, healing under these conditions for 7 days [41,42]. If the scions can grow new cotyledons, they will be considered as having survived. Otherwise, they will be not. Then, the grafting survival rate will be counted.

3. Results and Discussion

3.1. Simulation Results for the Seedling Adsorption

The effects of the number of suction holes and the number of suction holes on the adsorption properties of the rootstock adsorption mechanism were analyzed using fluid simulation. When the diameter of the adsorption hole is 2 mm, the distribution of the number of adsorption holes in the flow field is shown in Figure A7.

The adsorption force generated by a single adsorption hole acts on the rootstock seedling at the middle of the seedling stem, the average velocity of the airflow is 220 m/s, and the average pressure is −25,646 Pa. Since the rootstock seedling belongs to a non-rigid object, the two ends of the seedling stem may be shifted. Airflow dispersion due to double adsorption holes, the airflow speed is significantly lower than the single adsorption hole speed. The vacuum here compared to single adsorption holes is also reduced, the average speed of the airflow is 168 m/s, and the average pressure is −17,996 Pa. With adsorption holes of the same diameter, the total adsorption force for double adsorption holes is greater than single adsorption holes' adsorption force. The rootstock has two points of force with two holes, so both from the point of view of the size and force stability of the adsorption force, double adsorption holes are better than single adsorption holes. With three adsorption holes, the gas flow field distribution and the gas flow rate at both ends of the adsorption hole channel are the same, the average velocity of the gas flow at the exit is 97.6 m/s, and the average pressure is −5921 Pa. The gas flow rate for the middle adsorption hole channel is obviously larger than the gas flow rate of the two ends of the channel, the average velocity of the gas flow at the exit is 148.5 m/s, and the average pressure is −15,067 Pa. In the adsorption process for rootstock seedlings, the force between the middle and both ends of the seedling stems was different, and the two ends of the stems might buckle, which does not meet the requirements of stable adsorption. Therefore, the rootstock adsorption mechanism has a good adsorption stability under double adsorption hole conditions.

The effect of different adsorption hole diameters on the average flow rate and adsorption force of the adsorption holes under double adsorption hole conditions is shown in Figure A8. Calculations using Equation (8) show that as the diameter of the adsorption hole increases, the velocity of the airflow at the exit of the adsorption hole decreases gradually, and the total adsorption force increases gradually, which suggests that the effect of the diameter of the adsorption hole on the adsorption force is more significant than the velocity of the airflow, and when the diameter of the adsorption hole is 2.5 mm, the average flow rate of the airflow is 146 m/s, the average pressure is −16,281 Pa, and the magnitude of the adsorption force is 0.124 N, which signifies a good adsorption performance.

In summary, the optimal combination of parameters for the rootstock adsorption mechanism is double adsorption holes and an adsorption hole diameter of 2.5 mm. The length of the seedling stem after scion cutting is 10 mm to 15 mm, and the scion seedling stem is smaller, so the scion adsorption mechanism is designed as a single adsorption hole, and the diameter of the adsorption hole is 2.0 mm.

3.2. Spraying Simulation Results

The atomizing pressure of the spray valve determines the spray cone angle, and the glue supply pressure determines the mass flow rate for the adhesive spray. A velocity cloud for the gas-phase flow field of the spray valve is shown in Figure 16. With the nozzle exit contraction at the maximum speed, the airflow at the exit began to gradually diffuse. The flow field has an approximately trapezoidal distribution due to the nozzle spraying the beam of liquid under the influence of the airflow so that large-particle-size droplets are broken into smaller droplets to achieve a better diffusion of the atomization effect [43]. With an increase in the atomization pressure, the airflow field velocity increases significantly, and the flow field cone angle also increases slightly. With an atomizing pressure of 0.2 Mpa in the nozzle outlet near the range of the axis with an asymmetric distribution, the airflow produces uneven diffusion, with drift slightly to the left. At an atomization pressure of 0.25 Mpa/0.3 Mpa, the airflow field distribution is more uniform, so a 0.25 Mpa/0.3 Mpa atomization pressure is used to help achieve a better atomization effect. The direct factors affecting the distribution range of the adhesive are the spray cone angle and spray height. While the spray cone angle is affected by the atomization pressure, the glue supply pressure also has a certain effect on the spray cone angle, and the adhesive film thickness is affected by the mass flow rate (glue supply pressure).

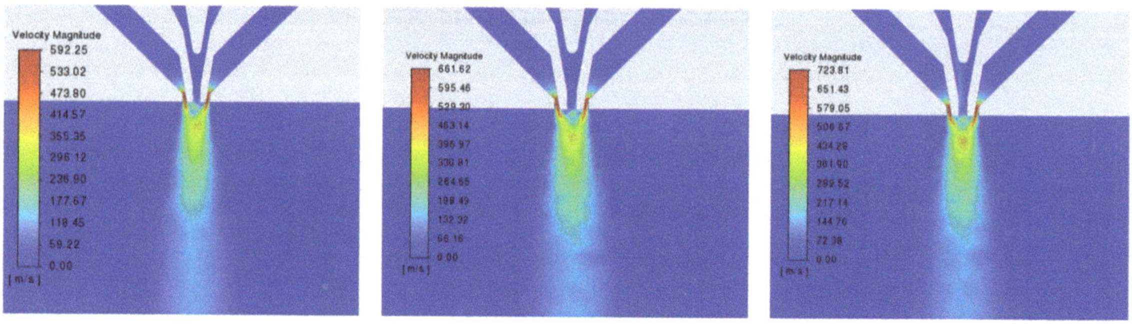

(a) (b) (c)

Figure 16. Velocity flow field distribution of atomized gas. (**a**) 0.20 Mpa; (**b**) 0.25 Mpa; (**c**) 0.30 Mpa.

Nine groups of orthogonal simulation tests were arranged according to the test factors in Table 2 to analyze the liquid film coverage of the spraying model, as shown in Figure 17. All nine groups of tests could satisfy the value of the adhesive coverage length L in Figure 6, in which the maximum value was 8.34 mm and the minimum value was 6.07 mm.The mass flow rate of the adhesive is mainly related to the glue supply pressure, the 1st, 4th, and 7th group; the 2nd, 5th, and 8th group; and the 3rd, 6th, and 9th group were three kinds of similar-quality adhesives under the condition of the spraying results. The smaller the range of the adhesive film and the greater the value of the thickness, the higher the effective spraying rate of the adhesive; on the contrary, if the range of the film is greaterthe thickness is smallerand the effective spraying rate is less. Therefore, under the condition of the same flow rate quality, groups 1, 2, and 6 have the highest effective spraying rate, but due to the small coverage area, the poor wrapping of the incision may lead to the actual grafting effect being less desirable. Meanwhile, groups 3, 5 and 7 have the lowest effective spraying rate, which will result in the waste of too much adhesive. Groups 4, 8, and 9 not only ensure that there is no unnecessary waste due to a low-efficacy spraying rate but also make the coating film have a better coverage.

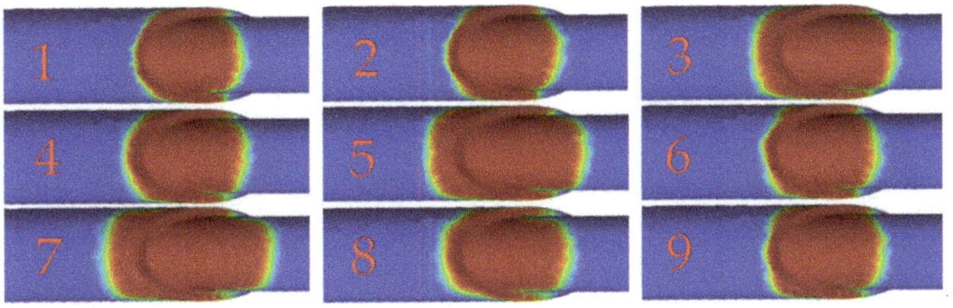

Figure 17. Film-forming coverage.

3.3. *Grafting Test Results*

3.3.1. Grafting Success Rate

The experimental program and the result statistics are seen in Table 3. For the nine groups, the experimental success rate reached more than 50%. Among the main reasons for grafting failure were as follows: ① After curing the adhesive film, the area is larger on the seedling incision, and the wrapping performance is better, but in terms of the adhesive spraying mass flow rate under the same conditions, the larger the spraying area, the thinner the adhesive film's thickness, resulting in a reduction in the adhesive's performance; ② After curing, the film area for the adhesive is small, which is not enough to cover and bond the incision effectively, resulting in an unsatisfactory bonding effect and grafting failure.

Table 3. Experimental program and results.

Test Number	A (Mpa)	B (Mpa)	C (mm)	Grafting Success Rate (%)
1	1 (0.20)	1 (0.15)	1 (5)	56.70
2	1 (0.20)	2 (0.20)	2 (10)	63.33
3	1 (0.20)	3 (0.25)	3 (15)	70.00
4	2 (0.25)	1 (0.15)	2 (10)	100.00
5	2 (0.25)	2 (0.20)	3 (15)	90.00
6	2 (0.25)	3 (0.25)	1 (5)	76.67
7	3 (0.30)	1 (0.15)	3 (15)	63.33
8	3 (0.30)	2 (0.20)	1 (5)	83.33
9	3 (0.30)	3 (0.25)	2 (10)	100.00

The grafting success rate for both test groups 4 and 9 reached 100%. Comparing the two groups of test parameter matching, the parameter matching scheme for group 4 was $A_2B_1C_2$, and the parameter matching scheme for group 9 was $A_3B_3C_2$. The level of factor C (a spraying height of 10 mm) for the two groups of schemes was the same, being in the middle, whereas the levels of factor A and factor B were higher in group 9 than in group 4, which leads to a higher amount of adhesive and higher cost. In addition, it is easy to affect the normal growth of the seedlings.

The analysis of variance of the graft success rate is shown in Table 4. It was based on the magnitude of the *p* values. The three factors influencing the grafting success rate in order of priority are A (atomization pressure) > C (spraying height) > B (glue supply pressure),under the same experimental results, the lower the factor level value, the better the scheme. The fourth group of tests not only ensured a success rate of 100% grafting but also made the amount of adhesive used in the grafting reach the minimum value. Therefore, the parameters of the fourth group of test conditions are considered to be the best combined parameters for spray grafting, and the parameters are $A_2B_1C_2$ (atomization pressure of 0.25 Mpa, glue supply pressure of 0.15 Mpa, spraying height of 10 mm). The optimal schemes obtained using the simulations were $A_2B_1C_2$, $A_3B_2C_1$, and $A_3B_3C_2$. It is obvious that the optimal parameter combination conforms with the results of the simulation parameters.

Table 4. Analysis of variance for success rate.

Source	Square Sum	Degrees of Freedom	Mean Square	F	*p*
A	1053.450	2	526.725	2.143	0.318
B	120.716	2	60.358	0.246	0.803
C	424.257	2	212.129	0.863	0.537
Error	491.657	2	245.829		
Total	57,058.446	9			
Total after amendments	2090.080	8			

3.3.2. Adhesive Film Length and Effective Spray Rate

The results for the adhesive spray grafting film-forming length L and effective spraying rate in the above nine groups of tests are shown in Figure 18. Among them, the optimal parameter combination was group 4 ($A_2B_1C_2$), with a simulated adhesive film-forming length of 7.51 mm, an actual adhesive film-forming length of 7.86 mm, and an effective adhesive spraying rate of 83.03%. The maximum relative deviation between the actual film-forming length and the simulated film-forming length was 4.69%, indicating that the simulated and actual values were in good agreement. The minimum film-forming length for the adhesive was seen in the first group test ($A_1B_1C_1$), for which the simulated value and the actual value were 6.15 mm and 6.33 mm, respectively, and the effective spraying rate was 90.69%. The maximum film-forming length for the adhesive was seen the seventh group test ($A_3B_1C_3$), for which the simulated value and the actual value were 8.34 mm and 8.60 mm, respectively, and the effective spraying rate was 68.31%. The effective spraying rate was negatively correlated with the adhesive film-forming length. This is due to the fact that the larger the spray's spraying area, the larger the adhesive film length for the grafted seedling, but at the same time, the dose of adhesive sprayed into the air is also larger.

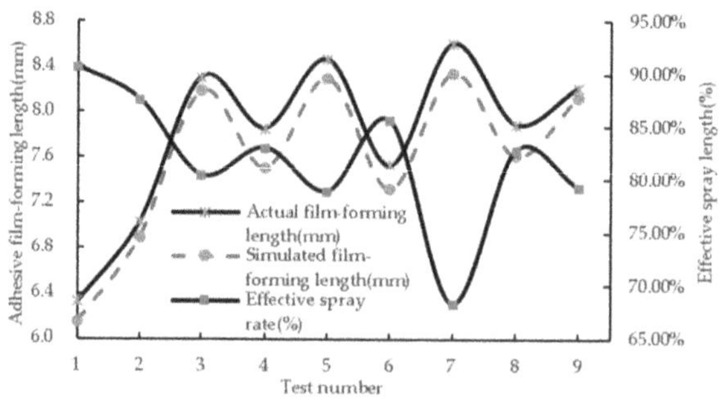

Figure 18. Adhesive film formation length and effective spray rate.

3.3.3. Grafting Survival Rate

The grafting survival rate was verified under the conditions of a 0.25 Mpa atomization pressure in the spray valve, a 0.15 Mpa glue supply pressure, and a 10 mm spraying height. The number of completed watermelon grafts with broken roots was 200, and the grafts were replanted into 50-hole cavity trays and placed in an NK system growth cabinet for healing treatment (model: LPH-1-CT, Nippon Medical & Chemical Instruments Co., Ltd. 3–9 Tamatsukuri Motomachi, Tennoji-ku, Osaka 543-0014, Japan), as shown in Figure A9. Observation: A total of 11 seedlings showed scion dislodgement, with a grafting survival rate of 94.5%. Among them, the reasons for scion shedding were as following: ① The scion and rootstock incision did not reach each other in a close fit, resulting in the rootstock being unable to deliver nutrients to the scion; ② Part of the scion seedling stem was too large, or the adhesive in the scion incision butt joint area was too small. With the growth of the rootstock, the scion wounds did not yet heal, and the adhesive film came off. For the remaining surviving seedlings, some of them have grown new leaves.

3.4. Analysis of the Grafting Costs and Ecological Benefits

The grafting device in this research is simple in structure. The feeding clip mechanism of a traditional grafting machine is replaced with spraying glue and a curing mechanism. The structures of the grafting machine feeding clip mechanisms developed by both ISEKI & Co., Ltd. (GR800-U) in Matsuyama city, Japan and Helper Robotech Co., Ltd. (GR-800CS) in Gimhae-si, Gyeongsangnam, South Korea are complex, and the feeding clip work is completed by pushing the cylinder and the feeding clip cylinder. The plastic clamp becoming stuck often occurs, and the machine needs to be shut down to take it out so that the machine can resume normal operation, which shortens the effective working time. In addition, a vibration feeding device is required to separate, sort, and output the plastic clamp. Reverse ordering of the plastic clip will directly lead to the failure of the feeding clip, resulting in a waste of grafted seedlings. ISO Group Co., Ltd. (Graft 1200) in Middelkampseweg 9, 5311 PC, Gameren, Netherlands developed a grafting feeding clip mechanism based on silicone clips, which exists in the form of the whole volume supply. The operation process includes conveying, cutting, and the feeding clip, so the structure and process are more complicated.

The UV adhesive grafting method proposed in this research relies on a spray valve and a point light source curing machine for the implementation process. Compared with the above feeding clip mechanism, the structure of the spraying and curing mechanism is simple, and the operation stability and success rate are greatly improved. The grafting success rate of the traditional grafting machine is 90~98%, and the grafting survival rate is 90~95%, while the results of this study have achieved a 100% grafting success rate and a 94.5% grafting survival rate, which has completely surpassed the success rate, and the grafting survival rate has reached the leading level as well. Therefore, the results of this paper are expected to be applied to the development of new spray glue grafting robot technology, changing the traditional machine grafting production mode.

At present, the splice grafting methods for both cucurbits and Solanaceae use plastic clips or silicone clips, while producers in the grafting nursery and seedling planting link cannot recover the plastic clips fast enough. Usually, each plastic clip weighs approximately 1 g. Taking a seedling nursery enterprise with an annual output of 10 million grafted seedlings as an example, it needs to consume 10,000 kg of ABS raw materials, and each person can recycle about 10,000 plastic clips per day, so it takes 100 people to work for 10 days to complete the recycling task. In addition, the plastic clips on each grafted seedling should be removed one by one, and the process of removing the plastic clips requires careful operation to avoid damage to the grafted seedlings. With each worker's daily wage of $14~17, a total of $14,000 to $17,000 of labor costs needs to be invested. The cost of the plastic clips is only about $0.007 per piece. If plastic clip recycling is implemented, it not only wastes time but there is also a huge difference between the labor cost input and output. Therefore, seedling nurseries are reluctant to spend this investment to complete the recycling of plastic clips. The above problems have had an adverse effect on the green and efficient development of the vegetable nursery industry, and new and innovative methods are urgently needed by producers and management to alleviate the traditional grafting problems.

Using the UV adhesive grafting method proposed in this research, the grafting cost is calculated to be only $0.0011/plant, which is 16% of the grafting cost using traditional plastic clips, and this can greatly reduce the grafting costs. In 2021, China's demand for vegetable grafted seedlings was as high as 50 billion plants. If the technology can be successfully applied, within the grafting costs, more than $277 million per year can be saved, and the market prospects for its application are very promising. In addition, the adhesive film formed on the surface of the grafted seedlings can be automatically dislodged, with gradual thickening of the stem of the grafted seedling. The adhesive film to fix the incision will be gradually weakened until it is dislodged. Due to the UV adhesive reaching the level of medical grade and having biodegradable characteristics, it can appropriately alleviate the problem of pollution of the agricultural planting environment, and its ecological benefits are very significant, with significant potential for its application.

4. Conclusions

In response to the current global use of plastic clips or silicone clips for vegetable seedling grafting, which has the problem of serious pollution of the planting environment caused by the difficulty of recycling the plastic clips, this research proposes a grafting method based on a biodegradable UV adhesive, which forms a stationary adhesive film around the stems of the grafted seedlings in place of plastic clips. In this research, a combination of theoretical modeling, structural design, and experimental validation was used to systematically study the atomization parameters of the UV adhesive and curing film, and the experimental results were a 100% grafting success rate, an adhesive film length of 7.86 mm, a spray rate of 83.03%, and a grafting survival rate of 94.5%, which meet the technical requirements of a watermelon grafting nursery. In terms of the grafting success rate and survival rate, this research, compared with the existing grafting machines, is clearly superior in its grafting success rate, and its grafting survival rate is comparable. In terms of the cost input, adhesive grafting represents 16% of the cost of grafting with plastic clips, which can significantly reduce the production cost of grafting in seedling nursery enterprises, which is also the main advantage of the popularization and application of this technology. In terms of the ecological benefits, the adhesive has biodegradable and self-shedding properties, which are important for planting environment protection and ecological environment improvement. Therefore, the results of the research not only provide a theoretical basis for the development of a new type of spray glue grafting robot but also are expected to change the traditional plastic clip grafting seedling mode and realize the transformation and upgrading of the seedling industry to green and environmentally protective and friendly.

5. Research Deficiencies and Future Prospects

In existing grafting production, due to a large number of grafting plastic clips causing planting pollution problems, a UV adhesive atomization spray grafting method was formulated based on the feasibility of the tests used to verify it. However, due to a lack of time and inexperience, a follow-up will be carried out covering the following aspects of the study.

(1) Seedling adaptability. The design of this paper adopts the seedling positioning method for negative pressure adsorption. The characteristics of tomato seedlings and watermelon seedlings are different, so it cannot be adapted to tomato seedlings and other tomato seedling grafting practices. In addition, in actual grafting operations, some of the seedling stems are twisted or bent and form in other non-ideal growth states. In the incision butt joining process, there are scion incision contact areas that are too small or even do not make contact and other problems, which leads to a lower grafting quality.

(2) Grafting efficiency. The grafting test bench is a linear mobile working mode, which does not include the cutting process, and seedling cutting and then seedling supply have to be completed with the help of other auxiliary cutting devices. The institutions devices to be reset sequentially after the grafting is completed, which increases the amount of ineffective time in the grafting process, resulting in a low grafting efficiency. The design and research of multi-station rotary grafting devices including cutting mechanisms will be carried out in the future.

(3) The shedding characteristics of the adhesive film are still unclear. It is necessary to further explore the shedding characteristics of the UV film, analyze and test the mechanical properties of the film in terms of the fixation of grafted seedlings, track the process of film shedding after the healing and planting of the grafted seedlings, and clarify the environmental conditions and the shedding cycle of the film.

Author Contributions: Conceptualization, X.Z., L.K. and K.J.; methodology, L.K. and K.J.; software, L.K., K.J. and H.L.; validation, L.K. and K.J.; formal analysis, X.Z., Q.Z. and T.L.; investigation, L.K. and K.J. resources, K.J., Q.F., H.L. and Q.Z.; data curation, X.Z. and K.J.; writing—original draft preparation, L.K.; writing—review and editing, Q.F. and K.J.; visualization, T.L. and K.J.; supervision, X.Z., Q.F. and K.J.; project administration, K.J.; funding acquisition, K.J. All authors have read and agreed to the published version of the manuscript.

Funding: This research was supported by the National Nature Science Foundation of China (Grant No. 32171898), the BAAFS Innovation Ability Project (KJCX20220403), the China National Agricultural Research System (CARS-25-07), and the Beijing Key Laboratory of Agricultural Intelligent Equipment Technology (PT2024-44).

Institutional Review Board Statement: Not applicable.

Informed Consent Statement: Not applicable.

Data Availability Statement: Data are contained within the article.

Conflicts of Interest: The authors declare no conflicts of interest.

Appendix A

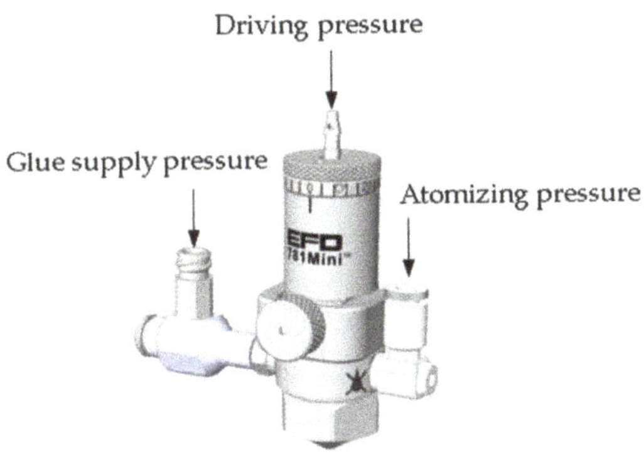

Figure A1. The 781mini spray valve.

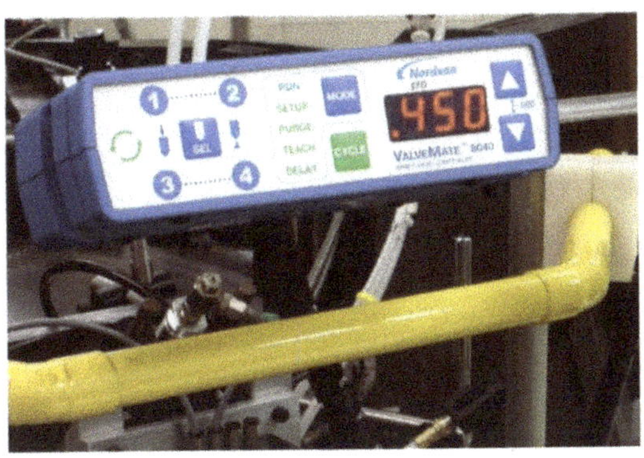

Figure A2. Controller operation panel.

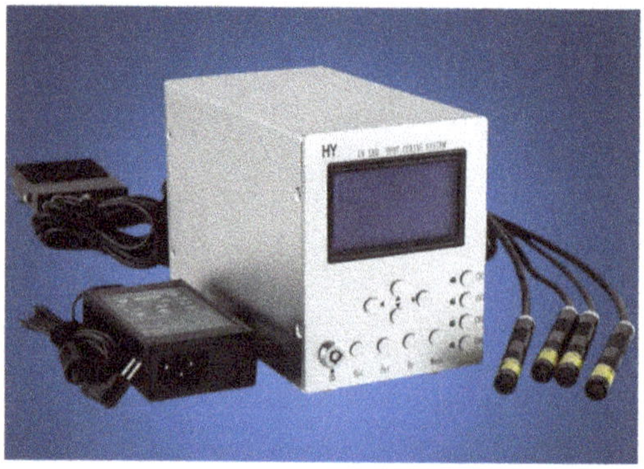

Figure A3. Point source curing machine.

Figure A4. Nozzle flow field meshing.

Figure A5. Grafting device.

Figure A6. Grafting situation. (**a**) Grafting success; (**b**) grafting failure.

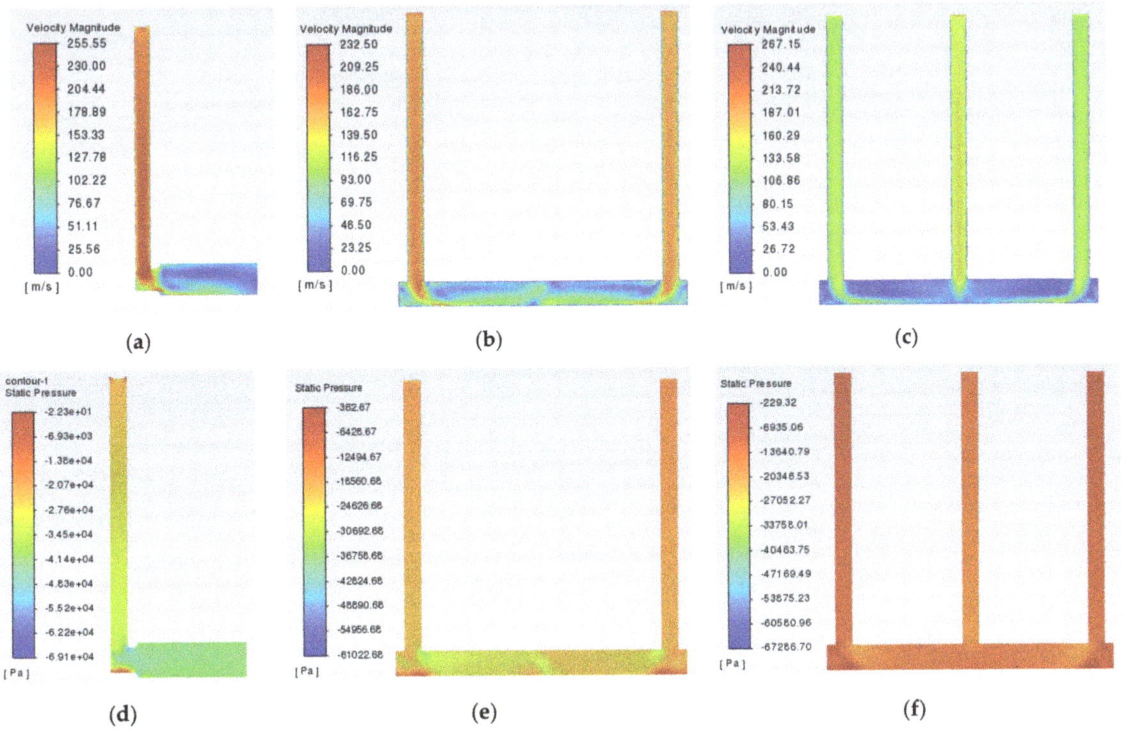

Figure A7. Airflow field distribution with different numbers of adsorption holes. (**a**) Single-hole velocity cloud chart; (**b**) two-hole velocity cloud chart; (**c**) three-hole velocity cloud chart; (**d**) single-hole pressure cloud chart; (**e**) two-hole pressure cloud chart; (**f**) three-hole pressure cloud chart.

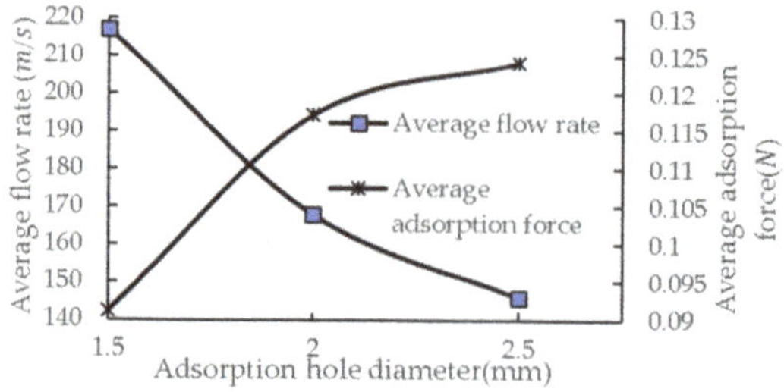

Figure A8. Effect of diameter of double adsorption holes on mean flow rate and adsorption force.

(**a**) (**b**)

Figure A9. Verification of graft survival rate. (**a**) Watermelon grafting seedlings; (**b**) growth cabinet.

References

1. Li, B.; Liu, W. Research progress on genetic map construction and gene mapping for watermelon. *China Cucurbits Veg.* **2019**, *32*, 1–6.
2. Guan, W.; Zhao, X.; Hassell, R.; Thies, J. Defense mechanisms involved in disease resistance of grafted vegetables. *Horticulture* **2012**, *47*, 164–170. [CrossRef]
3. Xie, Z.; Gu, S.; Chu, Q.; Li, B.; Fan, K.; Yang, Y.; Yang, Y.; Liu, X. Development of a high-productivity grafting robot for Solanaceae. *Int. J. Agric. Biol. Eng.* **2020**, *13*, 82–90. [CrossRef]
4. Maurya, D.; Pandey, A.K.; Kumar, V.; Dubey, S.; Prakash, V. Grafting techniques in vegetable crops: A review. *Int. J. Chem. Stud.* **2019**, *7*, 1664–1672.
5. Kyriacou, M.C.; Rouphael, Y.; Colla, G.; Zrenner, R.; Schwarz, D. Vegetable grafting: The implications of a growing agronomic imperative for vegetable fruit quality and nutritive value. *Front. Plant Sci.* **2017**, *8*, 732–741. [CrossRef] [PubMed]
6. Colla, G.; Rouphael, Y.; Cardarelli, M.; Salerno, A.; Rea, E. The effectiveness of grafting to improve alkalinity tolerance in watermelon. *Environ. Exp. Bot.* **2010**, *68*, 283–291. [CrossRef]
7. Ulas, F.; Aydın, A.; Ulas, A.; Yetisir, H. Grafting for sustainable growth performance of melon (*Cucumis melo*) under salt stressed hydroponic condition. *Eur. J. Sustain. Dev.* **2019**, *8*, 201–210. [CrossRef]
8. Chen, L.; Jiang, K.; Zhang, Q.; Guo, W.; Zheng, W. Design and experiment on scion cutting mechanism of grafting robot for cucurbit. *Int. J. Agric. Biol. Eng.* **2020**, *13*, 99–106. [CrossRef]
9. Lewis, M.; Kubota, C.; Tronstad, R.; Son, Y.J. Scenario-based Cost Analysis for Vegetable Grafting Nurseries of Different Technologies and Sizes. *HortScience* **2014**, *49*, 917–930. [CrossRef]
10. Tong, J.; Ding, Y.; Wu, C.; Yu, Q.; Pan, J.; Sun, L. Design and experiment of key mechanism for semi-automatic vegetable grafting machine. *Trans. Chin. Soc. Agric. Mach.* **2018**, *49*, 65–72.
11. Chen, S.; Liang, H.; Zhang, Q.; Feng, Q.; Li, T.; Chen, L.; Jiang, K. Melon Robotic Grafting: A Study on the Precision Cutting Mechanism and Experimental Validation. *Agriculture* **2023**, *13*, 2139. [CrossRef]
12. Bie, Z. The impact of COVID-19 on intensive vegetable nursery farms and suggestions. *China Veg.* **2020**, *03*, 1–4.
13. Zhao, S.; Liu, J.; Jin, Y.; Bai, Z.; Liu, J.; Zhou, X. Design and Testing of an Intelligent Multi-Functional Seedling Transplanting System. *Agronomy* **2022**, *12*, 2683. [CrossRef]
14. Fu, W.; Gao, J.; Zhao, C.; Jiang, K.; Zheng, W.; Tian, Y. Detection Method and Experimental Research of Leafy Vegetable Seedlings Transplanting Based on a Machine Vision. *Agronomy* **2022**, *12*, 2899. [CrossRef]
15. Wang, J.; Zhang, M.; Gao, C.; Shang, S.; Wang, D. Design and Test of Key Components of Vegetable Grafting Robot for Plug Seedlings. *Trans. Chin. Soc. Agric. Mach.* **2023**, *54*, 38–45.
16. Jiang, K. Study on Mechanism and Experimental Device of Splice Mechanical Grafting of Cucurbit. Ph.D. Thesis, Northeast Agricultural University, Harbin, China, 2019.

17. Kang, C.; Han, G.; Noh, T.; Choi, H.; Youn, K. Splice Grafting Robot for Fruit and Vegetable Plants. Patent WO2005089532A1, 29 December 2004.
18. Yan, G.; Feng, M.; Lin, W.; Huang, Y.; Tong, R.; Cheng, Y. Review and Prospect for Vegetable Grafting Robot and Relevant Key Technologies. *Agriculture* **2022**, *12*, 1578. [CrossRef]
19. Yang, L.; Liu, C.; Zhang, T. Design and experiment of vegetable grafting machine with double manipulators. *Trans. Chin. Soc. Agric. Mach.* **2009**, *40*, 175–181.
20. Chu, Q.; Jiang, K.; Liu, K.; Gu, S. Experimental study on 2JC-600 automatic grafting machine. *Agric. Mech. Res.* **2011**, *33*, 83–85.
21. Gu, S.; Yang, Y.; Zhang, Y.; Qiao, X. Development status of automated equipment systems for greenhouse vegetable seedlings production in Netherlands and its inspiration for China. *Trans. Chin. Soc. Agric. Eng.* **2013**, *29*, 185–194.
22. ISO Group, Inc. ISO Graft 1200 [S/OL]. Available online: http://www.iso-group.nl/en/machines/iso-graft-1200.html (accessed on 26 February 2024).
23. Kang, D.H.; Lee, S.Y.; Kim, J.K.; Park, M.J.; Son, J.K.; Yun, S.W. Development of an Automatic Grafting Robot for Fruit Vegetables using Image Recognition. *Prot. Hortic. Plant Fact.* **2019**, *28*, 322–327. [CrossRef]
24. Xu, P.; Zhang, T.; Chen, L.; Huang, W.; Jiang, K. Study on the Method of Matched Splice Grafting for Melon Seedlings Based on Visual Image. *Agriculture* **2022**, *12*, 929. [CrossRef]
25. EC21 Inc. Supreme Precision Grafting Robot [S/OL]. Available online: https://www.ec21.com/ (accessed on 26 February 2024).
26. Jiang, K.; Chen, L.; Zhang, Q. Design and experiment on flexible clamping and conveying mechanism of vegetable grafting robot. *Trans. Chin. Soc. Agric. Mach.* **2020**, *51*, 63–71.
27. Liang, H.; Zhu, J.; Ge, M.; Wang, D.; Liu, K.; Zhou, M.; Sun, Y.; Zhang, Q.; Jiang, K.; Shi, X. A Comparative Analysis of the Grafting Efficiency of Watermelon with a Grafting Machine. *Horticulturae* **2023**, *9*, 600. [CrossRef]
28. Lin, M.; Zhou, F.; Cai, K.; Pei, Y.; Wang, P.; Zhang, J.; Tian, H.; Tao, Z. An Efficient Horizontal Automatic Grafting Machine. Patent CN115088494B, 21 July 2023.
29. Liu, C.; Shi, Y.; Sun, X. New progress in biological evaluation of medical devices. *Chin. J. Med. Instrum.* **2021**, *1*, 72–75+80.
30. Chen, B.; Hu, G.; Liu, W.; Sun, S.; Sun, C.; Xiao, M. Alternate Automatic Seedling Picking and Dropping Mechanism Based on Symmetrically Arranged Seedling Trays for Automatic Vegetable Transplanters. *Trans. Chin. Soc. Agric. Mach.* **2022**, *53*, 131–139+151.
31. Zhang, Q.; Lü, Y.; Chu, Q.; Li, B.; Wang, Y.; Yang, Y.; Gu, S. Design and experiment on flexible seedling seat of pneumatic pick-up for seedlings. *Trans. Chin. Soc. Agric. Eng.* **2017**, *33*, 69–75.
32. Cao, X.; Ma, X.; Li, H.; Wen, Z.; Li, Z.; Wang, X. Design and experiments of pneumatic roller type precision seed-metering device for rapeseed plug seedlings. *Trans. Chin. Soc. Agric. Eng.* **2021**, *37*, 51–60.
33. Lou, J.; Li, J.; Zhu, P.; Lü, G.; Wang, M. Optimization of Suction Head of Scion Clamping Mechanism for Vegetable Grafting Machine. *Trans. Chin. Soc. Agric. Mach.* **2013**, *44*, 63–68.
34. Liu, C.; Zhou, L.; Lei, F. Overview on numerical simulations of primary atomization. *Rocket Propuls* **2014**, *40*, 10–17.
35. Li, B. Study on Pneumatic Atomization Characteristics of Coaxial Three Channel Nozzle for High Viscosity Liquid. Ph.D. Thesis, China Academy of Launch Vehicle Technology, Beijing, China, 2021.
36. Chen, Y.; Hu, J.; Zhang, G.; Chen, W.; Pan, H.; Lou, B. Research on Characteristics of Paint Deposition on Spherical Surface. *J. Hunan Univ. (Nat. Sci.)* **2019**, *46*, 37–46. [CrossRef]
37. Li, X.; Du, J.; Wang, L.; Fan, J.; Peng, X. Effects of different nozzle materials on atomization results via CFD simulation. *Chinese J. Chem. Eng.* **2020**, *28*, 362–368. [CrossRef]
38. Chen, Y.; Chen, S.; Chen, W.; Hu, J.; Jiang, J. An Atomization Model of Air Spraying Using the Volume-of-Fluid Method and Large Eddy Simulation. *Coatings* **2021**, *11*, 1400. [CrossRef]
39. Liang, H.; Li, A.; Wang, D.; Zhu, J.; Ge, M.; Zhou, M.; Shi, X. Studies on rootstock and Scion age suitable for mechanized grafting of watermelon. *China Veg.* **2022**, *6*, 72–78.
40. Liang, H.; Jiang, K.; Shi, X.; Zhu, J.; Liu, J.; Wang, D.; Ge, M.; Zhou, M.; Shi, F. An Experimental Study on the Effect of Cutting Angle on the Growth of Grafted Watermelon Seedlings Using the One-Cotyledon Grafting Method. *Agronomy* **2023**, *13*, 250. [CrossRef]
41. Yang, J.; Guo, Z.; Hang, Y.; Gao, S.; Jin, K.; Wu, X.; Yang, J. Early classification and detection of grafting union state of melon by hyperspectral imaging. *Spectrosc. Spectr. Anal.* **2022**, *42*, 2218–2224.
42. Nie, W.; Wen, D. Study on the Applications and Regulatory Mechanisms of Grafting on Vegetables. *Plants* **2023**, *12*, 2822. [CrossRef]
43. Jin, W.; Xiao, J.; Ren, H.; Li, C.; Zheng, Q.; Tong, Z. Three-dimensional simulation of impinging jet atomization of soft mist inhalers using the hybrid VOF-DPM model. *Powder Technol.* **2022**, *407*, 117622. [CrossRef]

horticulturae

MDPI

Article

Non-Destructive Detection of Cerasus Humilis Fruit Quality by Hyperspectral Imaging Combined with Chemometric Method

Bin Wang [1], Hua Yang [1], Lili Li [1] and Shujuan Zhang [2,*]

1 College of Information Science and Engineering, Shanxi Agricultural University, Jinzhong 030801, China; wangbin1759@126.com (B.W.)
2 College of Agricultural Engineering, Shanxi Agricultural University, Jinzhong 030801, China
* Correspondence: sxndgxyzsj@126.com; Tel.: +86-0354-6288339

Abstract: Cerasus Humilis fruit is susceptible to rapid color changes post-harvest, which degrades its quality. This research utilized hyperspectral imaging technology to detect and visually analyze the soluble solid content (SSC) and firmness of the fruit, aiming to improve quality and achieve optimal pricing. Four maturity stages (color turning stage, coloring stage, maturity stage, and fully ripe stage) of Cerasus Humilis fruit were examined using hyperspectral images (895–1700 nm) alongside data collection on SSC and firmness. These samples were divided into a calibration set and a validation set with a ratio of 3:1 by sample set partitioning based on the joint X-Y distances (SPXY) method. The original spectral data was processed by a spectral preprocessing method. Multiple linear regression (MLR) and nonlinear least squares support vector machine (LS-SVM) detection models were established using feature wavelengths selected by the successive projections algorithm (SPA), competitive adaptive reweighted sampling (CARS), uninformative variable elimination (UVE), and two combined downscaling algorithms (UVE-SPA and UVE-CARS), respectively. For SSC and firmness detection, the best models were the SNV-SPA-LS-SVM model with 18 feature wavelengths and the original spectra-UVE-CARS-LS-SVM model with eight feature wavelengths, respectively. For SSC, the correlation coefficient of prediction (Rp) was 0.8526, the root mean square error of prediction (RMSEP) was 0.9703, and the residual prediction deviation (RPD) was 1.9017. For firmness, Rp was 0.7879, RMSEP was 1.1205, and RPD was 2.0221. Furthermore, the optimal model was employed to retrieve the distribution of SSC and firmness within Cerasus Humilis fruit. This retrieved information facilitated visual inspection, enabling a more intuitive and comprehensive assessment of SSC and firmness at each pixel level. These findings demonstrated the effectiveness of hyperspectral imaging technology for determining SSC and firmness in Cerasus Humilis fruit. This paves the way for online monitoring of fruit quality, ultimately facilitating timely harvesting.

Keywords: Cerasus Humilis fruit; soluble solid content; firmness; visualization; hyperspectral imaging

Citation: Wang, B.; Yang, H.; Li, L.; Zhang, S. Non-Destructive Detection of Cerasus Humilis Fruit Quality by Hyperspectral Imaging Combined with Chemometric Method. *Horticulturae* **2024**, *10*, 519. https://doi.org/10.3390/horticulturae10050519

Academic Editor: Riccardo Lo Bianco

Received: 12 April 2024
Revised: 14 May 2024
Accepted: 15 May 2024
Published: 17 May 2024

1. Introduction

Cerasus Humilis (*Prunus humilis* Bunge) fruit is highly favored by most consumers due to its sweet and unique taste and rich nutrition. In evaluating Cerasus Humilis fruit quality, soluble solid content (SSC) and firmness are two important internal quality indicators. These properties are also two key indices for consumers to judge maturity and harvest time [1]. Detecting SSC and firmness indicators in Cerasus Humilis fruits is important for identifying ripeness, determining optimal harvest times, enhancing product value, and increasing fruit farmers' income. Traditionally, the SSC of Cerasus Humilis fruit is tested by the refractometer method, which requires the fruit to be extruded into juice and then titrated, and the firmness is tested with the fruit texture analyzer (TA-xt-plus) [2]. Obviously, the methods for determining SSC and firmness of Cerasus Humilis fruit are different. Generally, to obtain a set of data that can actively reflect SSC and firmness of the whole sample, multiple measurements are necessary and the average value is regarded as the representative data. Though traditional testing methods ensure the accuracy of individual sample measurements, they are limited to sampling tests, which are inefficient and time-consuming, and cannot be performed for all samples. Tested samples are destroyed during the testing process and cannot be sold or consumed, resulting in many drawbacks. Additionally, due to the mutual influence of the two indicators, traditional chemical methods cannot simultaneously obtain both indicators [3]. This is cumbersome, time-consuming, and impractical for large-scale sample testing. Consequently, developing nondestructive techniques to determine the SSC and firmness of Cerasus Humilis fruits is essential.

In recent years, many nondestructive testing techniques have been used to evaluate fruit quality. Among them, visible and near infrared (Vis-NIR) spectral sensing technology can effectively evaluate the quality indicators in fruit (SSC, firmness, moisture content, etc.). However, this technique is a single point measurement for the sample, and can only obtain the spectral information of the sample, which cannot reflect the spatial information of the internal quality of the sample, and there is no visualization. Therefore, the technology has limitations. As a comparison, the hyperspectral imaging (HIS) technology combines the advantages of both spectroscopy and image, which can fulfill the task of visualization and has the advantages of simplicity, speed, non-destructiveness, and accuracy [4]. When the molecules in a substance transform from ground state to high energy level, vibrations containing hydrogen bonds (C-H, O-H, N-H, etc.) generate spectral information [5]. The position and intensity of spectral absorption peaks vary depending on the molecules and content in the material, leading to changes in spectral characteristics [6]. Therefore, combining hyperspectral imaging technology with chemometrics can analyze the component content of the material organic compounds (for soluble sugars and organic acids). In recent years, this technology has achieved significant success in fruit maturity detection and quality analysis [7–9]. For example, Benelli et al. [10] employed hyperspectral imaging to predict grape SSC during harvest, achieving a correlation coefficient of prediction (Rp) of 0.77 and a root mean square error of prediction (RMSEP) of 0.779 °Brix for their established partial least squares regression (PLSR) model. Similarly, Pullanagari et al. [11] utilized hyperspectral imaging with characteristic wavelengths to predict total soluble solid concentration (TSS) and flesh firmness (FF) in cherry fruit. Their Gaussian process regression (GPR) model yielded the best results with Rp values of 0.88 and 0.60 for TSS and firmness, respectively, along with RMSEP values of 0.43% and 0.38 N, and RPD values of 3.04 and 2.54. Additionally, recent studies have also shown that HSI is feasible for visualizing the distribution of internal quality in fruits, such as SSC and PH distribution of cherries, mandarins and kiwifruit [12–14], and firmness of peach [15]. Beyond these examples, research using hyperspectral imaging and chemometrics has shown promising results in maturity analysis and quality evaluation of blueberries, persimmons, bananas, and other fruits [16–19]. The above researches mainly focus on the prediction of single indicators of SSC and hardness during the sample maturity stages. However, there are few reports about detection of SSC and firmness in Cerasus Humilis fruits at different maturity stages by using HIS technology.

Therefore, the purpose of this study was to explore the application of hyperspectral imaging technology in SSC and firmness of Cerasus Humilis fruit at different maturity stages. In this study, fresh fruits at four maturity stages (color turning, coloring, maturity, and fully ripe) were collected, and their hyperspectral information was captured. We then established prediction models to relate spectral information to both SSC and firmness. The advantages and disadvantages of these models were compared to determine the optimal approach. Additionally, a visual analysis of fruit quality was performed. This research provides a theoretical foundation for developing an online testing system for the comprehensive quality evaluation of Cerasus Humilis fruit.

2. Materials and Methods

2.1. Samples

Following the established production protocols, the ripening process of Cerasus Humilis fruit was stratified into four distinct stages: color turning, coloring, maturity, and fully ripe. Figure 1 illustrates samples of Cerasus Humilis fruit at different maturity stages.

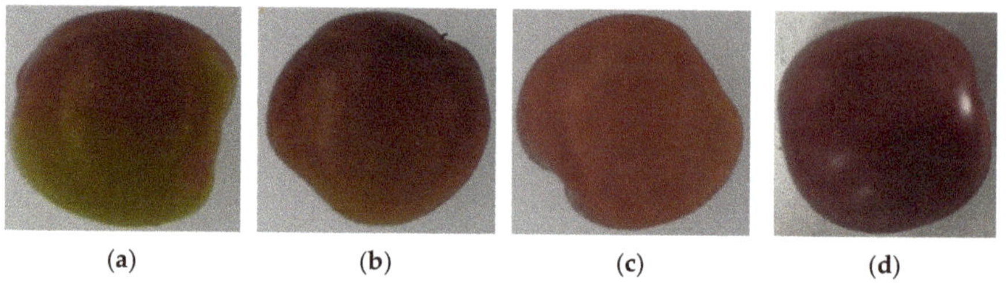

Figure 1. Samples of Cerasus Humilis fruit at different maturity stages: (**a**) color turning stage; (**b**) coloring stage; (**c**) maturity stage; (**d**) fully ripe stage.

This study focused on "Nongda 6" Cerasus Humilis fruit harvested from the Jinzhong Agricultural High-Tech Industry Demonstration Zone planting base between July and August 2023. With the guidance of experienced fruit growers, 110 samples were collected at four specific points in the growth period: day 75 (color turning stage), day 85 (coloring stage), day 95 (maturity stage), and day 100 (fully ripe stage).

Following harvest, the samples were transported to the laboratory on the same day in a low-temperature preservation box. From the initial set of 110 samples per maturity stage, a total of 320 samples were meticulously selected based on stringent criteria. These criteria ensured the samples were free from defects, bruises, and diseases, and possessed relatively uniform shapes.

2.2. Hyperspectral Image Acquisition

2.2.1. Hyperspectral Imaging System

The hyperspectral imaging system used in this study is depicted in Figure 2. The system comprises an imaging spectrograph (GaiaSorter, Beijing, China), light sources, a data acquisition and processing unit, a CCD camera, electronically controlled mobile platforms, a computer, and a darkroom [20]. The system boasts a spectral range of 895–1700 nm (encompassing 254 wavelengths), with a spectral resolution of 5.0 nm, a wavelength interval of 3.19 nm, an effective slit length of 9.6 mm, an incident slit width of 30 μm, and a resolution of 320 × 256 pixels per image.

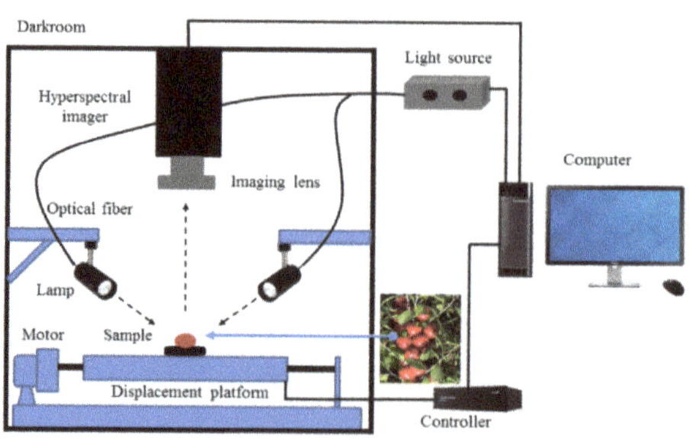

Figure 2. Experiment platform of hyperspectral imaging system.

2.2.2. Image Acquisition and Calibration

To minimize temperature's influence on the spectra and fruit quality, samples were equilibrated in the laboratory for 2 h at 25 °C and 40% RH before image collection. The hyperspectral imaging system was then warmed up for 30 min to ensure stability. For optimal image quality, we determined appropriate settings through repeated experiments. These included a sample-to-lens distance of 280 mm, a camera exposure time of 130 ms, and a mobile platform movement speed of 8.0 mm/s. Additionally, four tungsten halogen lamps positioned at a 45° angle above the sample reduced shadow effects.

A black and white correction was performed on the hyperspectral imaging system to compensate for the effects of light source fluctuations and camera dark current on the captured images [21]. Before measurements, the full black calibration image (I_{Dark}) was obtained with a black opaque cap completely covering the camera lens (0% reflectance), and a white reflection image (I_{White}) was recorded by acquiring a spectral image from a Teflon whiteboard (99.99% reflectance). The corrected image I_C was computed as follows:

$$I_C = \frac{I_{Raw} - I_{Dark}}{I_{White} - I_{Dark}} \times 100\%$$

where I_{Raw} is the original hyperspectral image acquired, I_{White} is a standard white reference image, and I_{Dark} is a dark reference image.

The ENVI 5.0 software was used to extract spectral data from regions of interest (ROIs) within the samples. For each fruit sample, a rectangular ROI measuring 60 × 70 pixels was manually extracted from the equatorial region. The average value of all spectral information within the ROI was then calculated to represent the corresponding reflected spectral value.

2.2.3. Chemical Analysis

Following the hyperspectral image acquisition of Cerasus Humilis fruit samples, a puncture test was conducted to assess firmness. The experiment utilized a Texture Analyzer (TMS-Pro, FTC, Washington, WA, USA) equipped with a 3 mm cylindrical probe. The force cell range was set to 100 N, loading speed to 0.3 mm/s, trigger force to 0.1 N, and maximum penetration depth to 10 mm. For firmness measurement, a 1 cm^2 section of peel was removed from the upper, middle, and lower positions of the ROI for each sample. The firmness value at each position was measured, and the average of the three measurements was used as the overall firmness value for the sample. After firmness measurement, pulp was extracted from the same three locations for juice extraction. A PAL-1 digital refractometer was then used to measure the SSC of the juice. The average of the three SSC measurements was taken as the SSC value for the Cerasus Humilis fruit sample. Figure 3 presents the average values of these physical and chemical indicators for Cerasus Humilis fruits at different maturity stages.

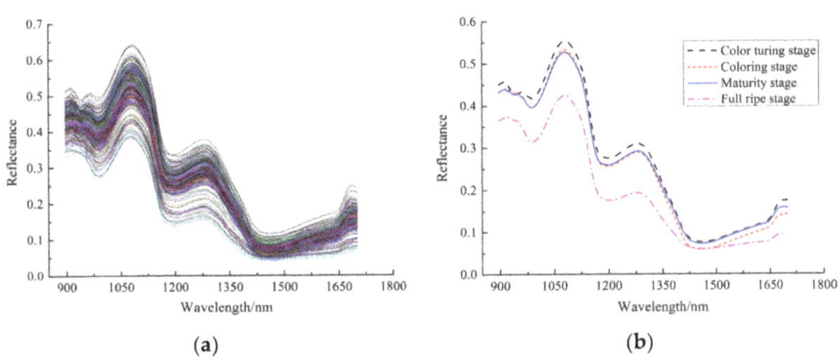

Figure 3. Raw reflectance spectra of Cerasus Humilis fruit samples at different maturity stages: (**a**) original spectral curve; (**b**) average spectrum.

2.3. Data Processing

2.3.1. Sample Partitioning

Prior to establishing a prediction model, the samples were divided into two sets: a calibration set and a prediction set. The model was trained using the calibration set data, followed by performance testing on the prediction set data. This sample division employed the Sample Set Partitioning Based on Joint X-Y Distance (SPXY) algorithm. The algorithm used spectral values as the X variable and physicochemical indicators (SSC and firmness) as the Y variable, achieving a 3:1 ratio between the calibration and prediction sets [22]. To assess the validity of this division, the maximum, minimum, mean, and standard deviation of spectral data were calculated for both the calibration and prediction sets.

2.3.2. Spectral Preprocessing

The collected spectral information comprised not only the internal structure and component information of the tested sample, but also interference caused by various factors such as light scattering, sample heterogeneity, temperature fluctuations, and instrument noise [23]. To enhance model stability, preprocessing the original spectrum is crucial. This study employed five spectral preprocessing methods: Savitzky–Golay (S-G), Standard Normal Variate (SNV), Multiplicative Scatter Correction (MSC), Baseline Correction (BC), and De-trending (De-T). Subsequently, Partial Least Squares (PLS) regression was used to evaluate the performance of the data processed with these different methods.

2.3.3. Selection Methods of Characteristic Wavelengths

The full spectrum, encompassing 254 wavelength variables, presents a challenge due to potential information redundancy. A large number of collinear wavelengths can contribute to this redundancy, which in turn increases computational load, reduces modeling efficiency, and ultimately affects model performance. To address these limitations and streamline the modeling process, enhance computational speed, and improve prediction accuracy, dimensionality reduction of the raw spectral data becomes necessary. This study employed several feature selection methods for dimensionality reduction: the Successive Projections Algorithm (SPA), Competitive Adaptive Reweighted Sampling (CARS), and Uninformative Variable Elimination (UVE). Additionally, combined approaches, including UVE-CARS and UVE-SPA, were utilized. These methods aim to reduce the dimensionality of the spectral data by selecting characteristic wavelengths that effectively capture the essential information from the full spectrum.

The SPA is a forward loop feature selection method. It effectively filters out irrelevant information, significantly reduces the collinearity effect between data points, and utilizes root mean square error (RMSE) for evaluation [24]. CARS is another effective method for wavelength selection. It leverages the principles of "survival of the fittest" and regression coefficients. First, a PLS model was used to identify wavelengths with large regression coefficients. Then, 10-fold cross-validation was employed to select the wavelength subset with the smallest RMSE [25]. UVE also selected wavelengths based on PLS regression coefficients. This method added a specific number of random variables to the spectral matrix. Subsequently, a PLS model was established through cross-validation. Finally, the ratio of regression coefficient to standard deviation was calculated to identify and select effective spectral information [20].

2.3.4. Models Establishment

Partial least squares regression offers several advantages over traditional methods like set-correlation analysis, multiple linear regression (MLR), and principal component analysis (PCA) [13]. PLSR is a widely used quantitative analysis technique based on multivariate linear modeling. It performs curve fitting by minimizing the sum of squared deviations between predicted and actual values. This method effectively utilizes both spectral information from samples and their physicochemical properties to create an optimal calibration model. PLSR is used to search for potential linear combinations of wavelength variables (X) and chemical composition (Y). Its expression is as follows:

$$Y = XT + e \quad (1)$$

where T and e are the regression coefficient matrix and regression residual matrix, respectively.

MLR, a common modeling approach, establishes relationships between dependent and multiple independent variables [26]. While MLR boasts a simple algorithm and easy interpretation, it suffers from vulnerabilities to collinearity (high correlation) between variables. Additionally, the number of variables in an MLR model must be less than the number of samples for proper function. The equation is as follows:

$$Y = a_0 + a_1X_1 + a_2X_2 + \cdots + a_iX_i + \varepsilon \quad (2)$$

where Y represents the component content of the tested sample, a_0 represents the constant, a_i represents the regression coefficient, X_i represents the spectral data at different wavelength bands, and ε represents the regression bias.

Least squares support vector machine (LS-SVM) represents an improvement over traditional support vector machines (SVM). LS-SVM effectively addresses both linear and non-linear problems by finding linear equations [27]. It achieves this by solving a set of linear equations with equality constraints, as opposed to the complex quadratic optimization problems encountered in traditional SVMs. Consequently, LS-SVM boasts improved computational efficiency and enhanced model prediction capabilities. The choice of kernel function directly determines the performance of the LS-SVM model. There are many types of kernel functions (such as linear, polynomial, and radial basis function (RBF)), among which the RBF has the characteristics of low complexity and few parameters. Therefore, RBF was chosen as the kernel function of LS-SVM in this study. The grid search algorithm and leave-one-out cross validation (LOOCV) method were used to determine the optimal parameters of gamma (γ) and sigma2 (σ^2) of the RBF kernel [13]. The LS-SVM model can be expressed as:

$$Y(x) = \sum_{k=1}^{N} a_k k(x, x_k) + b \quad (3)$$

where $k(x, x_k)$ is the kernel function, x_k is the input vector, a_k is the Langland multiplier (also known as the support vector), and b is the model bias.

3. Results and Discussion

3.1. Spectral Analysis

Following the removal of head and tail noise, spectral feature analysis was conducted on the 230 bands within the 945–1675 nm range. This analysis yielded the spectral curves and average spectral curves for all Cerasus Humilis fruit samples, as presented in Figure 3.

Figure 3 illustrates a degree of spectral overlap between samples from distinct maturity stages, although the overall spectral curve trends exhibit significant similarity. The spectra in Figure 3a display prominent absorption peaks at 980 nm, 1195 nm, and 1460 nm, attributed to the second overtone of O-H in water and carbohydrates (around 980 nm and 1195 nm) [28,29] and the first overtone of the O-H bond (near 1460 nm), respectively. As depicted in Figure 3b, the average spectral curves for the various maturity stages (color turning, coloring, maturity, and fully ripe) demonstrated generally consistent trends. However, the spectral reflectance of the samples exhibited a significant difference as the maturity period progressed, with the fully ripe stage exhibiting the lowest reflectance values. Notably, the range of 1180–1300 nm revealed a relatively substantial difference in fruit reflectivity, which then tended to decrease gradually. This observation might be linked to the alterations in the chemical composition of Cerasus Humilis fruits at different maturity stages. Nevertheless, it was not feasible to isolate the content of a specific compound based solely on these wavelengths, as each wavelength reflects intricate compositional information. Consequently, data analysis methods have become crucial for uncovering the underlying relationships between spectra and the measured physicochemical properties (SSC and firmness).

3.2. Statistics of SSC and Firmness

Figure 4 reveals significant differences in SSC and firmness across the various maturity stages of Cerasus Humilis fruit, reflecting the changes in internal quality during fruit development. The SSC content exhibits a continuous upward trend throughout the maturity stages, reaching a maximum value of 17.93 °Brix at the fully ripe stage. Conversely, the firmness value displays a continuous decline throughout development. The highest firmness (13.69 kg/cm^2) is observed during the color turning stage, while the fully ripe stage exhibits the lowest firmness (9.12 kg/cm^2). These findings align with the established research on the internal physical and chemical changes in fruits at different maturity stages.

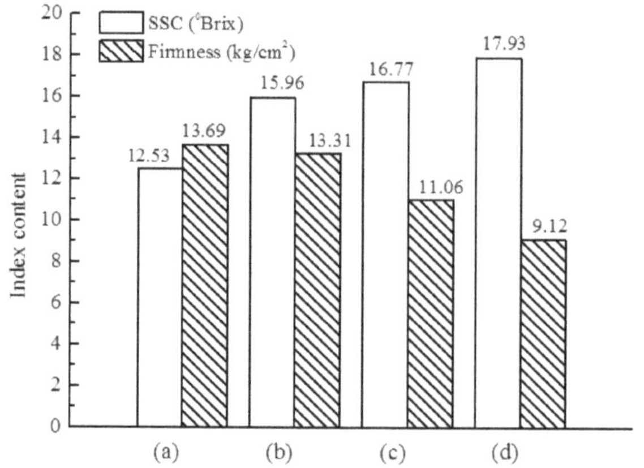

Figure 4. Average physical and chemical indexes of Cerasus Humilis fruit at different stages: (**a**) color turning stage; (**b**) coloring stage; (**c**) maturity stage; (**d**) fully ripe stage.

All Cerasus Humilis fruit samples were divided into calibration set and prediction set based on SPXY method mentioned in Section 2.3.1. Table 1 presents the statistical results of this sample set division. As can been seen, the SSC in the calibration set ranged from 9.24 to 18.5 °Brix with an average of 14.65 °Brix. Firmness values in the calibration set ranged from 5.86 to 16.64 kg/cm^2 with an average of 11.45 kg/cm^2. Similarly, the prediction set exhibited an SSC range of 11.32 to 18.17 °Brix with an average of 14.91 °Brix, and a firmness range of 7.61 to 16.53 kg/cm^2 with an average of 12.45 kg/cm^2. This distribution achieved through the SPXY algorithm ensured that the calibration set encompasses the range of SSC and firmness values present in the prediction set. Those properties were very important to establish a robust and stable prediction model.

Table 1. Statistical for the SSC and firmness of Cerasus Humilis fruit.

Index	Sample Set	No. of Samples	Minimum	Maximum	Average Value	Standard Deviation
SSC (°Brix)	Calibration set	240	9.24	20.85	14.65	1.88
	Prediction set	80	11.32	18.17	14.91	1.85
	Total samples	320	9.24	20.85	14.71	1.87
Firmness (kg/cm^2)	Calibration set	240	5.86	16.64	11.45	2.57
	Prediction set	80	7.61	16.53	12.45	2.27
	Total samples	320	5.86	16.64	11.76	2.53

3.3. Spectral Pretreatment

Given that the number of latent variables (Lvs) in PLSR significantly impacts model performance, this study employed 10-fold cross validation to determine the optimal Lvs for PLSR modeling of both SSC and firmness. Spectral data preprocessed using each method were then used as input variables for PLSR modeling, establishing corresponding prediction models for each quality parameter. Table 2 presents the PLSR modeling results obtained with different pretreatment methods.

Table 2. Results of full wavelength PLSR model based on different preprocessing methods.

Parameter	Preprocessing Methods	Lvs	Calibration Set		Prediction Set		
			Rc	RMSEC	Rp	RMSEP	RPD
SSC	Original spectra	12	0.7625	1.2148	0.8368	1.0289	1.7934
	S-G	10	0.7365	1.2703	0.8056	1.1123	1.6589
	SNV	11	0.7709	1.1963	0.8596	0.9865	1.8705
	MSC	9	0.7589	1.2228	0.8315	1.0528	1.7527
	BC	10	0.7611	1.2181	0.8308	1.0431	1.7689
	De-T	9	0.7497	1.2429	0.8229	1.0447	1.7662
Firmness	Original spectra	11	0.6737	1.4674	0.7836	1.2536	1.8074
	S-G	11	0.6738	1.4946	0.7607	1.2867	1.7609
	SNV	12	0.6728	1.4968	0.6689	1.3784	1.6438
	MSC	12	0.6724	1.4976	0.6586	1.3801	1.6418
	BC	10	0.6456	1.5527	0.6482	1.4079	1.6093
	De-T	12	0.6759	1.4852	0.6949	1.3123	1.7266

Table 2 reveals that for the SSC index of Cerasus Humilis fruit, the PLSR model established using SNV pretreatment on the original spectra achieved the best prediction performance compared to models based on other preprocessing methods. This model yielded the highest correlation coefficients (Rc = 0.7709, Rp = 0.8596), the highest RPD value (1.8705), and the lowest root mean square errors for calibration and prediction (1.1963 and 0.9865, respectively).

Interestingly, for the firmness index, the PLSR model based on the original spectra outperformed models based on all five preprocessing methods (S-G, SNV, MSC, BC, and DE-T). The model using the original spectra exhibited increased correlation coefficients (Rc = 0.6737, Rp = 0.7836), decreased RMSEC (1.4674) and RMSEP (1.2536) values, and the maximum RPD value (1.8074). These results suggest that, for firmness prediction, the original spectra contained more relevant information compared to the preprocessed data. It is highly conceivable that some preprocessing algorithms might have eliminated noise or scattering effects that were actually correlated with firmness.

3.4. Effective Wavelength Selection

3.4.1. Successive Projections Algorithm

In the SSC and firmness characteristic wavelength screening of Cerasus Humilis fruit, the minimum and maximum values of the wavelength variable were 2 and 30, respectively. Figure 5 illustrates the distribution of RMSE for the SPA algorithm with different numbers of selected variables. The RMSE values reached a minimum at 18 and 4 wavelength variables, being 1.2176 and 1.9416, respectively. Figure 6 depicts the distribution of the effective wavelengths chosen by SPA for predicting SSC and firmness. For SSC prediction, the 18 characteristic wavelengths were 1208, 1514, 1558, 1177, 1004, 1457, 1157, 1657, 1663, 956, 1014, 1402, 976, 1135, 1596, 947, 1393, and 1084 nm, respectively. These wavelengths accounted for approximately 3.91% of the total analyzed wavelengths, with the importance of each wavelength decreasing sequentially. For firmness prediction, the four characteristic wavelengths were 1100, 1202, 1364, and 1007 nm, constituting only 1.74% of the total wavelengths.

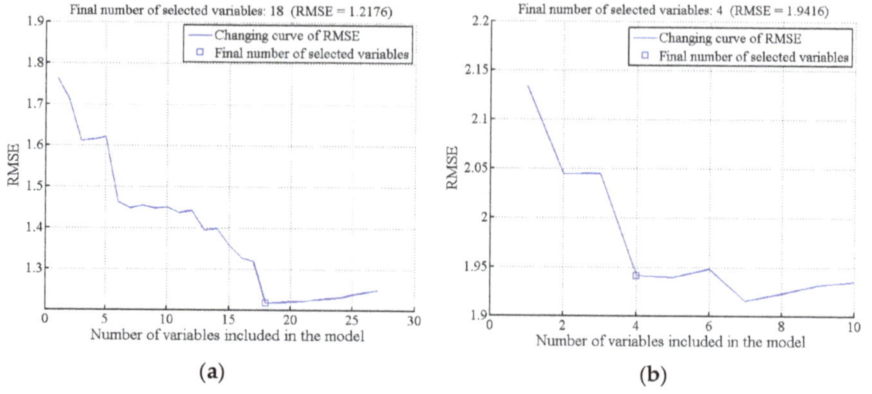

Figure 5. RMSE plots obtained by SPA: (**a**) SSC; (**b**) firmness.

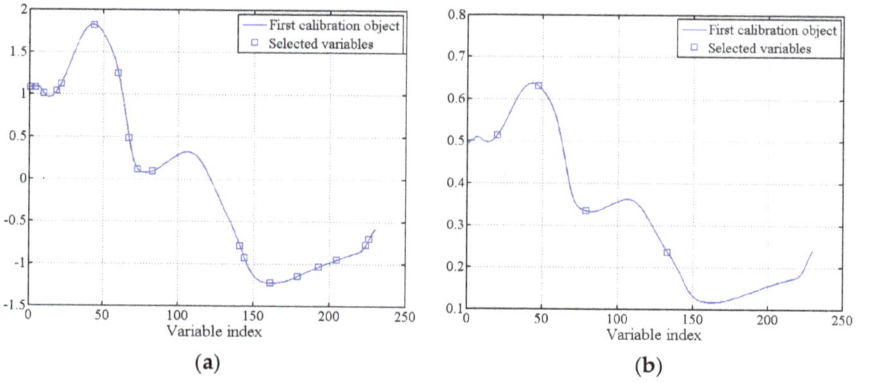

Figure 6. Optimal characteristic wavelength distribution selected by SPA: (**a**) SSC; (**b**) firmness.

3.4.2. Competitive Adaptive Reweighted Sampling

The CARS algorithm employed 50 Monte Carlo sampling iterations and a 5-fold cross-validation method for characteristic wavelength selection. Figure 7 illustrates the selection process. Figure 7a–c depict the trends observed during each CARS run as the number of Monte Carlo samples increased. These trends included the number of sampled variables (a), the RMSECV values (b), and the regression paths for each variable (c). In Figure 7c, each curve represents the change in regression coefficients for a spectral variable as the number of sampling runs increases. The vertical line marked "*" indicates the point where the RMSECV reaches a minimum value (1.1355) with 33 sampling runs. At this minimum, 10 characteristic wavelengths were selected: 960, 1081, 1119, 1154, 1215, 1298, 1383, 1622, 1625, and 1672 nm, constituting approximately 4.35% of the total analyzed wavelengths. Similarly, for firmness prediction, the minimum RMSECV value (1.9134) was reached with 31 sampling iterations. This resulted in the selection of eight characteristic wavelengths: 963, 1103, 1202, 1205, 1208, 1272, 1377, and 1520 nm, representing 3.48% of the total wavelengths.

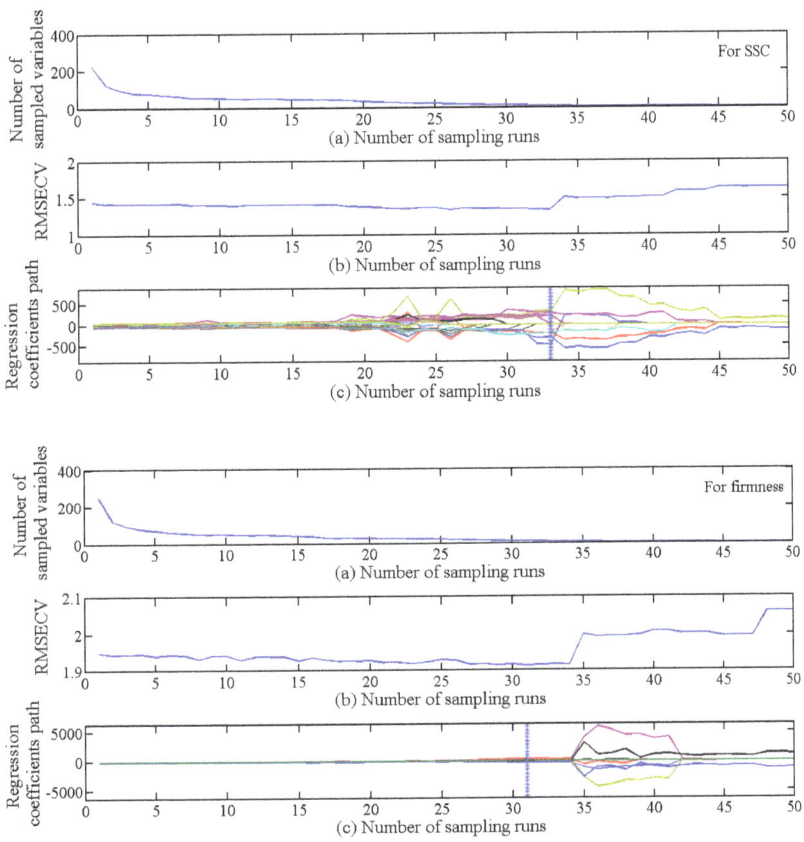

Figure 7. Process of selecting wavelength variables by CARS for SSC and firmness: (**a**) number of sampled variables; (**b**) variation of RMSECV under different sampling times; (**c**) regression coefficients of each wavelength variable.

3.4.3. Uninformative Variable Elimination

Taking the SSC feature wavelength selection of Cerasus Humilis fruit as an example, the UVE algorithm was used to eliminate the uninformative variables from the original 230 wavelengths. When using a five-fold cross-validation scheme and setting the number of principal components to 12, the minimum RMSECV value obtained was 1.3036, as illustrated in Figure 8a.

Figure 8b depicts the stability distribution results of the UVE algorithm for characteristic wavelength selection with 12 principal components. The curves on the left side of the vertical line represent the original 230 wavelength variables, while those on the right represent randomly introduced variables. The upper and lower dashed lines indicate the selection threshold for random variables (±41.03), set at 99% of the maximum stability value of the random variable. Information exceeding the absolute value of this threshold was considered valid. In simpler terms, information above and below the dashed lines was considered useful, while information between the lines was deemed irrelevant. Consequently, variables exceeding the threshold were preserved for further model analysis. This process resulted in the selection of 129 characteristic wavelengths, constituting approximately 56.09% of the total wavelengths.

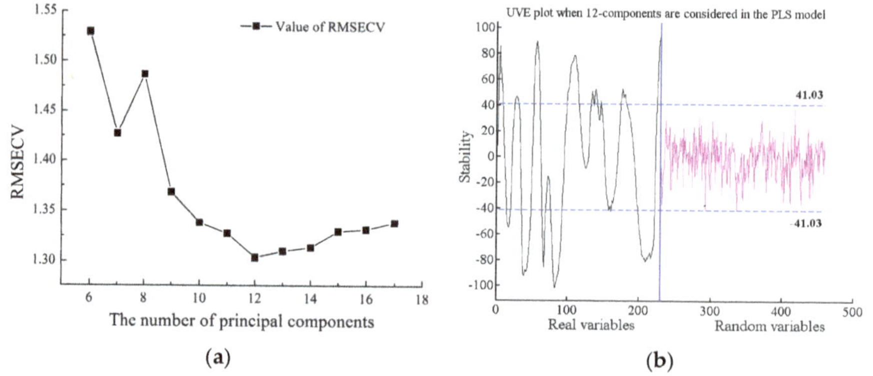

Figure 8. Process of selecting wavelength variables for SSC by UVE: (**a**) RMSECV distribution with a different number of principal components; (**b**) stability distribution of characteristic wavelength.

Similarly, the selection of characteristic wavelengths for firmness in Cerasus Humilis fruit yielded a minimum RMSECV value of 1.9097 with 10 principal components. The selection threshold in this case was ±35.96, leading to the extraction of 53 characteristic wavelengths, representing 23.04% of the total wavelengths. The analysis results are presented in Figure 9.

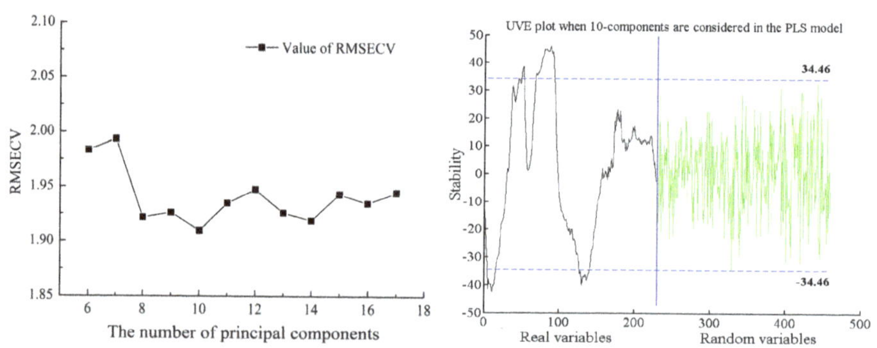

Figure 9. Process of selecting wavelength variables for firmness by UVE.

3.4.4. Secondary Effective Wavelength Selection

Dimensionality reduction using the three aforementioned methods (SPA, CARS, UVE) successfully eliminated irrelevant information and reduced the wavelength dimension. While the UVE algorithm yielded a higher number of characteristic wavelengths compared to SPA and CARS, it might introduce issues with redundant information and multicollinearity in the spectral data. Conversely, SPA and CARS algorithms bring the advantage of eliminating multicollinearity among spectral variables, effectively addressing the limitations of UVE. Therefore, SPA and CARS were employed for further screening.

For SSC prediction, the UVE-SPA algorithm was used to select characteristic wavelengths. The number of selected wavelengths was determined based on the RMSE value. Figure 10 illustrates the trend of RMSE with different variable numbers. As the number of variables increased, RMSE exhibited a decreasing trend. The minimum RMSE value (1.2518) was achieved with 15 variables (marked in blue squares). These 15 characteristic wavelengths were 1218, 1189, 969, 1151, 1453, 1577, 995, 1110, 1367, 1129, 953, 960, 1409, 1596, and 1647 nm, representing approximately 6.52% of the total wavelengths. Additionally, the UVE-CARS algorithm was used for characteristic wavelength selection. The minimum RMSECV value (1.3406) was obtained with 31 sampling runs. Figure 11 presents the results of effective wavelength selection. Ten characteristic wavelengths were identified: 995, 1036, 1075, 1081, 1119, 1122, 1151, 1208, 1383, and 1628 nm, constituting 4.35% of the original spectral wavelengths.

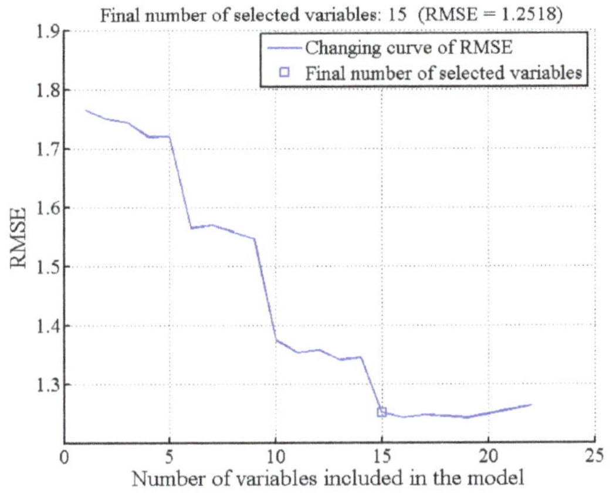

Figure 10. Characteristic wavelength selection results by UVE-SPA.

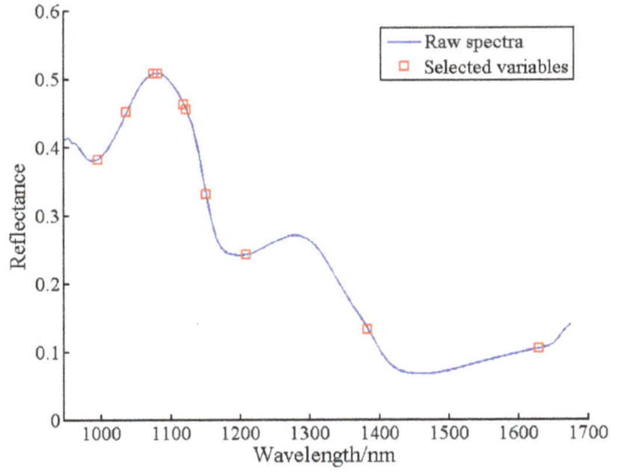

Figure 11. Distribution of the optimal characteristic wavelength selected by UVE-CARS.

Following the same approach, the UVE-SPA and UVE-CARS algorithms selected four and eight characteristic wavelengths for firmness prediction, respectively. These wavelengths accounted for 1.74% and 3.48% of the original spectral data, as detailed in Table 3.

Table 3. Characteristic wavelengths selected by secondary dimension reduction models.

Parameter	Variable Selection Methods	No. of Variables	Characteristic Wavelengths (nm)
SSC	UVE-SPA	15	1218, 1189, 969, 1151, 1453, 1577, 995, 1110, 1367, 1129, 953, 960, 1409, 1596, 1647
	UVE-CARS	10	995, 1036, 1075, 1081, 1119, 1122, 1151, 1208, 1383, 1628
Firmness	UVE-SPA	4	1106, 1180, 960, 1164
	UVE-CARS	8	966, 1106, 1198, 1202, 1205, 1208, 1237, 1377

3.5. *Model Analysis*

3.5.1. MLR Model

Table 4 demonstrates the effectiveness of dimensionality reduction using different characteristic wavelength extraction methods on the original spectral data. For the SSC index of Cerasus Humilis fruit, the UVE-SPA and UVE-CARS methods further reduced the number of characteristic wavelengths compared to other methods, leading to a significant improvement in model prediction performance. Notably, the MLR model established with UVE-SPA wavelengths achieved the best results (Rp = 0.8326, RMSEP = 0.9913, RPD = 1.8614). This suggested that the combined UVE-SPA approach effectively eliminated redundant and multicollinear variable information, extracting the most relevant characteristic wavelengths for prediction. Additionally, the UVE-SPA-MLR model benefited from a lower number of variables (only 15), significantly reducing modeling time.

Table 4. Results of MLR model based on different characteristic wavelength extraction methods.

Parameter	Variable Selection Methods (No.)	Calibration Set		Prediction Set		
		Rc	RMSEC	Rp	RMSEP	RPD
SSC	SPA (18)	0.7231	1.2973	0.7844	1.1708	1.5760
	CARS (10)	0.7187	1.3058	0.7728	1.1975	1.5409
	UVE (129)	0.8981	0.8258	0.6207	1.7198	1.0729
	UVE-SPA (15)	0.8045	1.1024	0.8326	0.9913	1.8614
	UVE-CARS (10)	0.7821	1.1702	0.8179	1.0749	1.7166
Firmness	SPA (4)	0.6762	1.4887	0.7049	1.3677	1.6566
	CARS (8)	0.6916	1.4518	0.7287	1.2926	1.7529
	UVE (53)	0.7449	1.4958	0.6839	1.3945	1.6248
	UVE-SPA (4)	0.6537	1.5402	0.6694	1.4005	1.6179
	UVE-CARS (8)	0.7211	1.4529	0.7629	1.2545	1.8061

For firmness prediction models, the UVE-SPA-MLR model exhibited the weakest performance (Rp = 0.6694, RMSEP = 1.4005, RPD = 1.6179), given that the SPA algorithm, during secondary screening, excluded some valuable spectral information identified by the UVE algorithm. In contrast, the CARS-MLR model achieved better results (Rp = 0.7287, RMSEP = 1.2926, RPD = 1.7529). However, the UVE-CARS-MLR model emerged as the best model for firmness prediction, with Rp = 0.7629, RMSEP = 1.2545, and RPD = 1.8061. Furthermore, UVE-CARS identified only eight characteristic wavelengths (3.48% of the total), demonstrating its efficiency in variable selection.

The MLR prediction model for the SSC index of Cerasus Humilis fruit based on UVE-SPA optimal characteristic wavelength was as follows:

$$Y_{SSC} = -18.449 - 173.894X_{1218} + 279.790X_{1189} - 378.044X_{969} + 202.343X_{1151} - 529.604X_{1453} - 158.681X_{1557} - 157.889X_{995} - 343.307X_{1110} + 193.646X_{1367} + 37.781X_{1129} + 391.801X_{953} - 21.612X_{960} + 105.39X_{1409} + 62.671X_{1596} + 239.260X_{1647}$$

The MLR prediction model for the firmness index of Cerasus Humilis fruit based on UVE-CARS optimal characteristic wavelength was as follows:

$$Y_{Firmness} = 7.378 - 148.275X_{996} + 103.483X_{1106} + 249.474X_{1199} - 78.415X_{1202} + 216.052X_{1205} + 158.313X_{1208} - 129.661X_{1237} - 96.598X_{1377}$$

3.5.2. LS-SVM Model

To investigate whether a nonlinear model outperforms the linear model in predicting the internal quality parameters of Cerasus Humilis fruit. Here, we established nonlinear LS-SVM prediction models for both SSC and firmness of Cerasus Humilis fruit. The predicted results from these LS-SVM models with different variables are presented in Table 5.

Table 5. Results of LS-SVM model based on different characteristic wavelength.

Parameter	Variable Selection Methods (No.)	(γ, σ^2)	Calibration Set		Prediction Set		
			Rc	RMSEC	Rp	RMSEP	RPD
SSC	SPA (18)	$(3.8888 \times 10^3, 660.6806)$	0.8146	1.0952	0.8526	0.9703	1.9017
	CARS (10)	$(4.4715 \times 10^3, 344.6078)$	0.7815	1.1737	0.8179	1.0681	1.7276
	UVE (129)	$(482.3338, 1.7519 \times 10^3)$	0.7952	1.1449	0.8514	0.9806	1.8817
	UVE-SPA (15)	$(4.1732 \times 10^3, 771.8445)$	0.8043	1.1224	0.8532	0.9650	1.9121
	UVE-CARS (10)	$(2.4422 \times 10^3, 782.9417)$	0.7631	1.2157	0.8244	1.0631	1.7357
Firmness	SPA (4)	$(318.8407, 2.0509 \times 10^3)$	0.6871	1.4635	0.7392	1.2879	1.7593
	CARS (8)	(561.8141, 210.9484)	0.7129	1.3991	0.7635	1.1889	1.9058
	UVE (53)	(262.0350, 193.0957)	0.7353	1.3390	0.7775	1.1333	1.9993
	UVE-SPA (4)	(42.8355, 12.0954)	0.6918	1.4525	0.6926	1.3060	1.7349
	UVE-CARS (8)	(143.1516, 29.2393)	0.7283	1.3578	0.7879	1.1205	2.0221

γ represents the penalty factor, σ^2 represents the nuclear parameter.

Table 5 demonstrates that, for the SSC index of Cerasus Humilis fruit, LS-SVM models based on all five characteristic wavelength extraction methods (SPA, CARS, UVE, UVE-SPA, and UVE-CARS) achieve consistently higher correlation coefficients (Rc and Rp) and lower root mean square errors (RMSEC and RMSEP). These findings indicated a strong overall prediction performance for the LS-SVM model across the various wavelength extraction techniques. The CARS-LS-SVM and UVE-CARS-LS-SVM models exhibited similar predictive capabilities, suggesting that both CARS and UVE-CARS were effective in extracting relevant variable information. While the SPA-LS-SVM model achieved comparable accuracy to the UVE-LS-SVM and UVE-SPA-LS-SVM models, it offered a distinct advantage: the SPA method eliminated the need for secondary feature wavelength screening (requiring only 18 variables). This translated to significant time savings and improved model fitting efficiency. Therefore, considering a balance between accuracy, stability, and model complexity, the SPA-LS-SVM model emerged as the optimal choice for predicting SSC in Cerasus Humilis fruit.

For the firmness index of Cerasus Humilis fruit, the LS-SVM model based on UVE-SPA characteristic wavelengths exhibited the weakest performance, achieving Rp of 0.6926, RMSEP of 1.3060, and RPD of 1.7349. This suggested that, during the secondary screening, the SPA algorithm might have eliminated some valuable spectral information identified by the UVE algorithm. The SPA-LS-SVM model achieved slightly better results (Rp = 0.7392, RMSEP = 1.2879, RPD = 1.7593). Notably, the CARS-LS-SVM model demonstrated superior performance (Rp = 0.7635, RMSEP = 1.1889, RPD = 1.9058). The UVE-LS-SVM model yielded results (Rc = 0.7353, RMSEC = 1.3390, Rp = 0.7775, RMSEP = 1.1333, RPD = 1.9993) slightly lower than the UVE-CARS-LS-SVM model (Rc = 0.7283, RMSEC = 1.3578, Rp = 0.7879, RMSEP = 1.1205, RPD = 2.0221). However, the UVE-CARS-LS-SVM model achieved this performance with a significantly smaller number of variables (only eight). In conclusion, among all LS-SVM models, the UVE-CARS-LS-SVM model emerged as the one with the best prediction performance for firmness.

3.5.3. Model Comparison and Analysis

Table 6 demonstrates that the established nonlinear LS-SVM models outperformed their linear MLR counterparts in predicting both SSC and firmness of Cerasus Humilis fruit. Notably, the SPA-LS-SVM and UVE-CARS-LS-SVM models emerged as the optimal models for SSC and firmness prediction, respectively. The LS-SVM model achieved exceptional prediction performance for firmness (RPD > 2.0), surpassing the performance for SSC. Nevertheless, the SSC prediction model's performance also meets acceptable standards. These findings collectively indicate that the nonlinear LS-SVM models yielded superior prediction accuracy and stability compared to the linear MLR models. The predicted and measured SSC and firmness values of Cerasus Humilis fruit by the optimal LS-SVM model were shown in Figure 12. The solid line is a regression line that corresponds to the ideal correlation between measured values and predicted values. From Figure 12, it can be seen that the predicted values were basically close to the corresponding actual values for both calibration and prediction sets.

Table 6. Prediction results of best MLR and LS-SVM models for SSC and firmness values of fruit.

Model	Parameter	Variable Selection Methods (No.)	Calibration Set		Prediction Set		
			Rc	RMSEC	Rp	RMSEP	RPD
MLR	SSC	UVE-SPA (15)	0.8045	1.1024	0.8326	0.9913	1.8614
	Firmness	UVE-CARS (8)	0.7211	1.4529	0.7629	1.2545	1.8061
LS-SVM	SSC	SPA (18)	0.8146	1.0952	0.8526	0.9703	1.9017
	Firmness	UVE-CARS (8)	0.7283	1.3578	0.7879	1.1205	2.0221

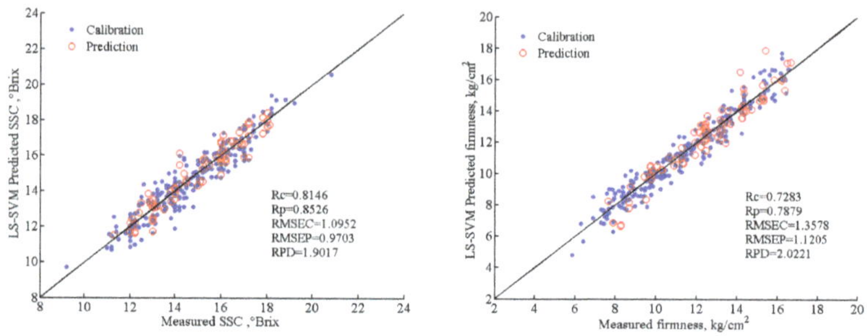

(**a**) (**b**)

Figure 12. Scatter plots of the measured and predicted various indexes of SSC and firmness of Cerasus Humilis fruit based on the optimal model: (**a**) SPA-LS-SVM model predicts SSC results; (**b**) UVE-CARS- LS-SVM model predicts firmness results.

As shown in Figure 13, the distribution of characteristic bands of SSC and firmness in the spectral curve. It can be seen that the specific characteristics of the first region (945–1100) may be caused by the vibration of hydrogen groups (O-H and N-H), which was the third overtone region of O-H and absorption band of water. The fruit contains acidic compounds, which usually contain hydrogen groups. However, the absorption peaks in the second region (1100–1210) and the third region (1220–1600) are mainly caused by vibrations of C=O (1160 nm) and C-H (1170 nm, 1194 nm), and these two chemical bonds are commonly present in soluble sugars, which are the major components of SSC [12]. In this figure, each triangle and pentagram marks the position of a feature band in the spectral curve.

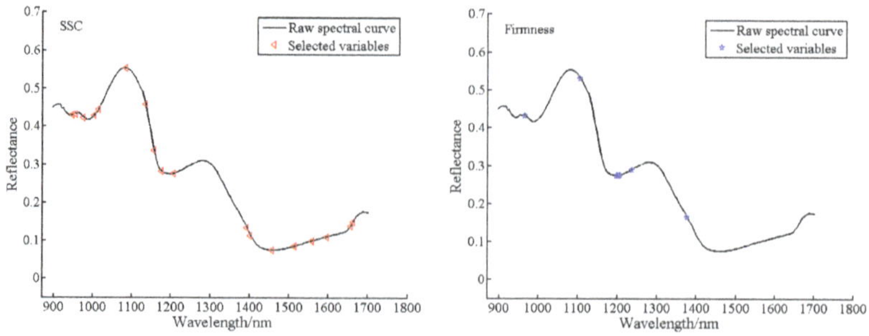

(**a**) (**b**)

Figure 13. The distribution of characteristic bands in spectral curves: (**a**) SSC; (**b**) firmness.

3.6. Visualization of SSC and Firmness Distribution

Distribution maps of SSC and firmness play a valuable role in online detection. This study employed SPA-LS-SVM and UVE-CARS-LS-SVM models to estimate the SSC and firmness values for each pixel within Cerasus Humilis fruit samples. Pseudo-color image processing technology was then utilized to generate the corresponding distribution maps. In these maps (Figure 14), different colors and shades represent varying magnitudes of SSC and firmness across the fruit at different maturity stages.

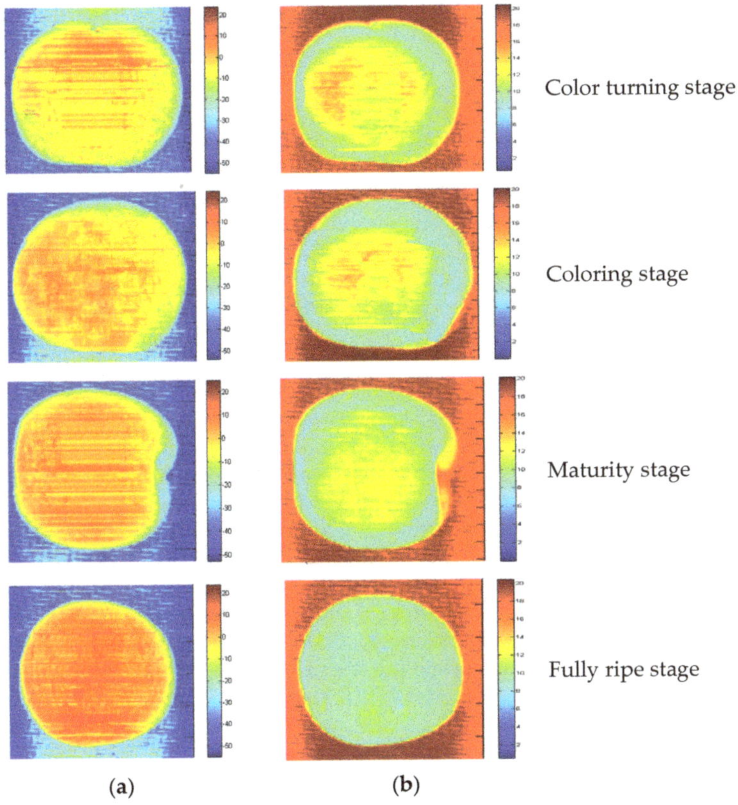

Figure 14. SSC and firmness distribution map of Cerasus Humilis fruit in different maturities: (**a**) SSC; (**b**) firmness.

Figure 14a depicts the SSC distribution during the fully ripe stage. The fruit appeared predominantly dark red, covering approximately 98% of the area, indicating an SSC concentration in the range of 17–20 °Brix. Compared to the fully ripe stage, the color density in Figure 14a gradually changed from red to yellow throughout the maturity, coloring, and color turning stages, reflecting a decrease in SSC.

Figure 14b shows the firmness distribution at the ripe stage. Here, the Cerasus Humilis fruit appeared primarily light green, covering almost the entire area, with firmness concentrated in the 9–10 kg/cm^2 range. In the coloring and ripening stages, the fruit color transitioned from dark yellow to light green, with the yellow pixels becoming lighter. This change suggested a gradual decrease in firmness value. Notably, the blue and red areas outside the fruit region in the SSC and firmness images, respectively, represented the background used for data collection.

The visualization of the distribution maps demonstrated an increase in SSC and a decrease in firmness as the ripening stage progressed. Furthermore, the maps revealed an uneven distribution of SSC and firmness across different locations within each sample. This unevenness likely stemmed from variations in the rates of compound conversion occurring within the Cerasus Humilis fruit during the ripening process. Therefore, SSC and firmness visualization distribution maps can be a valuable tool for consumers, enabling them to rapidly assess the quality of Cerasus Humilis fruit. This technology holds promise for online quality detection of Cerasus Humilis fruit.

4. Discussion

This study investigated the use of hyperspectral data for non-destructive, accurate, and rapid identification of SSC and firmness in Cerasus Humilis fruit. The study revealed a clear advantage of the LS-SVM model over the MLR model in predicting SSC and firmness under similar conditions. This finding suggested a potentially non-linear relationship between the spectral range of 895–1700 nm and the fruit's SSC and firmness. The superiority of LS-SVM compared to other models has been demonstrated in previous studies. The selected characteristic wavelengths were found to be distributed throughout the entire band, indicating that SSC is related to moisture (947, 956, and 976 nm) as well as carbohydrates such as soluble sugars and pectin (1104, 1014, 1084, 1135, 1157, 1177, 1208, 1393, 1402, 1457, 1514, 1558, and 1596 nm). The following studies draw the same conclusion. For example, Zhang et al. [30] employed hyperspectral imaging to measure the SSC of apples, achieving an optimal model (CARS-SPA-LS-SVR) with Rp = 0.917 and RMSEP = 0.453 °Brix. Similarly, Shao et al. [31] used Vis-NIR hyperspectral imaging to detect the SSC of winter jujube at different maturity stages, finding the SPA-LS-SVM model to be the most effective. It should be borne in mind that prediction results can vary depending on the research subject and specific conditions. At present, some research reports have conducted visualization studies on SSC, firmness, and anthocyanins content in fruit. For example, Li et al. [32] visualized anthocyanin content in mulberry fruit by using Vis-NIR hyperspectral imaging. Yang et al. [33] used near infrared hyperspectral imaging to visualize the SSC distribution in peaches. Similar studies include seed moisture content [34] and starch content of potato [35]. However, unlike most previous studies that focused on a single quality parameter, in this study, both the distribution of SSC and firmness were visually analyzed within Cerasus Humilis fruit at different maturity stages. This visualization capability provides valuable insights for consumers, allowing them to predict the internal quality of the fruit. To sum up, the feasibility and effectiveness of HSI for the determination of SSC and firmness in Cerasus Humilis fruit was confirmed.

5. Conclusions

This study explored the feasibility of determining the SSC and firmness in Cerasus Humilis fruit at different maturity stages by using hyperspectral imaging combined with chemometrics algorithms. The original spectral data were preprocessed. Five effective wavelength selection algorithms were employed to extract characteristic wavelengths. Subsequently, both linear MLR and nonlinear LS-SVM prediction models were established. The results indicate that the optimal model for SSC prediction was SPA-LS-SVM (Rp = 0.8526, RMSEP = 0.9703, RPD = 1.9017), while the optimal model for firmness prediction was UVE-CARS-LS-SVM (Rp = 0.7879, RMSEP = 1.1205, RPD = 2.0221). Finally, a color distribution map was constructed and visualized to represent the SSC and firmness of Cerasus Humilis fruit. The study indicates that the using of the hyperspectral imaging technology combined with chemometric algorithms provides a nondestructive determine method for determining the SSC and firmness of Cerasus Humilis fruit, evidencing the prospect of applying HSI to evaluating internal quality of fruits.

The method proposed in this study realizes the SSC and firmness detection of different maturity stages Cerasus Humilis fruits, which could be very helpful for real-time monitoring of Cerasus Humilis fruits quality by spectral imaging technique. It provides theoretical support for the development of portable equipment and on-line testing equipment. Even so, it should be noted that similar size and intact samples of Cerasus Humilis fruits were used at different maturity stages in this study. In practice, however, fruit quality is influenced by various factors, such as variety, origin, size, defects (cracks, pests, bruises, and rust spots), temperature, and humidity, etc. In future studies, the influence of the above factors should be considered to improve the robustness (universality) of the model, while adding new model algorithms (such as convolutional neural network). At the same time, the production line of Cerasus Humilis fruits quality detection will be developed and optimized based on hyperspectral imaging technology.

Author Contributions: Conceptualization, B.W.; methodology, B.W. and L.L.; software, B.W. and L.L.; data curation, B.W. and L.L.; writing—original draft, B.W.; writing—review and editing, B.W., S.Z., H.Y. and L.L. All authors have read and agreed to the published version of the manuscript.

Funding: This research was Supported by Fundamental Research Program of Shanxi Province (Grant NO. 202303021212120, No. 202203021212426). "Introduction of Talents and Scientific Research Initiation Project" of Shanxi Agricultural University (Project No. 2023BQ42, No. 2023BQ114).

Data Availability Statement: Data are contained within the article.

Acknowledgments: The authors thank the editor and anonymous reviewers for providing helpful suggestions for improving the quality of this manuscript. We thank Home for Researchers editorial team (www.home-for-researchers.com) for language editing service.

Conflicts of Interest: The authors declare no conflicts of interest.

References

1. Wang, B.; Yang, H.; Zhang, S.; Li, L. Detection of defective features in cerasus humilis fruit based on hyperspectral imaging technology. *Appl. Sci.* **2023**, *13*, 3279. [CrossRef]
2. Li, L.; Lu, L.M.; Zhao, X.H.; Hu, D.Y.; Tang, T.Y.; Tang, Y.L. Nondestructive detection of tomato quality based on multiregion combination model. *J. Food Process Eng.* **2022**, *45*, e14100. [CrossRef]
3. De Brito, A.A.; Campos, F.; dos Reis Nascimento, A.; Damiani, C.; da Silva, F.A.; de Almeida Teixeira, G.H.; Júnior, L.C.C. Non-destructive determination of color, titratable acidity, and dry matter in intact tomatoes using a portable Vis-NIR spectrometer. *J. Food Compos. Anal.* **2022**, *107*, 104288. [CrossRef]
4. Shang, Y.; Bao, L.; Bi, H.; Guan, S.; Xu, J.; Gu, Y.; Zhao, C. Authenticity discrimination and adulteration level detection of camellia seed oil via hyperspectral imaging technology. *Food Anal. Methods* **2024**, *17*, 450–463. [CrossRef]
5. Sun, J.; Shi, X.; Zhang, H.; Xia, L.; Guo, Y.; Sun, X. Detection of moisture content in peanut kernels using hyperspectral imaging technology coupled with chemometrics. *J. Food Process Eng.* **2019**, *42*, e13263. [CrossRef]
6. Baiano, A. Applications of hyperspectral imaging for quality assessment of liquid based and semi-liquid food products: A review. *J. Food Eng.* **2017**, *214*, 10–15. [CrossRef]
7. Sun, H.; Zhang, S.; Ren, R.; Xue, J.; Zhao, H. Detection of soluble solids content in different cultivated fresh jujubes based on variable optimization and model update. *Foods* **2022**, *11*, 2522. [CrossRef] [PubMed]
8. Wang, Z.; Wu, S.; Zuo, C.; Jiang, M.; Song, J.; Ding, F.; Pan, L. Exploring the variability and heterogeneity of apple firmness using visible and near-infrared hyperspectral imaging. *LWT-Food Sci. Technol.* **2024**, *192*, 115704. [CrossRef]
9. Feng, S.; Shang, J.; Tan, T.; Wen, Q.; Meng, Q. Nondestructive quality assessment and maturity classification of loquats based on hyperspectral imaging. *Sci. Rep.* **2023**, *13*, 13189. [CrossRef] [PubMed]
10. Benelli, A.; Cevoli, C.; Ragni, L.; Fabbri, A. In-field and non-destructive monitoring of grapes maturity by hyperspectral imaging. *Biosyst. Eng.* **2021**, *207*, 59–67. [CrossRef]
11. Pullanagari, R.R.; Li, M. Uncertainty assessment for firmness and total soluble solids of sweet cherries using hyperspectral imaging and multivariate statistics. *J. Food Eng.* **2021**, *289*, 110177. [CrossRef]
12. Li, X.; Wei, Y.; Xu, J.; Feng, X.; Wu, F.; Zhou, R.; He, Y. SSC and pH for sweet assessment and maturity classification of harvested cherry fruit based on NIR hyperspectral imaging technology. *Postharvest Biol. Technol.* **2018**, *143*, 112–118. [CrossRef]
13. Luo, W.; Zhang, J.; Liu, S.; Huang, H.; Zhan, B.; Fan, G.; Zhang, H. Prediction of soluble solid content in Nanfeng mandarin by combining hyperspectral imaging and effective wavelength selection. *J. Food Compos. Anal.* **2024**, *126*, 105939. [CrossRef]
14. Ma, T.; Xia, Y.; Inagaki, T.; Tsuchikawa, S. Non-destructive and fast method of mapping the distribution of the soluble solids content and pH in kiwifruit using object rotation near-infrared hyperspectral imaging approach. *Postharvest Biol. Technol.* **2021**, *174*, 111440. [CrossRef]
15. Xu, L.; Chen, M.; Wang, Y.; Chen, X.; Lei, X. Study on non-destructive detection method of kiwifruit sugar content based on hyperspectral imaging technology. *Spectrosc. Spectr. Anal.* **2021**, *41*, 2188–2195.
16. Choi, J.Y.; Kim, J.; Jeong, S.; Kim, M.; Park, S.; Moon, K.D. Hyperspectral imaging technigue for monitoring moisture content of blueberry during the drying process. *Korean J. Food Preserv.* **2021**, *28*, 445–455. [CrossRef]

17. Baek, M.W.; Tilahun, S. Prediction of tannin content and quality parameters in astringent persimmons from visible and near-infrared spectroscopy. *Front. Plant Sci.* **2023**, *14*, 1260644. [CrossRef]
18. Chu, X.; Miao, P.; Zhang, K.; Wei, H.; Fu, H.; Liu, H.; Ma, Z. Green banana maturity classification and quality evaluation using hyperspectral imaging. *Agriculture* **2022**, *12*, 530. [CrossRef]
19. Taghinezhad, E.; Rasooli Sharabiani, V.; Shahiri, M.; Moinfar, A.; Szumny, A. Predicting quality properties of pears during storage using hyper spectral imaging system. *Agriculture* **2023**, *13*, 1913. [CrossRef]
20. Wang, B.; He, J.; Zhang, S.; Li, L. Nondestructive prediction and visualization of total flavonoids content in Cerasus Humilis fruit during storage periods based on hyperspectral imaging technique. *J. Food Process Eng.* **2021**, *44*, e13807. [CrossRef]
21. Guo, Z.; Zhao, C.; Huang, W. Intensity correction of visualized prediction for sugar content in apple using hyperspectral imaging. *Trans. Chin. Soc. Agric. Mach.* **2015**, *46*, 227–232. [CrossRef]
22. Wang, X.H.; Xu, L.J.; Chen, H.; Zou, Z.Y.; Huang, P.; Xin, B. Non-destructive detection of pH value of kiwifruit based on hyperspectral fluorescence imaging technology. *Agriculture* **2022**, *12*, 208. [CrossRef]
23. Qin, J.; Chao, K.; Kim, K. Hyperspectral and multispectral imaging for evaluating food safety and quality. *J. Food Eng.* **2013**, *118*, 157–171. [CrossRef]
24. Tang, G.; Huang, Y.; Tian, K.; Song, X.; Yan, H.; Min, S. A new spectral variable selection pattern using competitive adaptive reweighted sampling combined with successive projections algorithm. *Analyst* **2014**, *139*, 4894–4902. [CrossRef] [PubMed]
25. Li, H.; Liang, Y.; Xu, Q.; Cao, D. Key wavelengths screening using competitive adaptive reweighted sampling method for multivariate calibration. *Anal. Chim. Acta* **2009**, *648*, 77–84. [CrossRef] [PubMed]
26. Meng, Q.; Shang, J.; Huang, R.; Zhang, Y. Determination of soluble solids content and firmness in plum using hyperspectral imaging and chemometric algorithms. *J. Food Process Eng.* **2021**, *44*, e13597. [CrossRef]
27. Gao, H.Z.; Wan, J.W.; Zhu, Z.Z.; Wang, L.B.; Nian, Y.J. Classification technique for hyperspectral image based on subspace of bands feature extraction and LS-SVM. *Pectrosc. Spectr. Anal.* **2011**, *31*, 1314–1317. [CrossRef]
28. Liu, Y.; Sun, X.; Zhang, H.; Aiguo, O. Nondestructive measurement of internal quality of Nanfeng mandarin fruit by charge coupled device near infrared spectroscopy. *Comput. Electron. Agr.* **2010**, *71*, S10–S14. [CrossRef]
29. Shinzawa, H.; Ritthiruangdej, P.; Ozaki, Y. Kernel analysis of partial least squares regression models. *Appl. Spectrosc.* **2011**, *65*, 549–556. [CrossRef] [PubMed]
30. Zhang, D.; Xu, Y.; Huang, W.; Tian, X.; Xia, Y.; Xu, L.; Fan, S. Nondestructive measurement of soluble solids content in apple using near infrared hyperspectral imaging coupled with wavelength selection algorithm. *Infrared Phys. Technol.* **2019**, *98*, 297–304. [CrossRef]
31. Shao, Y.; Ji, S.; Xuan, G.; Wang, K.; Xu, L.; Shao, J. Soluble solids content monitoring and shelf life analysis of winter jujube at different maturity stages by Vis-NIR hyperspectral imaging. *Postharvest Biol. Technol.* **2024**, *210*, 112773. [CrossRef]
32. Li, X.; Wei, Z.; Peng, F.; Liu, J.; Han, G. Non-destructive prediction and visualization of anthocyanin content in mulberry fruits using hyperspectral imaging. *Front. Plant Sci.* **2023**, *14*, 1137198. [CrossRef] [PubMed]
33. Yang, B.; Gao, Y.; Yan, Q.; Qi, L.; Zhu, Y.; Wang, B. Estimation method of soluble solid content in peach based on deep features of hyperspectral imagery. *Sensors* **2020**, *20*, 5021. [CrossRef] [PubMed]
34. Xu, Y.; Zhang, H.; Zhang, C.; Wu, P.; Li, J.; Xia, Y.; Fan, S. Rapid prediction and visualization of moisture content in single cucumber (*Cucumis sativus* L.) seed using hyperspectral imaging technology. *Infrared Phys. Technol.* **2019**, *102*, 103034. [CrossRef]
35. Wang, F.; Wang, C.; Song, S.; Xie, S.; Kang, F. Study on starch content detection and visualization of potato based on hyperspectral imaging. *Food Sci. Nutr.* **2021**, *9*, 4420–4430. [CrossRef] [PubMed]

horticulturae

MDPI

Article

A New Plant-Wearable Sap Flow Sensor Reveals the Dynamic Water Distribution during Watermelon Fruit Development

Runqing Zhang [1], Yangfan Chai [2], Xinyu Liang [1], Xiangjiang Liu [2], Xiaozhi Wang [3] and Zhongyuan Hu [1,4,*]

1 College of Agricultural and Biotechnology, Zhejiang University, Hangzhou 310058, China; 22116201@zju.edu.cn (R.Z.); 22216181@zju.edu.cn (X.L.)
2 College of Biosystems Engineering and Food Science, Zhejiang University, Hangzhou 310058, China; 12213070@zju.edu.cn (Y.C.); xjliu@zju.edu.cn (X.L.)
3 College of Information Science and Electronic Engineering, Zhejiang University, Hangzhou 310058, China; xw224@zju.edu.cn
4 Hainan Institute of Zhejiang University, Yazhou District, Sanya 572025, China
* Correspondence: huzhongyuan@zju.edu.cn

Abstract: This study utilized a plant-wearable sap flow sensor developed by a multidisciplinary team at Zhejiang University to monitor water distribution patterns in watermelon fruit stalks throughout their developmental stages. The dynamic rules of sap flow at different stages of fruit development were discovered: (1) In the first stage, sap flow into the fruit gradually halts after sunrise due to increased leaf transpiration, followed by a rapid increase post-noon until the next morning, correlating with fruit expansion. (2) In the second stage, the time of inflow sap from noon to night is significantly shortened, while the outflow sap from fruit is observed with the enhancement of leaf transpiration after sunrise, which is consistent with the slow fruit growth at this stage. (3) In the third stage, the sap flow maintains the diurnal pattern. However, the sap flow that inputs the fruit at night is basically equal to the sap flow that outputs the fruit during the day; the fruit phenotype does not change anymore. In addition, a strong correlation between the daily mass growth in fruit and the daily sap flow amount in fruit stalk was identified, validating the sensor's utility for fruit growth monitoring and yield prediction.

Keywords: sap flow sensor; watermelon; fruit development; water distribution

Citation: Zhang, R.; Chai, Y.; Liang, X.; Liu, X.; Wang, X.; Hu, Z. A New Plant-Wearable Sap Flow Sensor Reveals the Dynamic Water Distribution during Watermelon Fruit Development. *Horticulturae* **2024**, *10*, 649. https://doi.org/10.3390/horticulturae10060649

Academic Editor: Xuming Huang

Received: 10 May 2024
Revised: 31 May 2024
Accepted: 6 June 2024
Published: 17 June 2024

1. Introduction

Generally speaking, sap flow refers to the water flow within organs such as stems, leaf stalks, and fruit stalks that have transport tissues driven by water potential differences [1]. The water flow rate inside a plant is called the sap flow rate. Its measurement principle is to add a detection tracer to the water flowing inside the plant and calculate the sap flow rate by measuring the distance traveled by the detection tracer over a certain period of time. As an important water physiological indicator, the sap flow rate can reflect the transportation of water, minerals, and assimilates in plants. Therefore, monitoring the sap flow rate can provide clues for studying plant growth patterns [2,3]. In addition, compared to traditional phenotypic indicators, such as plant height, stem thickness, and leaf color, sap flow responds more quickly to environmental changes, providing a new perspective for studying the effects of environmental stresses such as water, temperature, light, and nutrients on plant growth and development [4,5]. Researchers often use temperature as a detection tracer; based on this, three methods have been developed to measure the sap flow rate, namely the thermal diffusion method, thermal equilibrium method, and thermal pulse method [6–8]. Traditional sap flow sensors require the probe to be inserted into the plant body to calculate the temperature difference between the two probes, which can cause certain damage to the plant. They are often used to measure the sap flow rate of tall woody plants and are not suitable for horticultural crops with a low biomass.

In 2020, the multidisciplinary team from Zhejiang University developed a new type of flexible wearable sensor for plant sap flow that was suitable for horticultural crops [9]. Chai et al. [10] deployed the new sap flow sensor on the surface of watermelon fruit stalks and conducted long-term and continuous monitoring. The plant condition was good, indicating that the sap flow sensor had good biocompatibility and could non-destructively monitor the plant sap flow rate. Compared with traditional sap flow sensors, the new sap flow sensor has excellent data collection capabilities and a simple and reliable structure, which is conducive to the high-throughput measurement of the plant sap flow rate in the field.

Watermelon (*Citullus lanatus*) is an annual herbaceous plant of the Cucurbitaceae family, originating from Africa [11]. It is widely cultivated worldwide and is one of the highest yielding fruits, known as the "king of summer fruits". The development period of watermelon fruit starts from pollination to the end of fruit harvest, which is the most important period in the entire growth period of watermelon [12]. With the assistance of the new sap flow sensor, long-term monitoring of watermelon fruit development in the field is conducted to explore the water distribution law during fruit development, investigate the physiological characteristics related to sap flow, provide new theoretical clues for the breeding of high-yield and high-quality watermelon varieties or the research and development of cultivation techniques, and contribute to the sustainable and healthy development of the watermelon industry.

2. Materials and Methods

2.1. Test Location and Materials

From February to June 2023, the experiment was conducted at the base of Zhejiang Wuwangnong Co., Ltd., Hangzhou, China (30.25° N, 120.25° E). The tested Zhemi 8 was a watermelon variety, which was bred by our laboratory and widely cultivated around Zhejiang, Jiangsu, and surrounding areas. After disinfection on February 17, the seeds were planted in hole trays. When watermelon seedlings had four true leaves on March 15, they were planted in plastic greenhouses. The greenhouse was 45 m long and 8 m wide, with four rows covered with silver-black dual-color plastic film and irrigated under the film. The planting distance was 40 cm, and the pruning method was single-vine pruning. When the plant entered the vine extension period, a strong main vine was selected and left. Daily agricultural operations were managed by farm workers. On April 27, plants generally grew to 10 true leaves and differentiated into a second female flower, beginning artificial pollination.

2.2. Method

2.2.1. Sap Flow Rate (mg/min) of Fruit Stalk and Leaf Stalk

On the 7th day after pollination, when the watermelon fruit grew to the size of an egg (with a transverse diameter range of 56 cm and a longitudinal diameter range of 6–7 cm), three individual plants with similar fruit sizes were selected as three biological replicates. Sap flow sensors were installed on the fruit stalks (Figure 1), with the upstream temperature sensor corresponding to the direction near the root system and the downstream temperature sensor corresponding to the direction away from the root system (please see Appendix A for a detailed description). The sap flow sensor program was set to make a measurement every 30 min from the beginning until fruit harvesting, so the sap flow rate (mg/min) represented the average value over 30 min.

Studies have shown that fruit growth was influenced by plant transpiration [13,14]. Because all sap flow passing through the leaf stalks would evaporate into the air through the leaves, sap flow sensors installed on the leaf stalks could be used to study transpiration [15–18]. In this study, the sap flow rate of the leaf stalk was utilized to reflect the intensity of leaf transpiration. The preliminary experiment measured the sap flow rate of leaf stalks at multiple nodes and showed similar variation patterns. Herein, the leaf stalks closest to the fruit, namely the leaf stalks of the fruit-setting node, were used to demonstrate their impacts on the sap flow of the fruit stalks. At the same time as the sap flow sensors were installed on the fruit stalks, sap flow sensors were also installed on the leaf stalks of the fruit-setting node. The installation method and measurement interval of the sensors were as described above. Unless otherwise specified, the subsequent plotting data represented the average of three biological replicates.

Figure 1. Field application of sap flow sensors to monitor fruit development. 'DAP' represents days after pollination.

2.2.2. Environmental Parameters

Data collection for total solar radiation (J/m^2/h) utilized photoelectric solar radiation sensor (RS4854, Vemsee Co., Ltd., Shandong, China), with a range of 0–2000 W/m^2 and an accuracy of ±5%. Measurements were automatically taken once every hour and transmitted to the cloud platform via a 4G signal. The sensor was mounted on the upper part of the greenhouse.

For air temperature (°C) data acquisition, external temperature and humidity sensor (ZL-TH10TP, CIMC Co., Ltd., Beijing, China) were employed, with a temperature measurement range of −40 °C to 80 °C and an accuracy of ±0.5 °C. Measurements were automatically taken once every half hour and sent to the cloud platform through a 4G signal. The sensor was positioned in the lower part of the greenhouse, near the fruits.

2.2.3. Daily Mass Growth (g) of Fruit

Destructive sampling was conducted every 24 h. Before sampling, the longitudinal and transverse diameters of the fruits containing sap flow sensors were measured using a tape measure with an accuracy of 0.1 cm. Subsequently, fruits matching their longitudinal and transverse diameters were selected and removed from the vine. These fruits were then weighed using an electronic balance, and the weights were recorded as Wx. The same procedure was repeated at the same time the following day to obtain the weight Wx + 1. Thus, the daily mass growth of fruit was calculated as Wx + 1 − Wx. This measurement was used to compare with the daily sap flow amount of fruit stalk during the same time period.

2.2.4. Daily Sap Flow Amount (g) of Fruit Stalk

The daily sap flow amount (g) was calculated based on the sap flow rate (mg/min), as shown in the following formula:

$$M = \sum(V \times 30/1000)$$

M represented the daily sap flow amount (g) of fruit stalk. V represented the sap flow rate (mg/min) of fruit stalk measured by the sap flow sensor. The sap flow sensor program was set to measure every 30 min from the beginning until fruit harvesting. Therefore, V represented the average sap flow rate (mg/min) over a 30 min period. (V × 30/1000) represented the estimated sap flow amount (g) for the 30 min interval. The estimated sap flow amount from 48 measurements in a day was then summed up to obtain the daily sap flow amount (g). As watermelon fruits primarily grew at night [10], the total of 48 measurements within a 24 h period from 12:00 on day x to 12:00 on day x + 1 was considered as the daily sap flow amount (g) for day x + 1. Consequently, the diameter and weight of the fruit were measured at 12:00 each day.

2.3. Data Processing and Statistical Analysis

In this experiment, we used the R language (4.2.1) to organize and clean the raw data measured by the sap flow sensors, calculate the sap flow rate, conduct descriptive statistics analysis and correlation analysis, and visualize the data using R language (4.2.1) and EXCEL (Microsoft Office 2016).

3. Results

3.1. Sap Flow Rate of Leaf Stalk

Figure 2 shows the variation pattern of the leaf stalk sap flow rate (equivalent to leaf transpiration rate, mg/min) with days after pollination (DAP) in watermelon. Total solar radiation and temperature exhibited similar trends, with the temperature in the greenhouse rising rapidly as the sun rose and then decreasing after noon as total solar radiation decreased. After conducting correlation analysis on the real-time data of total solar radiation and temperature, a highly significant positive correlation (R = 0.882 ***) was found. To avoid redundancy, only the temperature curve was added to the graph, while the total solar radiation curve was omitted.

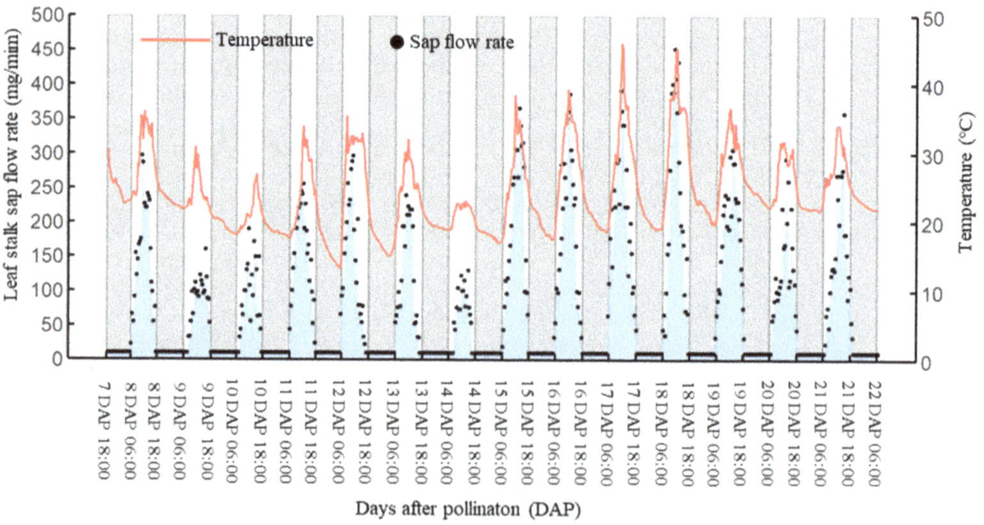

Figure 2. The variation of sap flow rate of leaf stalk at the fruit-setting node with the days after pollination. The black points represent the sap flow rate (mg/min) measured at 30 min intervals, and their values correspond to the *y*-axis on the left. The area marked by the blue shadow represents the sap flow amount (mg) passing through the leaf stalk between sequential time points. We will not consider the actual sunrise and sunset time here and use 6:00 as the sunrise time and 18:00 as the sunset time, which is shown as a black and white rectangular background in the figure. The red curve represents the temperature variation over time, and its value corresponds to the *y*-axis on the right.

As shown in the figure, the curve of the leaf transpiration rate and the temperature curve were in very good agreement, both in terms of days and the entire fruit development period. At 9 DAP, 10 DAP, and 14 DAP, under low-temperature and low-light conditions, the leaf transpiration rate was very low. During 15–19 DAP, when the temperature was high, the transpiration rate could even reach 450 mg/min. After Pearson's correlation analysis, a highly significant positive correlation (R = 0.903 ***) was found between the leaf transpiration rate and temperature.

The leaf transpiration rate exhibited a very distinct diurnal rhythm, regardless of weather conditions. Sap flow commenced around sunrise and peaked at noon due to the increase in total solar radiation. Subsequently, post-noon, as total solar radiation decreased, the transpiration rate weakened, approaching 0 mg/min around sunset, with no observed sap flow throughout the night. Following correlation analysis, the leaf transpiration rate and total solar radiation exhibited a highly significant positive correlation (R = 0.921 ***).

3.2. *Sap Flow Rate of Fruit Stalk*

According to the change rule of the sap flow rate in the fruit stalk and the time of appearance of negative-direction sap flow, the watermelon fruit development period was divided into three stages (stage 1 from 7 DAP to 20 DAP, stage 2 from 20 DAP to 29 DAP, and stage 3 from 29 DAP to 35 DAP):

The first stage began with the installation of sap flow sensors on the fruit stalks at 7 DAP and ended with the appearance of obvious negatively oriented sap flow at 20 DAP (Figure 3). The average value of the sap flow rate of fruit stalks in the first stage was 62 mg/min, and the total sap flow amount (area of blue shaded portion) was 1160 g.

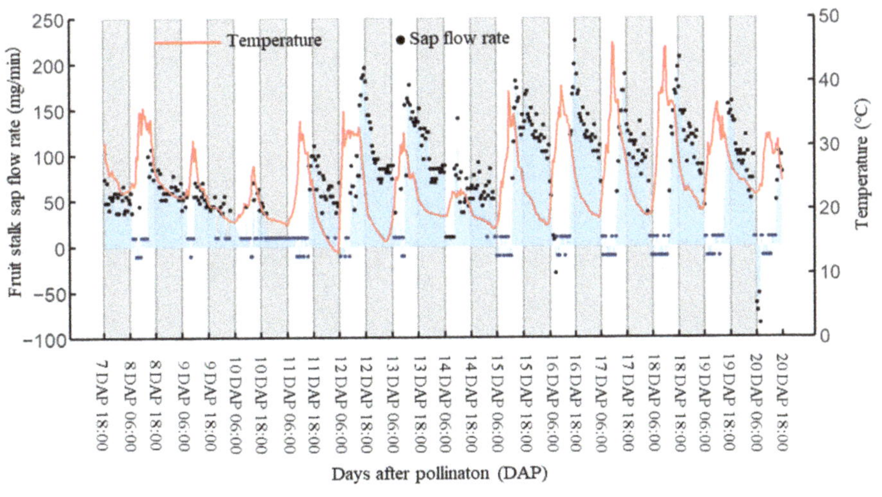

Figure 3. The variation of sap flow rate of watermelon fruit stalk versus the days after pollination in the first stage. The black points represent the sap flow rate (mg/min) measured at 30 min intervals, and their values correspond to the *y*-axis on the left. The area marked by the blue shadow represents the sap flow amount (mg) passing through the fruit stalk between sequential time points. The explanation of other information elements is the same as Figure 2.

The maximum value of the sap flow rate in this stage was 224 mg/min, which occurred around 18:00 at 16 DAP, and the sap flow rate remained high from 15 DAP to 19 DAP, with a daily average sap flow rate of 100–150 mg/min. Moreover, the area shaded in blue also showed that the sap flow into the fruit was very high during these days. At 10 DAP and 14 DAP, the overall sap flow rate as well as sap flow rate during these two days were significantly smaller in comparison to the two days prior and after, indicating that the low-temperature and low-sunshine weather would had a significant negative impact on fruit growth and development.

At this stage, the sap flow rate of fruit stalks showed a similar pattern of change nearly every day. Around 6:00 a.m., as the amount of light and temperature increased, the sap flow rate of fruit stalks rapidly approached 0 mg/min, and no sap flow was observed throughout the morning. After 12:00 a.m., as the temperature and total solar radiation decreased, sap flow reappeared between 13:00 and 16:00. The sap flow rate increased rapidly, reaching its maximum around sunset and gradually decreased after sunset, persisting until sunrise the next day.

During fruit development, the root system absorbed water from the soil, which was then transported to the fruit through the conductive tissues over a long distance, typically from the exterior to the interior of the fruit. However, the negative sap flow rate depicted in the figure indicated that water was flowing out of the fruit through the fruit stalk during certain periods. The sap flow rate calculations of the three biological replicates before 20 DAP also exhibited negative values, albeit with very small absolute magnitudes and infrequent occurrences, thus not being considered significant. Negative sap flow coincidentally occurred in all three biological replicates around 6:00 a.m. at approximately 20 DAP, with absolute sap flow rates exceeding 50 mg/min and lasting for about 3 h. This phenomenon persisted almost every morning from 20 DAP until the fruits were harvested.

The beginning of the second stage was marked by the appearance of significant negatively oriented sap flow at 20 DAP, and the end of the second stage was marked by the fact that the night-time sap flow amount into the fruit was essentially equal to the daytime sap flow amount out of the fruit at 29 DAP (first half of Figure 4). The average sap flow rate of fruit stalks in the second stage was 23.3 mg/min, and the total sap flow amount (area shaded blue above the x-axis minus the area shaded blue below the x-axis) was 335.3 g.

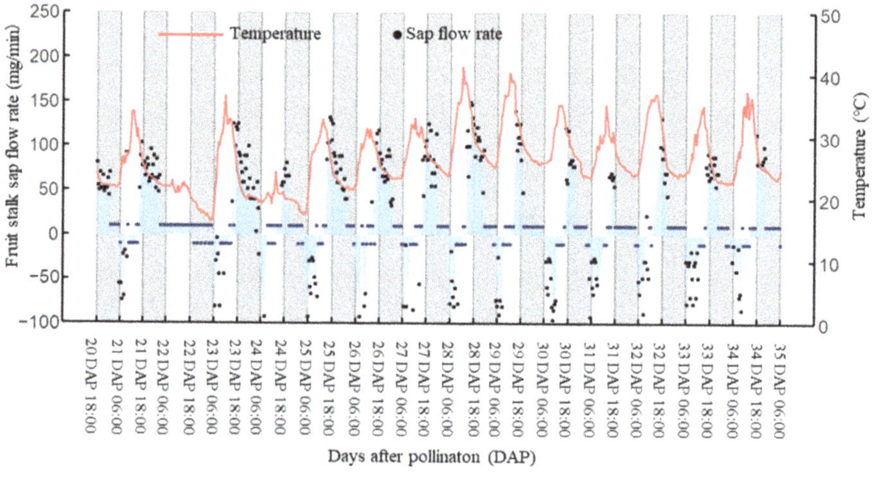

Figure 4. The variation of sap flow rate of watermelon fruit stalks versus the days after pollination in the second stage and third stage. The explanation of information elements is the same as Figure 3.

Compared to the first stage, the average temperature was higher in the second stage, but the sap flow rate was considerably lower, and the duration of positively oriented sap flow was significantly lower in this stage compared to the first stage, when water flowed into the fruit almost the entire night. In the first few days (20 DAP, 21 DAP, and 23 DAP), positive sap flow appeared around 16:00 and disappeared between 2:00 and 4:00. With the passing of developmental days, the disappearance of positive sap flow during the night began earlier and earlier, and the amount of water flowing into the fruit during the night decreased progressively (the blue shaded portion above the x-axis became smaller).

Negative-direction sap flow was observed almost every morning during this phase. On the initial days (20 DAP, 21 DAP, 23 DAP, and 24 DAP), the cessation of negative sap flow occurred around 9:00 a.m., with an approximate duration of 3 h from the onset to cessation. With the advancement of fruit development, the cessation of negative sap flow during daytime progressively delayed, accompanied by an increasing outflow of water from the fruit during daytime (the area shaded blue below the x-axis became larger and larger).

The beginning of the third stage was marked by the 29 DAP night-time sap flow into the fruit being essentially equal to the daytime sap flow out of the fruit, and the end of the stage was marked by the harvest of the fruit at 35 DAP (Figure 4, second half). The mean sap flow rate of fruit stalks in the third stage was 7.6 mg/min, and the total sap flow amount (the area shaded blue the x-axis minus the area shaded blue below the x-axis) was 60.6 g.

Positive sap flow in this phase occurred only around sunset and lasted less than 6 h. Negative sap flow occurred during the day for about 6 h and disappeared after 12:00 a.m. as temperature and total solar radiation decreased. The duration of positive-direction sap flow at night was basically equal to that of negative-direction sap flow during the day, and the amount of sap flow into the fruit at night was basically equal to the amount of sap flow out of the fruit during the day (the area shaded blue above the x-axis was basically equal to the area shaded blue below the x-axis), and the daily sap flow amount tended to be 0 g.

3.3. *Daily Sap Flow Amount of Fruit Stalk and Daily Mass Growth in Fruit*

As shown in Figure 5, the daily sap flow amount and daily mass growth showed an increasing and then decreasing trend versus the number of days of fruit development after pollination, and the difference between the daily sap flow amount and daily mass growth on the same day was not significant. Except for 10 DAP, 11 DAP, and 15 DAP, which were affected by low temperature and rain, the daily mass growth from 8 DAP to 20 DAP was more than 50 g, and the daily mass growth at 13 DAP, 16 DAP, and 17 DAP was even more than 150 g. From 20 DAP onwards, the daily mass growth gradually decreased, and the daily mass growth for the 6 days before harvest was around 0 g. By summing the daily sap flow amount, a total sap flow amount of 1479 g was obtained for the entire fruit development period. Additionally, summing the daily mass growth resulted in a mass increase of 1511 g during the same period. This represents that there was a small difference between the two values.

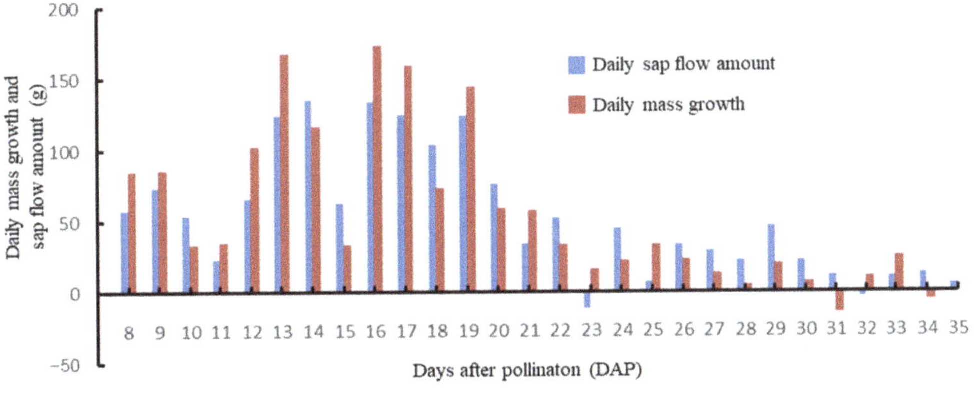

Figure 5. Changes in daily sap flow amount and daily mass growth versus the days after pollination.

As shown in Figure 6 (left), the longitudinal and transverse diameters of watermelon fruits after pollination conformed to an S-shaped growth curve. Initially, the longitudinal diameter was greater than the transverse diameter. As the fruits grew and developed, the gap between the two gradually decreased. When growth ceased, the transverse diameter averaged 16.7 cm, and the longitudinal diameter averaged 16 cm. The longitudinal diameter ceased to grow around 22 DAP, while the transverse diameter ceased around 24 DAP. At 10 DAP and 15 DAP, there were two days of cold and rainy weather, resulting in significantly smaller increases in longitudinal and transverse diameters compared to the dates before and after.

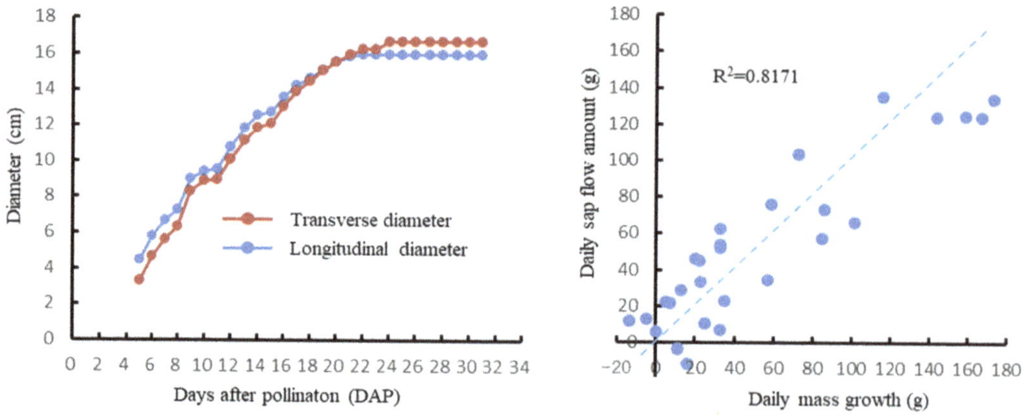

Figure 6. Changes in fruit longitudinal and transverse diameters versus the days after pollination (**left**); comparison of daily sap flow amount and daily mass growth with y = x straight line (**right**).

The daily sap flow amount over these 28 days was compared and analyzed against the daily mass growth. As shown in Figure 6 (right), the daily mass growth and the daily sap flow amount were evenly distributed above and below the y = x line, with an R^2 value of 0.817, indicating a good fit. After field verification, the sap flow sensor accurately predicted the sap flow rate of fruit stalk, which was essentially equivalent to the real-time growth rate of fruit mass. The daily sap flow amount of fruit stalk corresponded closely to the daily mass growth of the fruits.

4. Discussion

4.1. The Sap Flow Rate Is Sensitive to Environmental Factors and Can Reflect the Growth Status of Fruits

Water transport, traditionally focused on the plant stem and leaf, has been well developed to understand the soil–plant–atmosphere continuum (SPAC) [19]. However, research on water transport during fruit development has been limited, largely due to the lack of suitable technical means. Methods such as microscopy, high-pressure flow meters, microCT, and others [20–23], were often destructive and time-consuming, with significant drawbacks. In our study, we installed a new sap flow sensor on watermelon fruit stalk to investigate the sap flow patterns during watermelon fruit development. Previous studies have indicated that water accounted for more than 90% of the fresh weight in many fleshy fruits, and, hence, water content largely determined fruit size [24,25]. Our research results also confirmed this point, as the daily sap flow amount of the fruit stalk was essentially equal to daily fruit mass growth. Furthermore, we found that sap flow was highly sensitive to environmental changes: when encountering low-temperature and low-light conditions, the sap flow of the fruit stalk was notably weak, while under favorable weather conditions, the sap rate flowing into the fruit increased rapidly. Therefore, the sap flow rate of the fruit stalk can reflect the real-time growth rate of the fruit, and the new sap flow sensor can be used to monitor fruit growth status.

4.2. Employing the New Sap Flow Sensor to Examine Diurnal Shifts in Water Allocation during Watermelon Fruit Development

The plant root system relies on two different mechanisms for water uptake: one is passive water uptake through leaf transpiration, and the other is active water uptake that is dependent on the ion concentration in the reservoir [26]. In either case, the water potential difference is the fundamental driving force for water transportation in the plant, and water always passes from a higher water potential to a lower one [27]. Previous studies have shown that at different stages of fruit development, the partition of vascular flows in fleshy fruits varied [28–31]. According to the change rule of the fruit stalk sap flow rate and the duration of negative-direction sap flow, the watermelon fruit development period was divided into three stages. Moreover, based on the fundamental cause of water transportation and the synergistic changes of the sap flow rate of fruit stalks and leaf stalks, fruit mass, as well as the longitudinal and transverse diameters of the fruit observed in the experiments, the water distribution laws of watermelon fruit development in each stage are as follows (Figure 7):

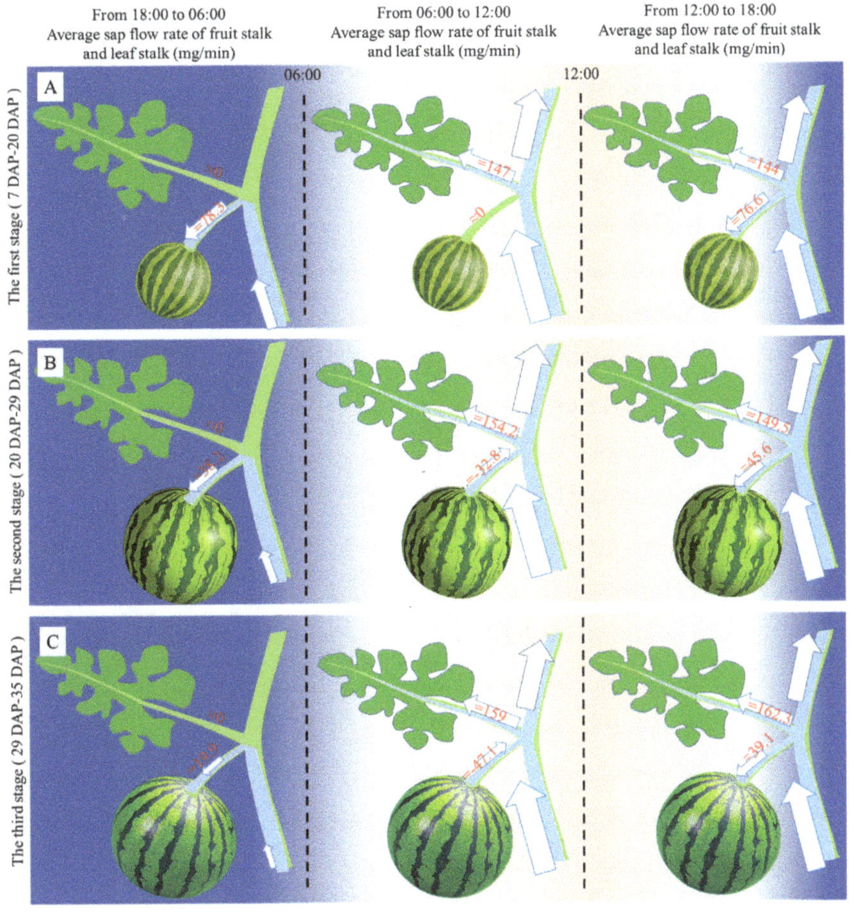

Figure 7. Water distribution patterns during watermelon fruit development. Rows ((**A–C**) in figure) represent different stages of fruit development, while columns represent different time periods within the same day. The values next to the leaf stalks and fruit stalks represent the average sap flow rate (mg/min) of the leaf stalks and fruit stalks during the corresponding time period, respectively.

4.2.1. The First Stage (7 DAP–20 DAP), as in Figure 7A

In the evening, transpiration is extremely weak, coupled with the transfer and accumulation of photosynthetic products from source to reservoir during the day. At this time, the water potential difference between the inside and outside the fruit is the greatest, and the sap flow rate of the fruit stalk reaches its maximum. Subsequently, with the inflow of water, the water potential inside the fruit increases, and the water potential difference between the inside and outside of the fruit decreases. The sap flow rate also decreases, but until the next morning, the sap flow rate does not reach 0 mg/min, indicating continuous water inflow into the fruit throughout the night. Morphologically, watermelon fruits expand significantly, and the longitudinal diameter, transverse diameter, and mass measured every other day increase greatly compared with those of the previous day.

After sunrise, with the increase in temperature and total solar radiation, leaf transpiration becomes active. The water potential outside of the fruit decreases, and the difference between the internal and external water potentials decreases until it reaches 0. The sap flow rate of the fruit stalk rapidly decreases to 0 mg/min, water stops flowing into the fruit, and the sap flow in the positive direction disappears. This phenomenon lasts until about 12:00 a.m.

After noon, as leaf transpiration weakens and the transpiration rate decreases, the water potential outside the fruit rises, allowing water to flow into the fruit. The sap flow sensor re-monitors the sap flow in the positive direction of the fruit stalk. With the continuous decline in the transpiration rate and the accumulation of assimilates in the fruit, the difference between the water potential inside and outside of the fruit increases. Finally, the sap flow rate of the fruit stalk reaches its maximum value around sunset.

4.2.2. The Second Stage (20 DAP–29 DAP), as in Figure 7B

The peak difference in water potential between the inside and outside of the fruit occurs in the evening, while the sap flow rate reaches its maximum value. Subsequently, as water flows in, the water potential inside the fruit increases, leading to a decrease in the water potential difference between the inside and outside of the fruit. The sap flow rate gradually decreases until it reaches 0 mg/min, and the water inflow into the fruit stops. Compared to the first stage, the water inflow is significantly reduced. Morphologically, watermelon phenotypic indexes exhibit slow growth. With the progression of developmental days, the positive sap flow disappears earlier and earlier. Night-time water inflow diminishes progressively. Morphologically, the longitudinal diameter, transverse diameter, and mass of fruits gradually cease to change.

After sunrise, as the temperature rises and the total solar radiation increases, leaf transpiration is initiated. The external water potential of the fruit begins to decrease, becoming lower than the internal water potential, resulting in water flowing out of the fruit. The sap flow sensor detects the sap flow in a negative direction, indicating the intense transpiration of the leaf "snatching" water from the fruit. As water flows out of the fruit, the water potential inside the fruit decreases, reducing the difference between the internal and external water potentials. After a few hours, the difference in water potentials tends to be 0, and the sap flow rate tends to be 0 mg/min. Water stops flowing out of the fruit, and the negative sap flow disappears. With the increase in developmental days, the sap flow in the negative direction disappears later and later, and the outflow amount becomes bigger and bigger.

After noon, the pattern of change of sap flow in the fruit stalk is the same as that of the first stage.

4.2.3. The Third Stage (29 DAP–35 DAP), as in Figure 7C

The diurnal variation in water allocation in this stage is the same as that of the second stage. However, the occurrence of positive-direction sap flow at night is basically equal to the occurrence of negative-direction sap flow during the day. The sap flow into the fruit at night is basically equal to the sap flow out of the fruit during the day, resulting in the daily sap flow amount approaching 0 g. The phenotype of watermelon fruits remains unchanged.

A comparison between the first and second stages reveals that even with high morning temperatures and total solar radiation in the first stage, there is almost no negative sap flow. However, in the second stage, the phenomenon of water outflow from the fruit occurs almost every morning. How can we explain this difference? During the pre-developmental period of watermelon fruits, it has been demonstrated that cell division and cell expansion are vigorous [32]. Watermelon fruits rapidly decompose assimilates into various components to enable cellular metabolism. The cytoplasm has a strong water-binding capacity, making it difficult for transpiration outside the fruit to "snatch" water [33]. As fruit expansion gradually stops, large amounts of sucrose, fructose, and other soluble sugars accumulate in the late stages of fruit development. Consequently, the free water content within the cells increases, facilitating transpiration from outside the fruit to "snatch" water.

4.3. It Was Firstly Discovered That Sap Flows out the Watermelon Fruit through the Fruit Stalk

It was firstly found that the direction of water transport during watermelon fruit development did not only occur from the outside to the inside of the fruit, but also from the inside of the fruit to the outside through the fruit stalk due to the intense transpiration of the leaves. Negative sap flow implies that watermelon fruit growth not only increases but also decreases, possibly resulting in a decrease in mass or diameter at certain times. However, capturing this small phenotypic change with conventional instruments is challenging.

In recent years, researchers have also observed similar phenomena in other crops using high-precision sensors. The linear variable displacement transducer (LVDT) is a distance sensor capable of measuring micrometer-level variations, and W. Conejero et al. mounted it on the trunk of a peach tree [34]. The trunk diameter was found to be smallest at noon, and the maximum daily trunk shrinkage was significantly positively correlated with the noon air temperature and transpiration rate. The maximum daily trunk shrinkage in the water deficit stress treatment was also significantly higher than that in the control group and returned to normal after normal irrigation. Sun Qing et al. installed linear variable displacement sensors on apple fruits, together with the Internet of Things, forming a fruit diameter dynamic monitoring system. After observing two growing seasons, it was found that the fruit diameter from young fruit to mature fruit exhibited obvious diurnal variation rules, generally peaking in the early morning and reaching a minimum in the evening, with a maximum daily shrinkage of 200 μm [35]. Through a real-time weighing system, Wang Dandan et al. found that changing tomato plants in solar greenhouses throughout the day causes an increase in plant growth at night. After sunrise, due to transpiration, plant weight begins to decline; after sunset, plant weight rises again [36]. In conclusion, the negatively oriented sap flow of watermelon fruits can be used as a new indicator to study the responses of watermelon fruits to water-deficit stress and high-temperature stress during their developmental period.

5. Conclusions

With the assistance of the new sap flow sensor, this study elucidated the dynamic changes in sap flow during watermelon fruit development, filling the gap in the research on water distribution during watermelon fruit development: (1) In the first stage, sap flow into the fruit gradually halts after sunrise due to increased leaf transpiration, followed by a rapid increase post-noon until the next morning, correlating with fruit expansion. (2) In the second stage, the time of inflow sap from noon to night is significantly shortened, while the outflow sap from fruit is observed with the enhancement of leaf transpiration after sunrise, which is consistent with the slow fruit growth at this stage. (3) In the third stage, the sap flow maintains a diurnal pattern. However, the sap flow into the fruit at night is basically equal to the sap flow out of the fruit during the day, and the fruit phenotype does not change anymore. In addition, the daily sap flow amount of the fruit stalk was essentially equal to the daily fruit mass growth, and the sap flow rate of fruit stalk was essentially equal to real-time fruit mass growth, indicating the sensor's potential in fruit growth monitoring and yield prediction.

Author Contributions: Conceptualization, R.Z. and Z.H.; data curation, R.Z., Y.C. and X.L. (Xinyu Liang); methodology, R.Z. and Y.C.; software, X.L. (Xiangjiang Liu) and X.W.; validation, R.Z., Z.H. and Y.C.; formal analysis, R.Z.; investigation, R.Z., Y.C. and X.L. (Xinyu Liang); resources, Y.C., X.L. (Xiangjiang Liu) and X.W.; visualization, R.Z.; writing—original draft preparation, R.Z.; writing—review and editing, R.Z., Y.C. and Z.H.; supervision, Z.H.; seedling management, X.L. (Xinyu Liang); funding acquisition, Z.H. All authors have read and agreed to the published version of the manuscript.

Funding: This work was supported by the Earmarked Fund for China Agriculture Research System (CARS25); the Project of Sanya Yazhou Bay Science and Technology City (SCKJ-JYRC-2022-18); and the Science and Technology Innovation Platform for Watermelon and Melon Breeding, Reproduction, and Spreading of Zhejiang Province (2020-KYY-NSFZ-0314).

Data Availability Statement: The data that support the findings of this study are available from the first author, upon reasonable request.

Conflicts of Interest: The authors declare no conflicts of interest.

Appendix A

The detection principle of the new sap flow sensor is shown in Figure A1. When sap flows from the upstream temperature sensor to the downstream temperature sensor, it is heated by the thermistor, resulting in an increase in temperature. Thus, the temperature value measured by the downstream temperature sensor is higher than the temperature value measured by the upstream temperature sensor. As the thermistor continues to heat up, the heat is continuously transported from the upstream to downstream sensor, and the temperature difference ($\Delta T = T_{downstream} - T_{upstream}$) between the two sensors becomes larger and larger.

It has been shown that the sap flow rate was significantly negatively correlated with the time to reach the maximum ΔT: the larger the sap flow rate, the shorter the time to reach the maximum ΔT; the smaller the sap flow rate, the longer the time to reach the maximum ΔT. Therefore, a prediction model was developed to calculate the sap flow rate by inputting the time to reach the maximum ΔT [10].

When sap flows from the upstream temperature sensor to the downstream temperature sensor, the sap flow rate is positive, and we define this direction as the positive direction of sap flow. When sap flows from the downstream temperature sensor to the upstream temperature sensor, the sap flow rate is negative, and we define this direction as the negative direction of sap flow. Generally, the installation direction of the sap flow sensor on the surface of the plant material is such that the upstream temperature sensor corresponds to the direction near the root system, and the downstream temperature sensor corresponds to the direction away from the root system.

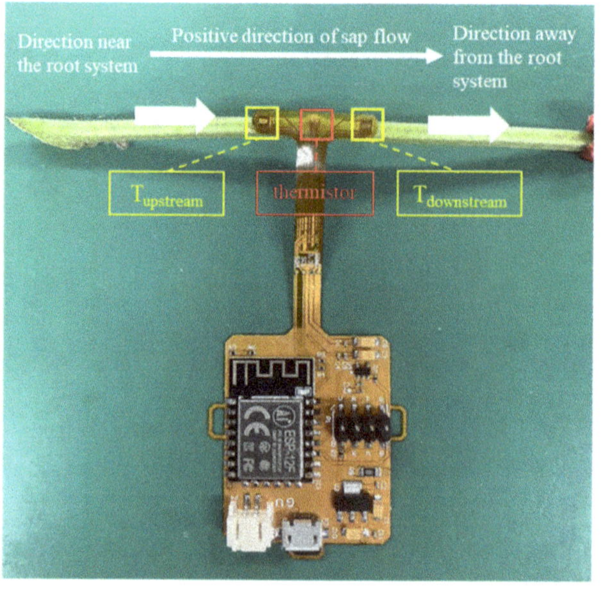

Figure A1. Detection principle of the new sap flow sensor. There are three core elements of the sensor: the upstream temperature sensor, the thermistor, and the downstream temperature sensor.

References

1. Smith, D.M.; Allen, S.J. Measurement of sap flow in plant stems. *J. Exp. Bot.* **1996**, *47*, 1833–1844. [CrossRef]
2. Busby, P.E.; Soman, C.; Wagner, M.R.; Friesen, M.L.; Kremer, J.; Bennett, A.; Morsy, M.; Eisen, J.A.; Leach, J.E.; Dangl, J.L. Research priorities for harnessing plant microbiomes in sustainable agriculture. *PLoS Biol.* **2017**, *15*, 134–156. [CrossRef] [PubMed]
3. Forster, M.A. How significant is nocturnal sap flow. *Tree Physiol.* **2014**, *34*, 757–765. [CrossRef] [PubMed]
4. Miner, G.L.; Ham, J.M.; Kluitenberg, G.J. A heat-pulse method for measuring sap flow in corn and sunflower using 3D-printed sensor bodies and low-cost electronics. *Agric. For. Meteorol.* **2017**, *246*, 86–97. [CrossRef]
5. Fang, W.; Lu, N.; Zhang, Y.; Jiao, L.; Fu, B. Responses of nighttime sap flow to atmospheric and soil dryness and its potential roles for shrubs on the Loess Plateau of China. *J. Plant Ecol.* **2018**, *11*, 717–729. [CrossRef]
6. Čermák, J.; Deml, M.; Penka, M. A new method of sap flow rate determination in trees. *Biol. Plant.* **1973**, *15*, 171–178. [CrossRef]
7. Helfter, C.; Shephard, J.D.; Martínez-Vilalta, J.; Mencuccini, M.; Hand, D.P. A noninvasive optical system for the measurement of xylem and phloem sap flow in woody plants of small stem size. *Tree Physiol.* **2007**, *27*, 169–179. [CrossRef] [PubMed]
8. Cohen, Y. Accuracy of Sap Flow Measurement Using Heat Balance and Heat Pulse Methods. *Agron. J.* **1993**, *85*, 1080–1086. [CrossRef]
9. Chen, C.Y. Study of Non-Invaisve Plant Wearable Sap Flow Sensor. Master's Thesis, Zhejiang University, Hangzhou, China, 2020.
10. Chai, Y.; Chen, C.; Luo, X.; Zhan, S.; Kim, J.; Luo, J.; Wang, X.; Hu, Z.; Ying, Y.; Liu, X. Cohabiting Plant-Wearable Sensor In Situ Monitors Water Transport in Plant. *Adv. Sci.* **2021**, *8*, 2003642. [CrossRef]
11. Guo, S.; Zhang, J.; Sun, H.; Salse, J.; Lucas, W.J.; Zhang, H.; Zheng, Y.; Mao, L.; Ren, Y.; Wang, Z. The draft genome of watermelon (*Citrullus lanatus*) and resequencing of 20 diverse accessions. *Nat. Genet.* **2013**, *45*, 51–58. [CrossRef]
12. Yativ, M.; Harary, I.; Wolf, S. Sucrose accumulation in watermelon fruits: Genetic variation and biochemical analysis. *J. Plant Physiol.* **2010**, *167*, 589–596. [CrossRef] [PubMed]
13. Morandi, B.; Losciale, P.; Manfrini, L.; Zibordi, M.; Anconelli, S.; Pierpaoli, E.; Grappadelli, L.C. Leaf gas exchanges and water relations affect the daily patterns of fruit growth and vascular flows in Abbé Fétel pear (*Pyrus communis* L.) trees. *Sci. Hortic.* **2014**, *23*, 106–113. [CrossRef]
14. Carella, A.; Massenti, R.; Lo Bianco, R. Testing effects of vapor pressure deficit on fruit growth: A comparative approach using peach, mango, olive, orange, and loquat. *Front. Plant Sci.* **2023**, *20*, 1294195. [CrossRef] [PubMed]
15. Forster, M.A. The importance of conduction versus convection in heat pulse sap flow methods. *Tree Physiol.* **2020**, *40*, 683–694. [CrossRef] [PubMed]
16. Ma, Y.; Ren, R.; Fu, H.; Si, B.; Kinar, N.J.; Liu, G.; Steppe, K. Comparing dual heat pulse methods with Péclet's number as universal switch to measure sap flow across a wide range. *Tree Physiol.* **2023**, *43*, 1691–1703. [CrossRef] [PubMed]
17. Pradiko, I.; Rahutomo, S.; Farrasati, R.; Ginting, E.N.; Hidayat, F.; Syarovy, M. Transpiration of oil palm (*Elaeis guineensis* Jacq.) based on sap flow measurement: The relation to soil and climate variables. *J. Oil Palm Res.* **2023**, *35*, 168–184. [CrossRef]
18. Bayona-Rodríguez, C.J.; Romero, H.M. Estimation of transpiration in oil palm (*Elaeis guineensis* Jacq.) with the heat ratio method. *Agron. Colomb.* **2016**, *34*, 172–178. [CrossRef]
19. Hou, X.; Li, H.; Zhang, W. Water transport in fleshy fruits: Research advances, methodologies, and future directions. *Physiol. Plant.* **2021**, *172*, 2203–2216. [CrossRef] [PubMed]
20. Ma, S.; Li, Y.; Li, X. Phloem unloading strategies and mechanisms in crop fruits. *J. Plant Growth Regul.* **2019**, *38*, 494–500. [CrossRef]
21. Anderegg, W.R.; Venturas, M.D. Plant hydraulics play a critical role in Earth system fluxes. *New Phytol.* **2020**, *226*, 1535–1538. [CrossRef]
22. Guo, X.M.; Wang, G.X.; Gao, R.F. Vascular anatomy of kiwi fruit and its implications for the origin of carpels. *Front. Plant Sci.* **2013**, *16*, 61185. [CrossRef] [PubMed]
23. Rančić, D.; Quarrie, S.P.; Radošević, R.; Terzić, M.; Pećinar, I.; Stikić, R.; Jansen, S. The application of various anatomical techniques for studying the hydraulic network in tomato fruit pedicels. *Protoplasma* **2010**, *246*, 25–31. [CrossRef] [PubMed]
24. Knipfer, T.; Fei, J.; Gambetta, G.A. Water transport properties of the grape pedicel during fruit development: Insights into xylem anatomy and function using microtomography. *Plant Physiol.* **2015**, *168*, 1590–1602. [CrossRef] [PubMed]
25. Trifilò, P.; Raimondo, F.; Lo Gullo, M.A. Hydraulic connections of leaves and fruit to the parent plant in *Capsicum frutescens* (hot pepper) during fruit ripening. *Ann. Bot.* **2010**, *106*, 333–341. [CrossRef] [PubMed]
26. Dubbert, M.; Couvreur, V.; Kübert, A.; Werner, C. Plant water uptake modelling: Added value of cross-disciplinary approaches. *Plant Biol.* **2023**, *25*, 32–42. [CrossRef] [PubMed]
27. Choat, B.; Brodribb, T.J.; Brodersen, C.R. Triggers of tree mortality under drought. *Nature* **2018**, *558*, 531–539. [CrossRef]
28. Brüggenwirth, M.; Winkler, A.; Knoche, M. Xylem, phloem, and transpiration flows in developing sweet cherry fruit. *Trees* **2016**, *30*, 1821–1830. [CrossRef]
29. Morandi, B.; Manfrini, L.; Lugli, S.; Tugnoli, A.; Boini, A.; Perulli, G.D.; Bresilla, K.; Venturi, M.; Grappadelli, L.C. Sweet cherry water relations and fruit production efficiency are affected by rootstock vigor. *J. Plant Physiol.* **2019**, *237*, 43–50. [CrossRef] [PubMed]
30. Morandi, B.; Manfrini, L.; Losciale, P. Changes in vascular and transpiration flows affect the seasonal and daily growth of kiwifruit (*Actinidia deliciosa*) berry. *Ann. Bot.* **2010**, *105*, 913–923. [CrossRef]
31. Nordey, T.; Lechaudel, M.; Génard, M. The decline in xylem flow to mango fruit at the end of its development is related to the appearance of embolism in the fruit pedicel. *Funct. Plant Biol.* **2015**, *42*, 668–675. [CrossRef]

32. Chen, J.W.; Zhang, S.L.; Zhang, L.C. Sugar Transport, Metabolism, Accumulation and Their Regulation in Fruits. *J. Plant Physiol. Mol. Biol.* **2004**, *30*, 1–10.
33. Srivastava, M.K.; Dwivedi, U.N. Delayed ripening of banana fruit by salicylic acid. *Plant Sci.* **2000**, *158*, 87–96. [CrossRef] [PubMed]
34. Conejero, W.; Alarcón, J.J.; García-Orellana, Y.; Abrisqueta, J.M.; Torrecillas, A. Daily sap flow and maximum daily trunk shrinkage measurements for diagnosing water stress in early maturing peach trees during the post-harvest period. *Tree Physiol.* **2007**, *27*, 81–88. [CrossRef] [PubMed]
35. Sun, Q.; Zhao, Y.X.; Cheng, J.X.; Zeng, T.Y.; Zhang, Y. Fruit growth modelling based on multi-methods—A case study of apple in zhaotong, yunnan. *Sci. Agric. Sin.* **2021**, *54*, 3737–3751. [CrossRef]
36. Wang, D. Research on Irrigation Model for Tomatoes in Solar Greenhouse Bag Culture Based on Real Time Growth. Master's Thesis, Shenyang Agricultural University, Shenyang, China, 2017.

MDPI AG
Grosspeteranlage 5
4052 Basel
Switzerland
Tel.: +41 61 683 77 34

Horticulturae Editorial Office
E-mail: horticulturae@mdpi.com
www.mdpi.com/journal/horticulturae

www.ingramcontent.com/pod-product-compliance
Lightning Source LLC
Chambersburg PA
CBHW040312270926
51850CB00007BC/1151
9783725816958